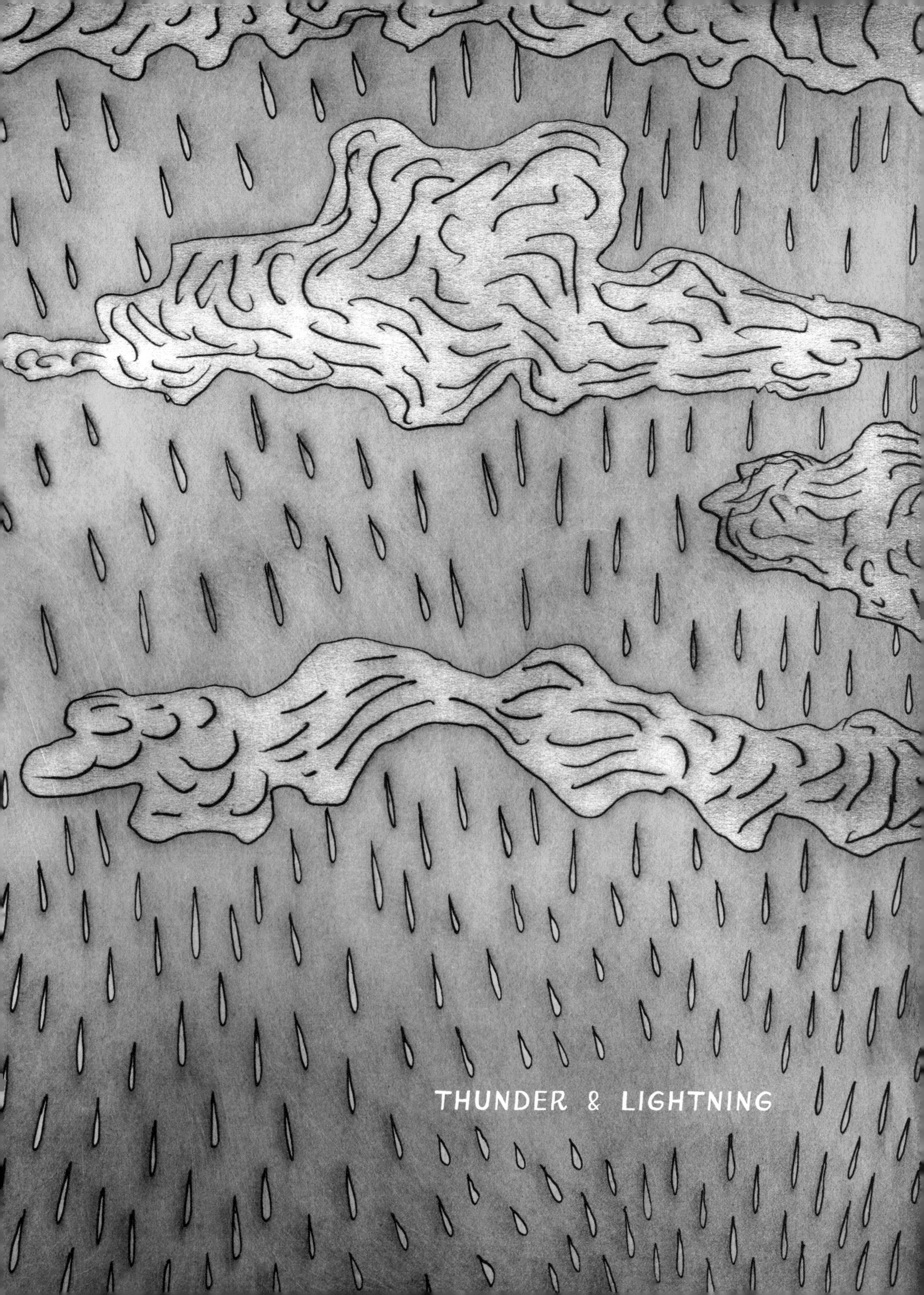
THUNDER & LIGHTNING

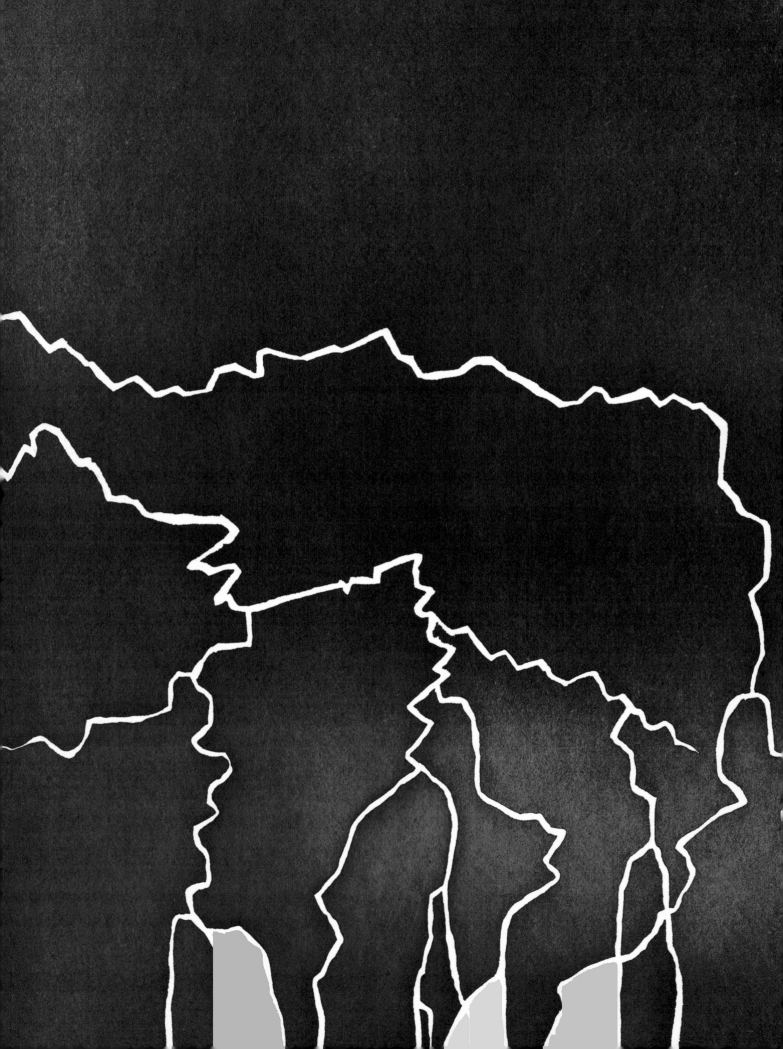

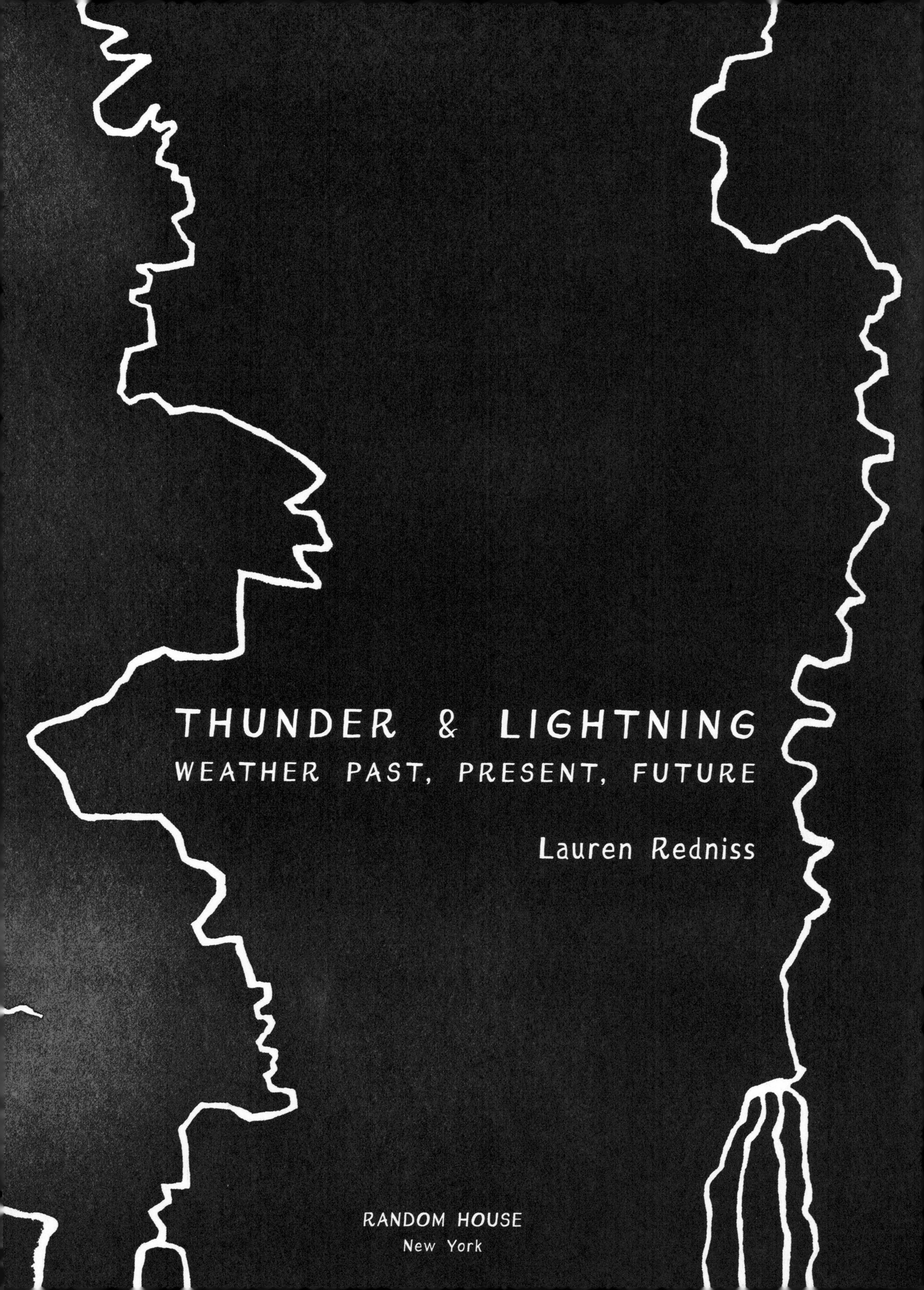

THUNDER & LIGHTNING
WEATHER PAST, PRESENT, FUTURE

Lauren Redniss

RANDOM HOUSE
New York

for
J & S & T

"Just as I was thinking I had better try to fill in with something about the weather, she spoke."

— **P.G. Wodehouse,** *The Code of the Woosters*

CONTENTS

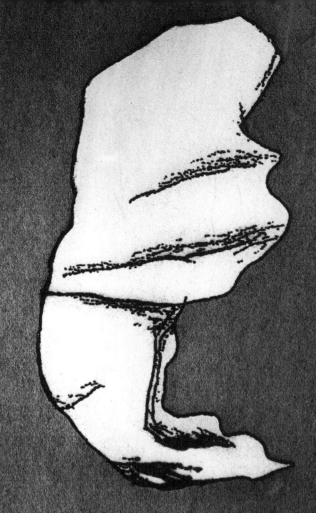

CHAPTER 1:

CHAOS

"Before,

it was beautiful and serene.

It sat up above things,

on this little plateau.

You could look across and see

mountains

all bordering it.

Flowers, bushes, shrubs. Just a

very pretty

cemetery."

Sue Flewelling is commissioner at Woodlawn Cemetery in Rochester, Vermont.

"Now, with that big hole ripped out, the cemetery is just like a rape victim. That's blunt — but that's the feeling of how it is now. I mean, it can be repaired, but the scar is still going to be there."

Hurricane Irene began in the Caribbean, in late August 2011, as an elongated area of low pressure — a "tropical wave." Clouds and thunderstorms formed. Winds picked up, gathering into a storm that hit St. Croix on August 20. Gaining strength over warm waters, Irene became a Category 3 hurricane that spun through the Bahamas with 120 mph winds, then weakened to a Category 1 before making landfall in North Carolina. Spilling torrents of rain, Irene continued northward. She struck Little Egg Inlet, New Jersey, just before dawn on August 28, Coney Island later that morning, and traveled across Vermont and New Hampshire that evening.

By the time Irene dissipated, 49 people were dead. Damage from the storm was estimated at nearly $16 billion. In Rochester, Vermont, roads were washed out. A bridge collapsed; houses were toppled. A culvert meant to drain water from Rochester's Nason Brook became clogged with debris. Instead of flowing through the pipe, storm waters surged around it, flooding Woodlawn Cemetery. Graves were unburied. The Vermont medical examiner's office was called in to identify bodies. Deputy chief medical examiner Dr. Elizabeth Bundock worked with local officials, including Flewelling.

Sue Flewelling: "That day we expected wind. That was the weather report. We prepared for the wind. We got hardly any wind. It was about water."

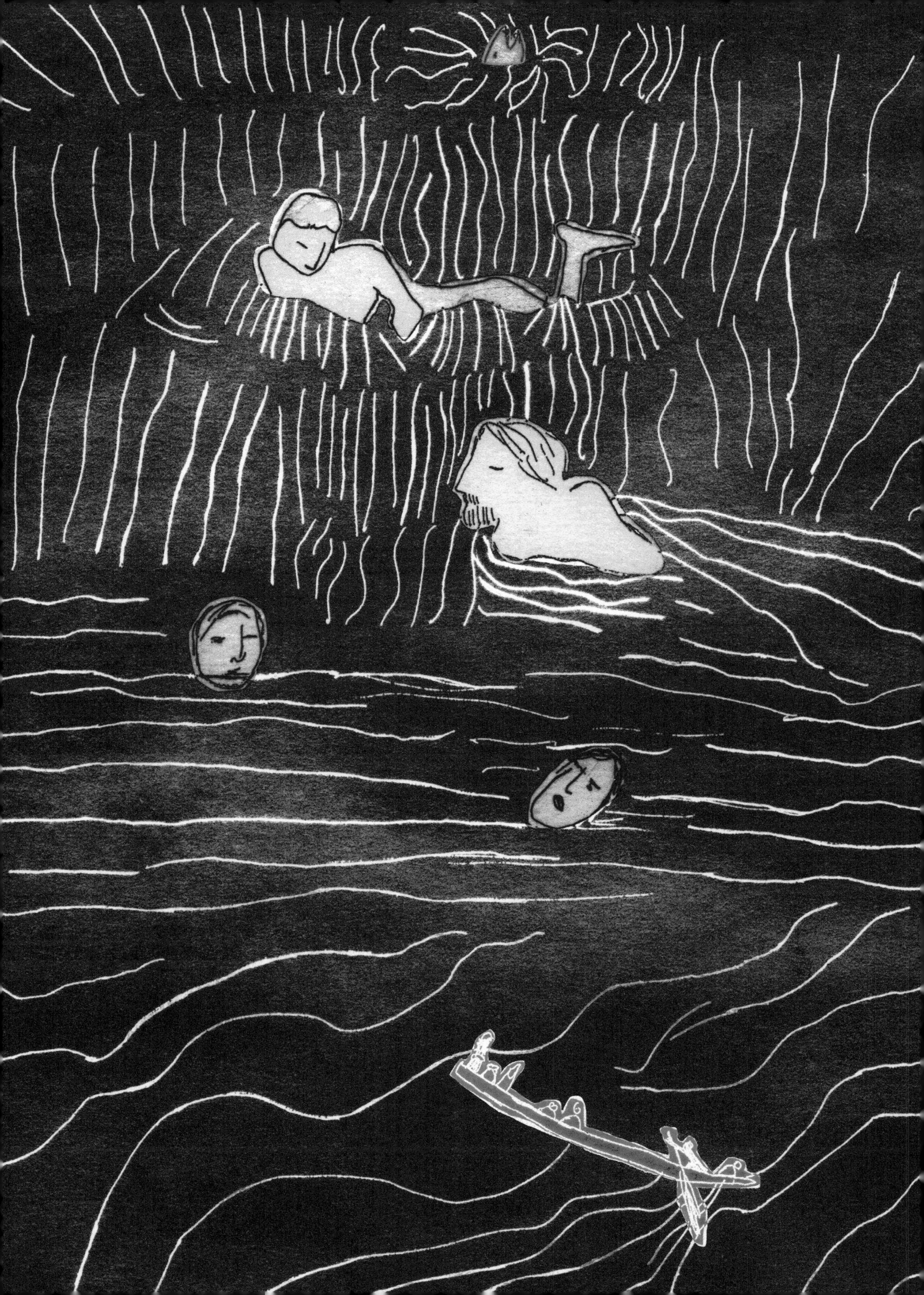

Elizabeth Bundock: "Part of the cemetery was washed out. The water carried with it dry bones which settled down onto the road and the field."

Sue Flewelling: "The cemetery is way up from the river — I mean, it's really a little tiny brook, one of the little tributaries that feed into the White River. We're talking about a brook that you step on stones to get across."

Elizabeth Bundock: "It sometimes dries up in the summer."

Sue Flewelling: "But the culvert failed, and when it failed, it took down big trees, which blocked the water and diverted it towards the cemetery."

Elizabeth Bundock: "Vaults, which are 800, 1000 pounds or so apiece, tumbled down. Caskets came out of their vaults. Some of the caskets were partially buried by silt and rock and boulders. Some of them were dented and broken open."

Sue Flewelling: "Some were cremations so we'll never find anything of them."

Elizabeth Bundock: "I have a list of the involved boxes. It's a process of elimination."

Sue Flewelling: "We lost, as far as we can tell, 54 people."

Elizabeth Bundock: "We have a list of 50 names, but we don't have 50 bodies. Half of those are over 50 years old. Back then they would have been put in wooden caskets, and by now they would be skeletonized. The caskets could have rotted. So for those, it's hard to even know what we would be looking for."

Sue Flewelling: "Phone calls were coming in from families inquiring whether their plots were washed out."

Elizabeth Bundock: "We did extensive searching for any kind of remains. But inevitably there's a lot of things we haven't found because there are huge piles of debris, logs, and vegetation potentially covering remains."

Sue Flewelling: "For some people all we'll find is bones and pieces."

Elizabeth Bundock: "We're using casket characteristics as well as personal features — whatever unusual characteristics someone might have — scars, clothing, jewelry, personal mementos, birthstone rings. Sometimes features of the person's body, whether they had a scar or amputation, or a unique facial feature like a really big nose or a very prominent, strong chin. Their hair can help. Curly hair, straight hair, long hair. A lot of people are buried with clothing and jewelry. Some people were buried with pictures or handwritten cards, cards from grandchildren, things like that. We have Masons and Shriners, so they had things related to those aspects of their lives. A pen or a fez."

Sue Flewelling: "They took DNA from the ones we couldn't positively identify and sent them off. There was one casket that had a little vial inside, and in that vial was a paper. It was damp but it didn't disintegrate. It rolls out and tells who we have, when they died, when they were buried, who they're related to. I think they paid $70 extra for that vial to go in the coffin."

Elizabeth Bundock: "I'm sure we'll never identify everyone."

Sue Flewelling: "We have unidentified bones that we have to put somewhere. We will dedicate one lot for all the ones that were unidentified. Put a stone there, for everyone that's missing. Some headstones washed away with the bodies. We find them in the rubble."

Elizabeth Bundock: "I found myself looking at this destruction, and I turn around in the other direction — 180-degree turnaround — and see the most beautiful, peaceful landscape with the Vermont hills in the background. And the sky, you know, it's just beautiful. You stand there in amazement of the power of all that water, how fast it must have been going to tumble thousand-pound concrete vaults down a riverbed. Most people expect their loved ones, once they're buried, to remain that way."

Sue Flewelling: "A cemetery, you know, is a little dot in the whole scheme of things. You need to help the living people."

CHAPTER 2:

COLD

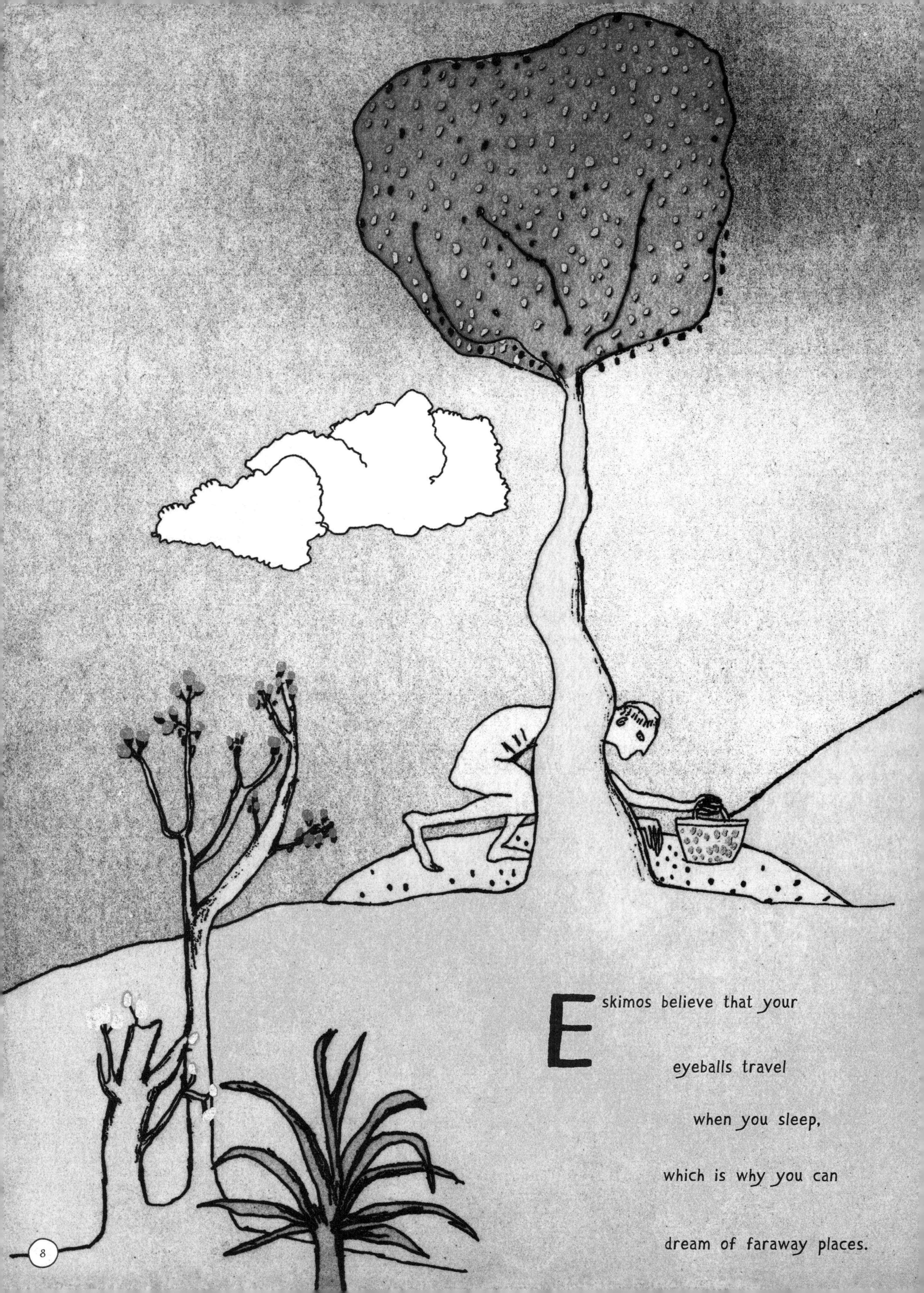

Eskimos believe that your
eyeballs travel
when you sleep,
which is why you can
dream of faraway places.

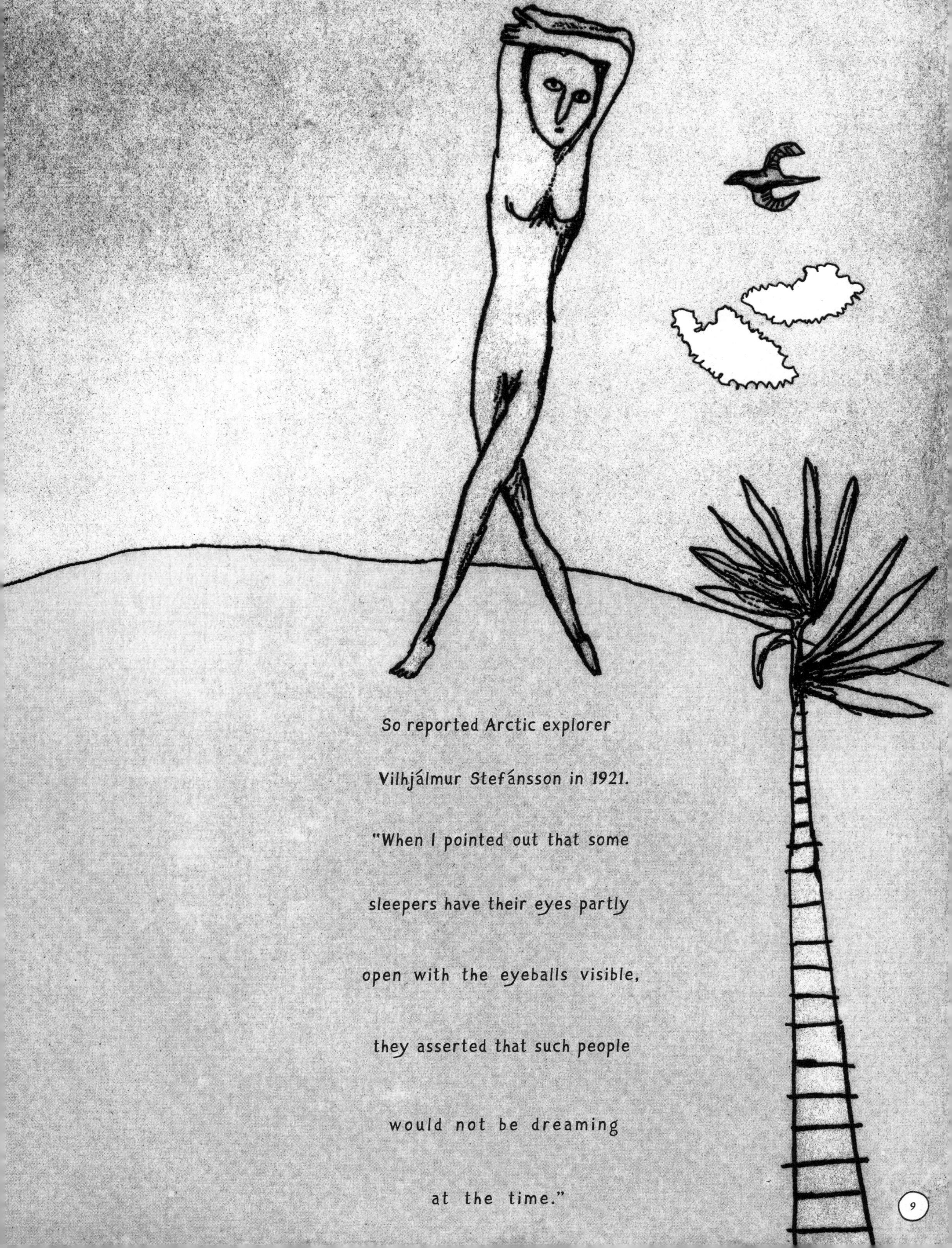

So reported Arctic explorer

Vilhjálmur Stefánsson in 1921.

"When I pointed out that some

sleepers have their eyes partly

open with the eyeballs visible,

they asserted that such people

would not be dreaming

at the time."

Stefánsson was born in Manitoba, Canada, to Icelandic parents. He studied anthropology at Harvard, and took off for the Canadian Arctic in 1906. Over the next dozen years he made three expeditions to the region. Stefánsson was seduced by the frozen landscape. In *The Friendly Arctic: The Story of Five Years in Polar Regions*, he describes the peculiarities of polar optics over the snow-covered terrain: "The daylight is negligible; and the moonlight, which comes to you first through clouds that are high in the sky and later through an enveloping fog, is a light which enables you to see your dog team distinctly enough, or even a black rock a hundred yards away, but is scarcely better than no light at all upon the snow at your feet." In the white-on-white expanse, details are erased, creating the disorienting illusion of perfect emptiness. "It looks as if there were nothing there and as if you were stepping into space each time you lift your foot." Stefánsson would toss his deerskin mittens in front of him to mark a path. "After throwing one of them about ten yards ahead, I would keep my eyes on it till I got within three or four yards and then throw the other, so that most of the time I could see the two black spots on the snow ahead of me separated by five or six yards of whiteness."

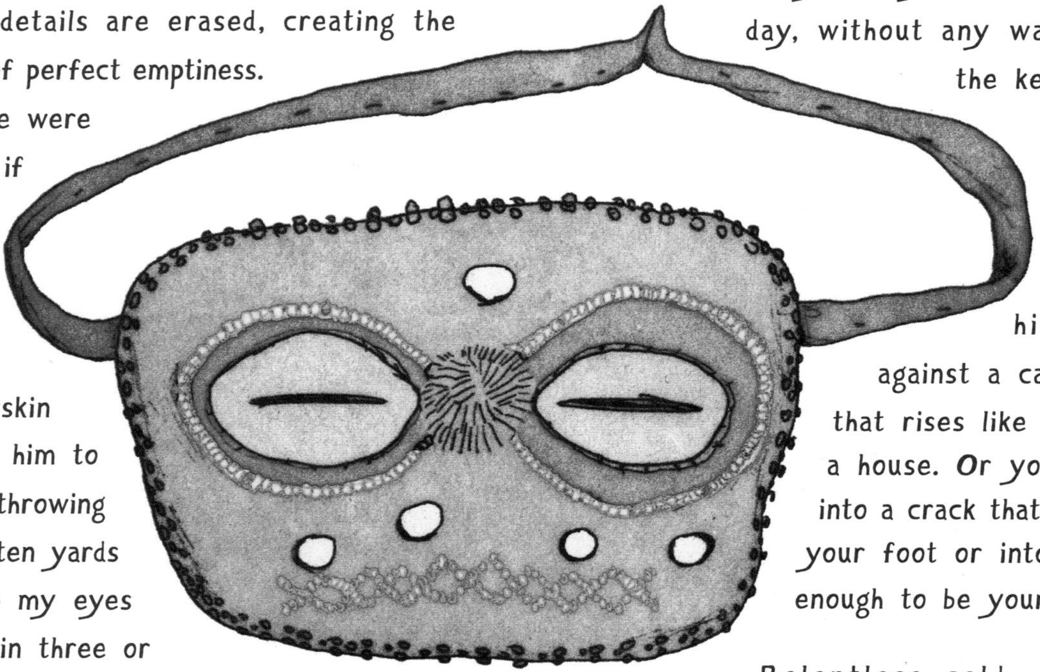

The trick did not circumvent all hazards. "All this would not be so bad if you really had the strength of mind to realize that your eyes are useless. But you are continually trying your best to see and the strain brings on the condition known as snowblindness."

Snowblindness (or photokeratitis, a temporary eye injury) occurs when the ultraviolet rays of the sun reflect against the brilliance of the snow and ice, burning the corneas of unprotected eyes. Arctic peoples have fought snowblindness by carving wood or caribou antlers into goggles with thin slits to let in just a sliver of light. Elaborately beaded goggles have been found in Siberia—one early 19th century pair held by the British Museum has slitted, convex brass "lenses" sewn into a reindeer leather mask. The wearer would place the soft, furry side against his face.

Stefánsson wore amber-tinted eyeglasses, yet still suffered occasionally: "It might be inferred that snowblindness is most likely to occur on days of clear sky and bright sun. This is not the case. The days most dangerous are those when the clouds are thick enough to hide the sun, but not heavy enough to produce what we call heavily overcast or gloomy weather. Then light is so evenly diffused that no shadows can be seen anywhere.... On the rough sea ice you may, on an unshadowed day, without any warning from the keenest eyes, fall over a chunk of ice that is knee high or walk against a cake on edge that rises like the wall of a house. Or you may step into a crack that just admits your foot or into a hole big enough to be your grave."

Relentless cold, months of darkness, lack of food, the threat of polar bear attack, loneliness, exhaustion, treacherous mirages: these have been essential elements in the heroic narratives of Arctic and Antarctic voyages, peril heightening glory. Not long after returning from his last polar expedition, Vilhjálmur Stefánsson wrote: "My first ambition, so far as I remember, was to be Buffalo Bill to kill Indians. That was when I was still a small boy.... [Then] my ambition shifted and my ideal became Robinson Crusoe.... Twenty years later when I discovered lands and stepped ashore on islands where human foot had never trod, I had in reality very much the thrills of my boyhood imagination when I dreamed of being a castaway on my own island."

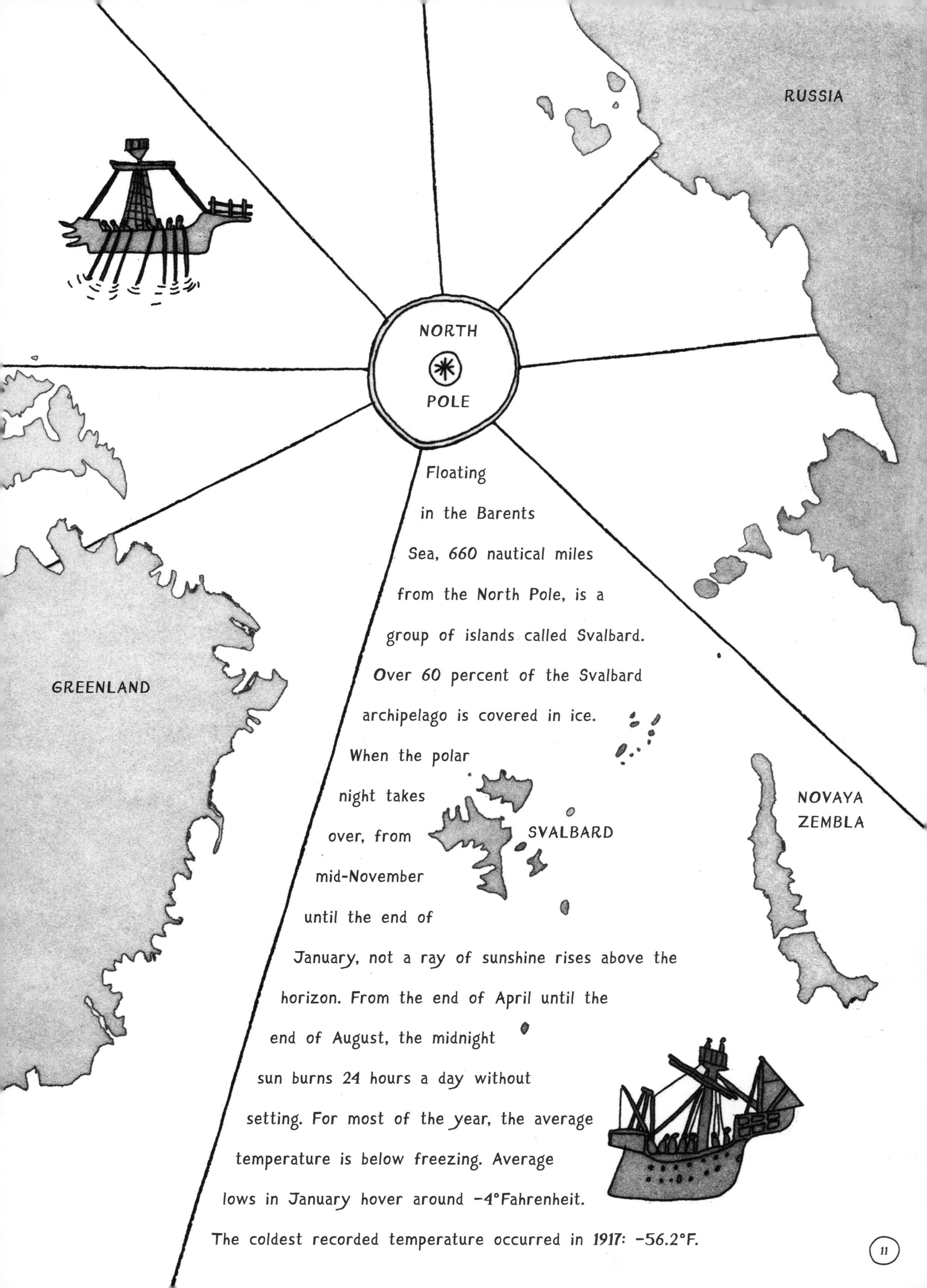
RUSSIA
NORTH POLE
GREENLAND
SVALBARD
NOVAYA ZEMBLA

Floating in the Barents Sea, 660 nautical miles from the North Pole, is a group of islands called Svalbard. Over 60 percent of the Svalbard archipelago is covered in ice. When the polar night takes over, from mid-November until the end of January, not a ray of sunshine rises above the horizon. From the end of April until the end of August, the midnight sun burns 24 hours a day without setting. For most of the year, the average temperature is below freezing. Average lows in January hover around −4°Fahrenheit. The coldest recorded temperature occurred in 1917: −56.2°F.

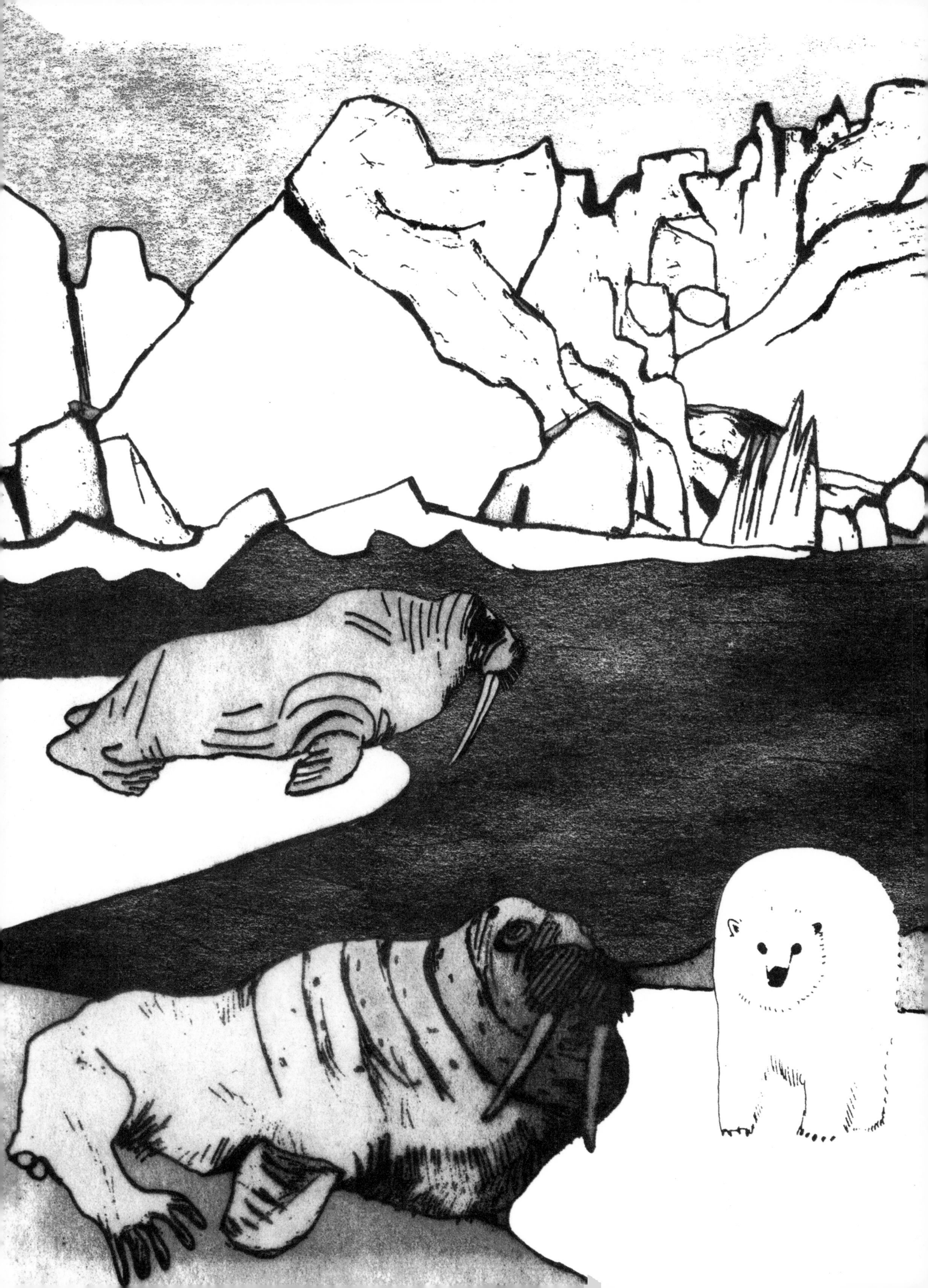

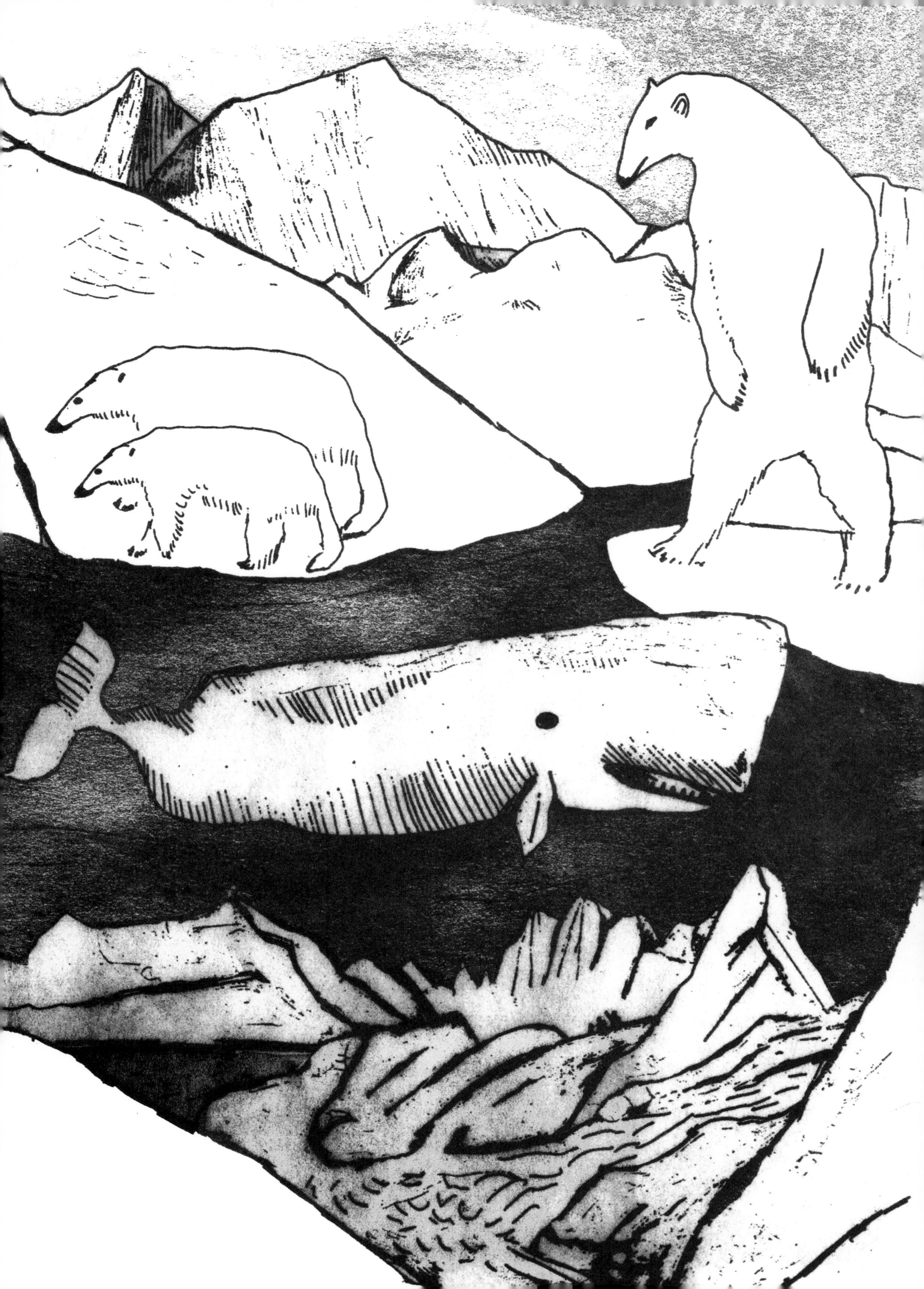

In Svalbard, gusting air blows powdery snow in swirling eddies over the frozen ground. There are no trees, no crops, no arable land. Today, Svalbard has a population of approximately 2000 people and 3000 polar bears. The land is covered by permafrost, soil that does not thaw over the course of the year. A thin so-called "active layer" on the surface is warmed enough during summer months to nourish small flowers and low-growing berries.

Dutch explorer Willem Barentsz is usually credited with the discovery of the archipelago in 1596, though it is speculated that Vikings arrived in the 12th century, and that not long after, the Pomors of northern Russia may have hunted there, bringing home furs and walrus tusks. In the 17th and 18th centuries, Svalbard became a destination for whaling. The center of operations was a settlement called Smeerenburg ("Blubbertown" in Dutch) until the whale population was hunted to near extinction. Industrial coal mining began in the early 20th century and is still a mainstay of the island's economy.

Life on Svalbard is challenging. Ingrid Urberg, a professor of Scandinavian studies at the University of Alberta, examined the 17th-century records of the London-based English and Russian Muscovy Company. "Concerned about safeguarding its whaling stations, the company tried to entice prisoners who had been sentenced to death to spend the winter on Svalbard by promising them both wages and freedom. Once the prisoners arrived, however, they grew so fearful they refused the offer and begged to be taken home. For them, death was preferable to spending a winter on Svalbard facing polar bears, frigid temperatures, and scurvy."

On Svalbard, even the dead cannot rest. An Austrian woman named Christiane Ritter traveled to the archipelago in the 1930s and became the "first European woman to spend a winter so far north." Ritter returned to Austria after a year, wrote a memoir, and lived to be 103 years old. In her book, *A Woman in the Polar Night*, she wrote: "The ground is frozen hard as steel, and for the first time I understand why the dead cannot be buried in [Svalbard] in wintertime, and why the hunters, to save them from the bears and foxes, keep their dead comrades with them in the hut right through the winter."

Caskets that have been buried in Svalbard are gradually unearthed: summer rains are absorbed by the ground, which then freezes in wintertime and expands, pushing the graves little by little towards the surface. Decades ago the small local cemetery stopped accepting new bodies. According to Liv Asta Ødegard, the governor's office communications advisor, "In a kind of ironic way we say it is illegal to die on Svalbard. The Norwegian government does not want this to be a place to be born or to die. There is one gynecologist at the hospital, but it is not sure that he is on duty every day, or if he is even on the island. We do not have any social welfare system here. If you are old and need help, you need to go away from Svalbard." In September 2012, Svalbard's English language weekly, *Ice People*, published a story about 80-year-old resident Anne Maeland facing pressure to leave. The article quotes a city councilman, Jon Sandmo: "We can sink into poverty due to 20 retired persons."

It would seem to make no sense to live in such a place — a place where it is too cold to be born, too cold to die.

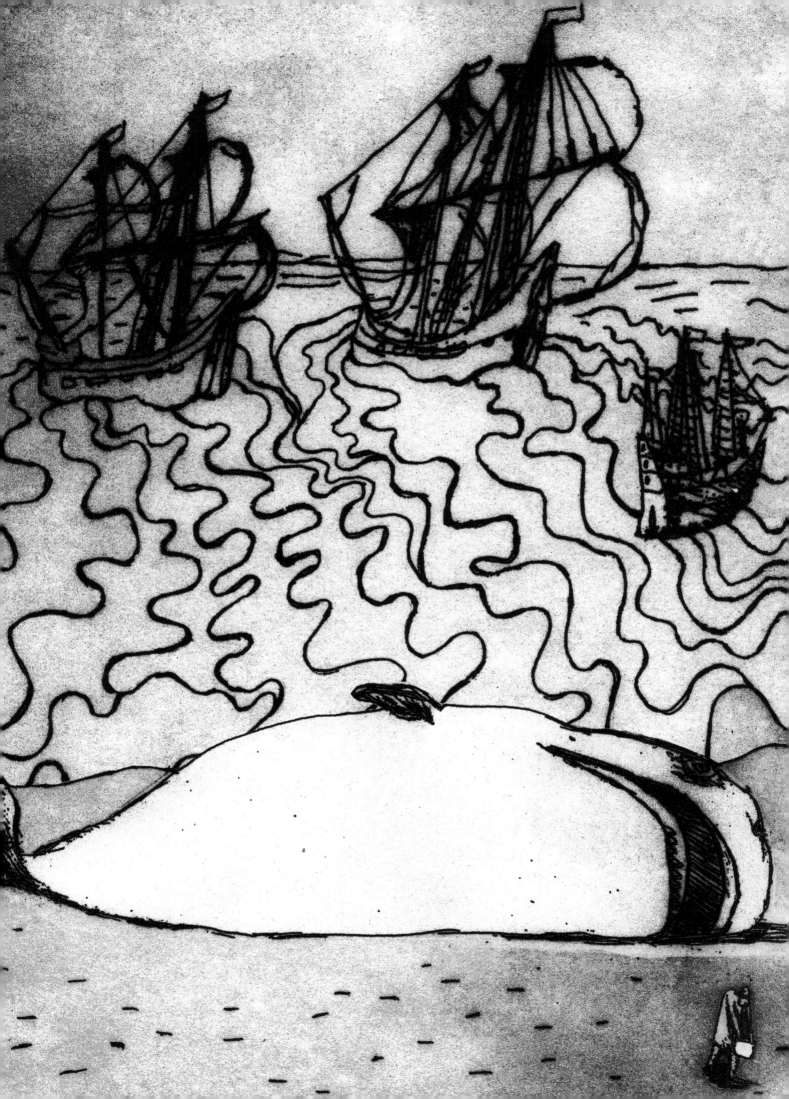

But the cold can also be a life-sustainer.

Freezing slows decay and retards the growth

of microorganisms, preserving vitality

far past what would be possible

under "normal" conditions. Cold can,

in a sense, derail time. We rely

on this principle in our home

freezers, wrapping up

last night's leftovers or

saving a slice of wedding

cake years into a marriage. Some

have dreamed that these ideas could be applied to the

prolongation of human life. The Alcor Life Extension Foundation is one of a number

of companies offering "speculative life support" services. Alcor uses cryonics — "the

low-temperature preservation of humans and animals" — to "save lives by using temperatures

so cold that a person beyond help by today's medicine might be preserved for decades or

centuries until a future medical technology can restore that person to full health."

In 2012, Russian scientists

reported that they had

regenerated a 30,000-year-old

flower from tissues preserved in the Siberian

permafrost: Ice Age squirrels had buried fruit and

seeds of the *Silene stenophylla* plant in a burrow in northeastern

Siberia. The squirrels lined the cavity with hay and animal fur. According to

Stanislav Gubin, an author of the study, it was "a natural cryobank." From fossilized

fruit tissue, the scientists were able to coax delicate, five-petaled white flowers to blossom.

Outside the settlement of Longyearbyen on Svalbard's Spitsbergen Island, a tunnel drives some 500 feet deep into the sandstone mountainside. A cement portal, walls tilting diagonally upwards, juts out into the cold air, sheltering the steel-reinforced entrance door. Behind the entryway, a tunnel of corrugated steel with bands of ice hardened into the grooves slopes downwards towards another locked door. That door opens into a space perpendicular to the tunnel, like a church nave crossed by its transept. Undulating rock walls have been coated with spray-on concrete and impregnated plastic. Entering this brilliant white cave, you stand facing three doors. The middle door, made of steel, glitters with ice crystals. The lock, too, is encrusted in frost. This is the Svalbard Global Seed Vault.

It's been called the "Fort Knox of seeds" and the "Doomsday Vault." The Global Seed Vault is a storage facility and an insurance policy for the world's agricultural diversity. Countries from all over the world deposit seeds here for safekeeping. Local seed banks are vulnerable to disasters: war, mismanagement, energy failure, financial instability, extreme weather, climate change. In recent years, seed banks in Iraq, Afghanistan, and Egypt were destroyed or looted. Svalbard—a place that is entirely hostile to agriculture, inhospitable to life in almost every form—turns out to be an ideal spot for protecting the world's harvest.

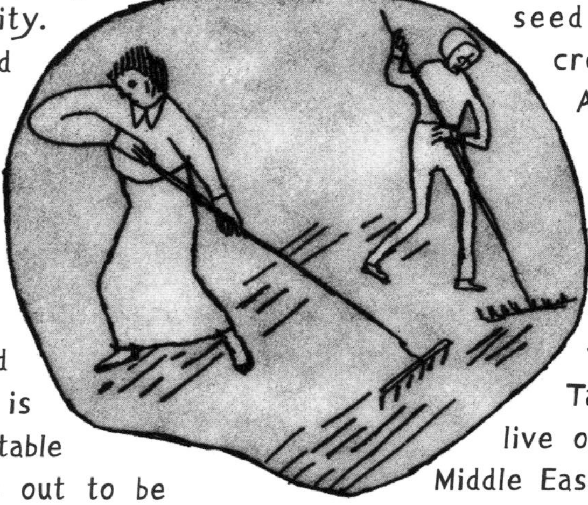

The Arctic permafrost naturally creates an interior environment in the vault of 21°F. Additional mechanical cooling brings the temperature down to just below 0°F, a level at which the U.S. Department of Agriculture says food "will always be safe," and a temperature that has been deemed appropriate for seed preservation. According to information issued by the three organizations that jointly run the vault, the Global Crop Diversity Trust, the Nordic Genetic Resource Center, and the Norwegian government, "Even given worst-case scenarios for global warming, the Seed Vault storage rooms will remain naturally frozen for up to 200 years." There are other natural security systems in place as well: "The region on Svalbard surrounding the Vault is remote, severe, and inhabited by polar bears."

At full capacity, the vault will store 2.25 billion seeds. Crops that contribute to "sustainable agriculture and food security" are given priority. The samples are dried, wrapped in small four-ply foil bags, and placed inside sealed boxes. There is barley and goat grass from Armenia; peas from Australia; cumin, flax, wild rye, alfalfa, and sunflowers from Canada; wheat from Israel; lentils from Ukraine; needlegrass, giant hyssop, monkshood, yarrow, marigold, foxtail, pigweed, hollyhock, saltbush, yellowrocket, snapdragon, chamomile, asparagus, and wild onion from Germany; goosegrass and sorghum from Uganda. There is mahogany from Kenya, clover from Ireland, mustard and chickpea from Pakistan, melon and morning glory from Taiwan. The United States' deposits include basil, mint, evening primrose, parsley, chicory, okra, blackberry, pear, watermelon, lawngrass, bluegrass. There is sesame, spinach, radish, peanut, tomato, wild carrot, and Job's Tears from South Korea. On a nearby shelf there are corn and rice seeds from North Korea.

As of 2014 the Global Seed Vault held seed samples representing the crops of some 230 countries. According to Cary Fowler, chair of the Global Seed Vault's International Advisory Council, "We have seeds from more countries than actually exist." In the vault, the Soviet Union and Tanganyika, today's Tanzania, live on. Territorial disputes in the Middle East do not affect the boxes of seeds labeled "Palestine." A large shipment arrived from Syria in the midst of that country's civil war in 2012. "We don't play politics here," says Fowler.

Until 1920, Svalbard was a nationless land, subject to no laws. During the negotiations at Versailles after World War I, the Spitsbergen Treaty made Svalbard a territory of Norway, though it remains distinct from the mainland in a number of ways. No residency permit, work permit, or visa is required to live on Svalbard. The treaty ensures that citizens of any signatory nation can exploit the land's natural resources and engage in commercial activity. Citizens of non-signatory nations can, too. "We don't discriminate," legal advisor Hanne Ingebrigsten told the *Asia Times* in 2007. Svalbard is not subject to customs regulations; shopping is duty free. In 2014, income tax was 27 percent on the Norwegian mainland; it dipped as low as 8 percent on Svalbard.

The island's largest settlement, Longyearbyen, is named for an American: John Munroe Longyear. Longyear, a Michigan politician, established the town when his Arctic Coal Company began mining operations in 1906. A century later, Svalbard remains attractive to economic migrants. While taxes are low, salaries, even for menial jobs, are high. Today the just over 2000 people that live on Svalbard come from some 44 countries. The list sweeps the globe: Iran, Botswana, Malaysia, India, China, Tunisia, Uruguay, Peru, Mexico, Colombia, Slovakia, Bosnia-Herzegovina, Azerbaijan, the Philippines, Russia, Lithuania, Hungary, the Netherlands, Germany, France, Great Britain, Finland, Denmark, Sweden, the United States, Argentina, Brazil, Chile, Vietnam. After Norwegians, Thai immigrants make up the largest segment of the population. Svalbard is forbidding terrain, but it is a land of opportunity.

Longyearbyen sits at the base of a valley, on the shoreline of a fjord. A collection of colorful, boxy houses like a Lego set are framed against the white mountains. The settlement has one commercial street. It is a pedestrian walkway, though it is not unusual to see people getting around on skis, or tugging children in sleds. Reindeer wander through town. Halfway down the street is the Fruene café, which sells sandwiches and cakes, knitting wool, mittens, and locally made women's dresses.

Tanyong Suwanboriboon has worked at Fruene for two years.

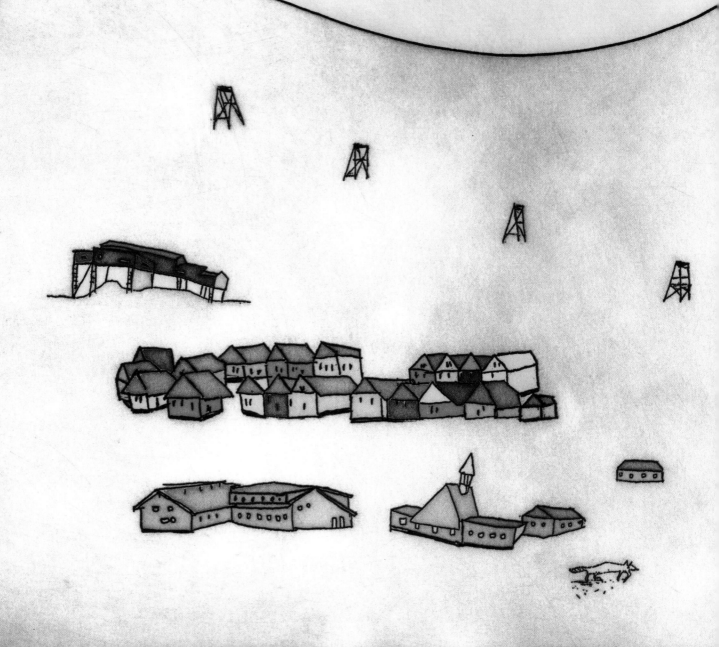

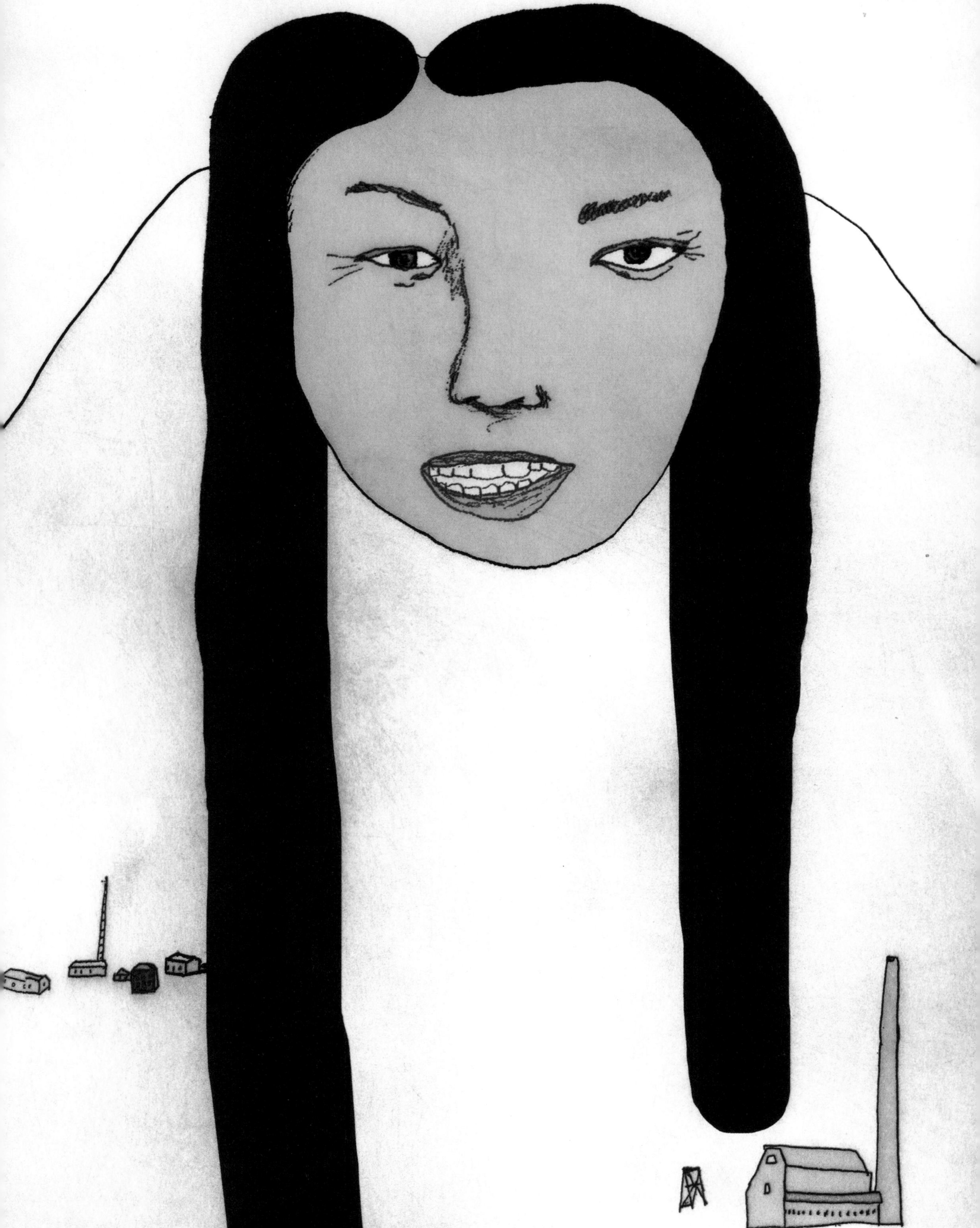

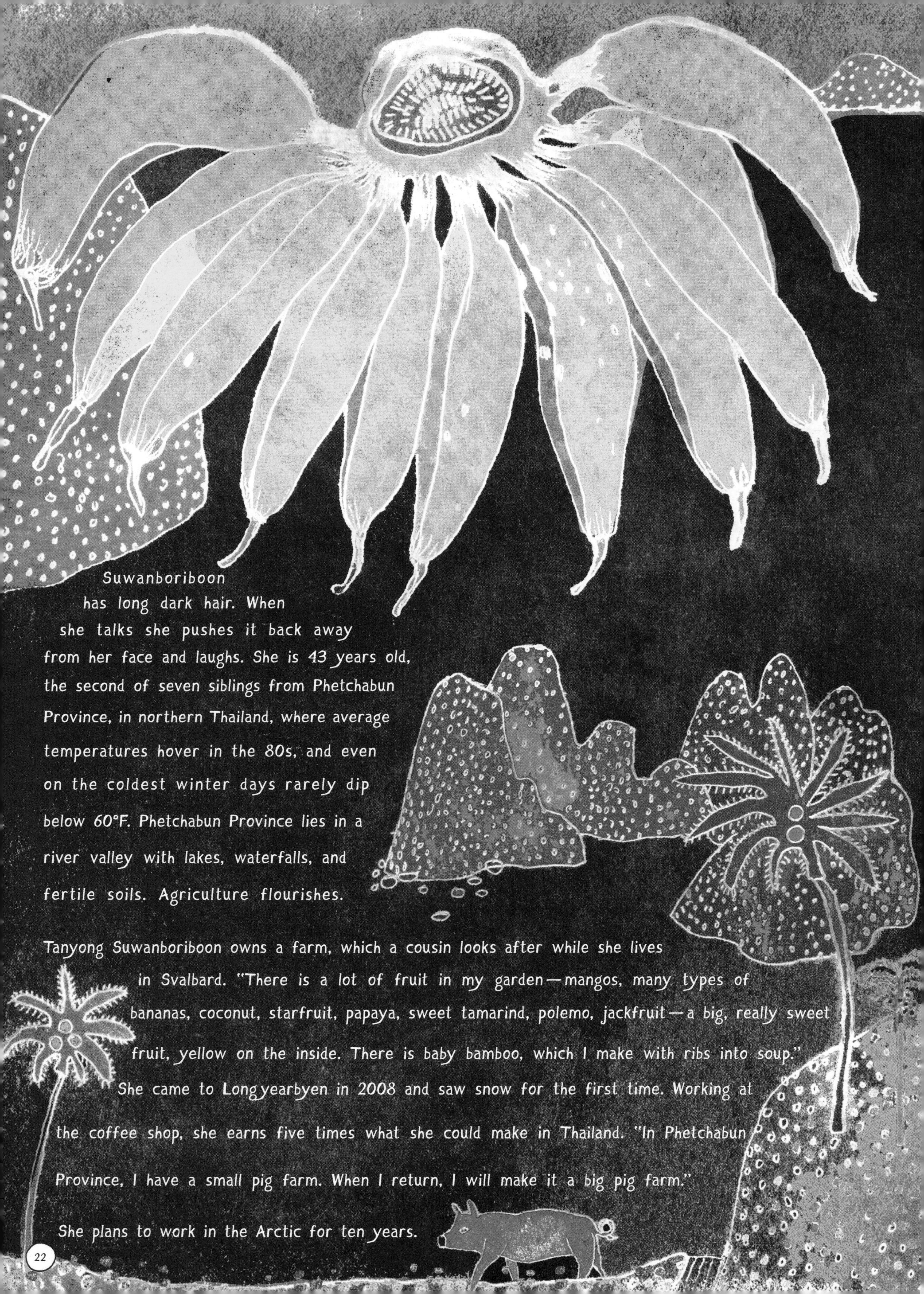

Suwanboriboon has long dark hair. When she talks she pushes it back away from her face and laughs. She is 43 years old, the second of seven siblings from Phetchabun Province, in northern Thailand, where average temperatures hover in the 80s, and even on the coldest winter days rarely dip below 60°F. Phetchabun Province lies in a river valley with lakes, waterfalls, and fertile soils. Agriculture flourishes.

Tanyong Suwanboriboon owns a farm, which a cousin looks after while she lives in Svalbard. "There is a lot of fruit in my garden—mangos, many types of bananas, coconut, starfruit, papaya, sweet tamarind, polemo, jackfruit—a big, really sweet fruit, yellow on the inside. There is baby bamboo, which I make with ribs into soup." She came to Longyearbyen in 2008 and saw snow for the first time. Working at the coffee shop, she earns five times what she could make in Thailand. "In Phetchabun Province, I have a small pig farm. When I return, I will make it a big pig farm."

She plans to work in the Arctic for ten years.

Suwanboriboon's world on Svalbard is tightly circumscribed. Roads from Longyearbyen's center dead-end just outside the settlement. Beyond the winter darkness and cold, mobility is further restricted by the polar bears. Anyone leaving the small settlement area is advised to carry a gun. Suwanboriboon is not bothered by these constraints. She is focused on work. "I don't think about outside," she says. "The polar night or the midnight sun—it doesn't make a difference. I don't think about the sun coming up or not. I concentrate on work." On her days off she cleans houses.

Suwanboriboon allows herself one complaint about her life on Svalbard: she doesn't like Norwegian food. "It has no nutrition." She recoils at the mention of classic Arctic fare—whale meat, reindeer, seal. In her kitchen there is an enormous chest freezer, the kind you would find in a New York City bodega holding varieties of ice cream.

Suwanboriboon's is filled with frozen shrimp and spring roll pastry dough and dozens of other ingredients for Thai dishes. In the adjoining room, a small foyer, are bags of potting soil from the local market. "In the summer I grow Thai herbs and spices on the windowsills," she says. She opens a Ziploc bag containing seed packets and shuffles them out over the sofa cushions. Coriander. Mustard seed. Sweet basil. Soy. Morning glory. Dill. Her family sends the seeds from Thailand. These plants could never take root outside, but they thrive indoors, photosynthesizing all night under the midnight sun. "I can always eat Thai food," Suwanboriboon says.

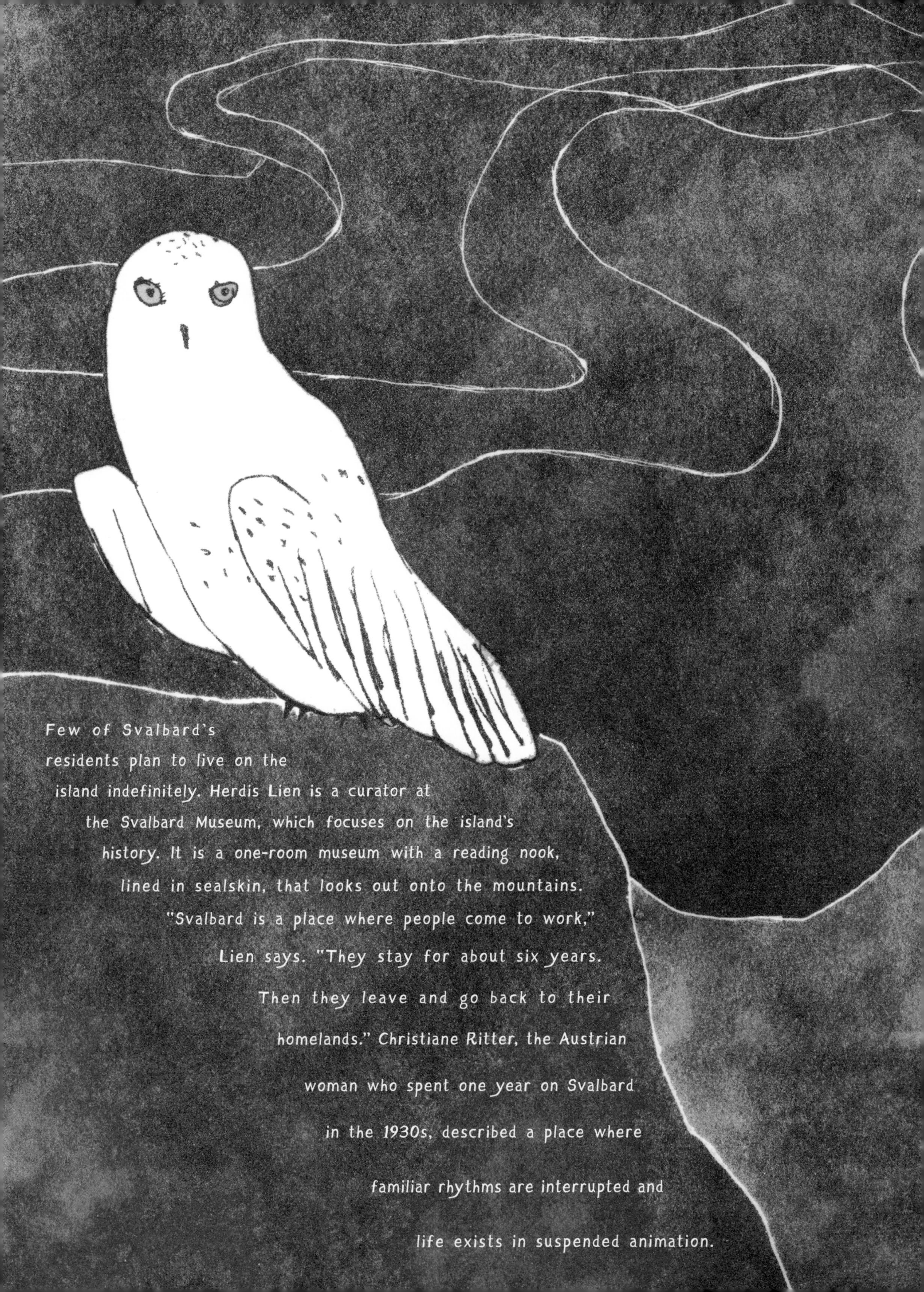

Few of Svalbard's residents plan to live on the island indefinitely. Herdis Lien is a curator at the Svalbard Museum, which focuses on the island's history. It is a one-room museum with a reading nook, lined in sealskin, that looks out onto the mountains. "Svalbard is a place where people come to work," Lien says. "They stay for about six years. Then they leave and go back to their homelands." Christiane Ritter, the Austrian woman who spent one year on Svalbard in the 1930s, described a place where familiar rhythms are interrupted and life exists in suspended animation.

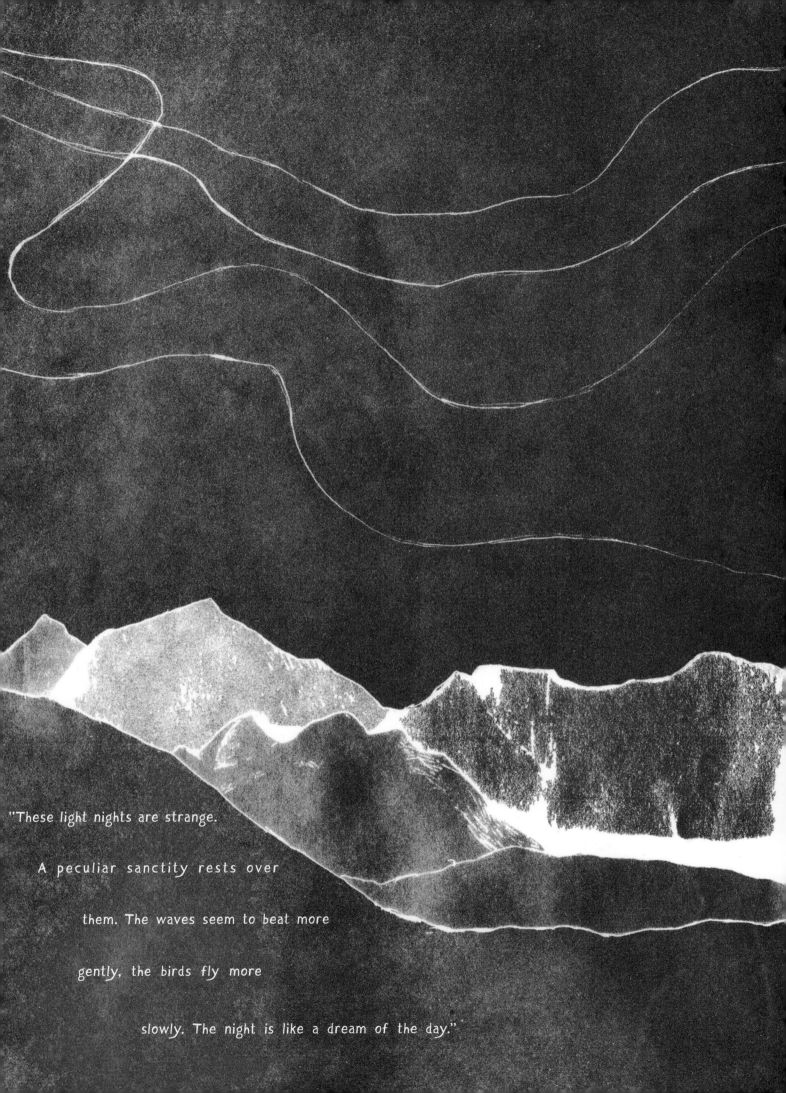

"These light nights are strange.

A peculiar sanctity rests over

them. The waves seem to beat more

gently, the birds fly more

slowly. The night is like a dream of the day."

CHAPTER 3:

RAIN

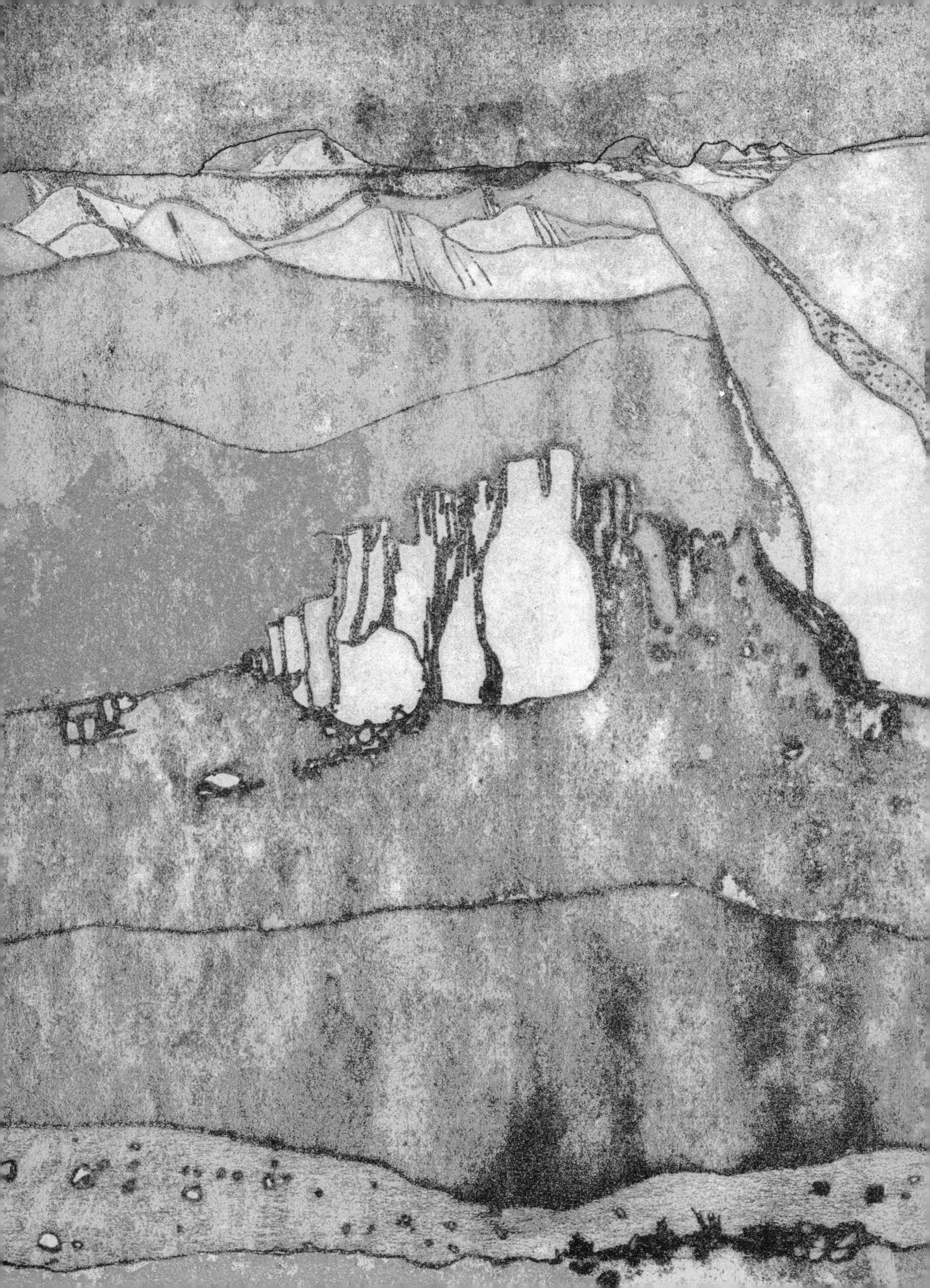

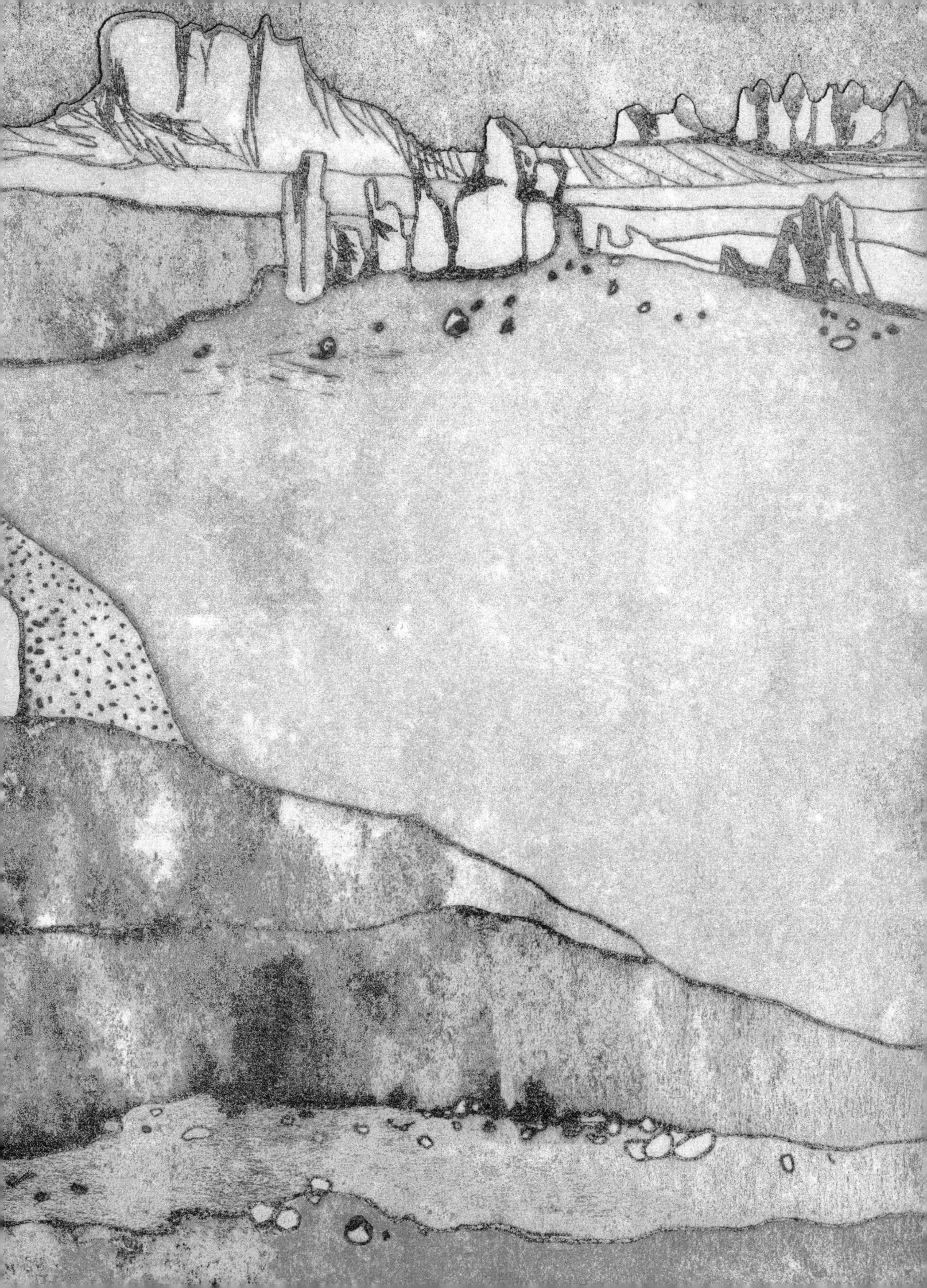

Just after midnight on October 13, 2010, after more than two months underground, 31-year-old Florencio Ávalos was lifted in a steel capsule into the dry air of northern Chile. One by one, through the night and the next day, the 32 other miners who had also been trapped in the collapsed San José copper mine were pulled to the surface. Wives, lovers, children, and cousins, along with Chilean president Sebastián Piñera and over a thousand journalists, waited at a makeshift camp to welcome the men. In Copiapó, the nearest city, people gathered in the main square, dancing and singing Chile's national anthem. Cars honked and passengers waved flags from open windows. Around the world, an estimated one billion television viewers watched the rescue.

The San José Mine is located in Chile's Atacama Desert. Chile is the world's largest producer of copper; in 2010, copper accounted for 20 percent of government revenues and, as of 2012, 15 percent of Chile's annual GDP. Chile's mineral deposits have amassed over millions of years due to a combination of geologic and climatic forces: volcanic activity and extreme aridity.

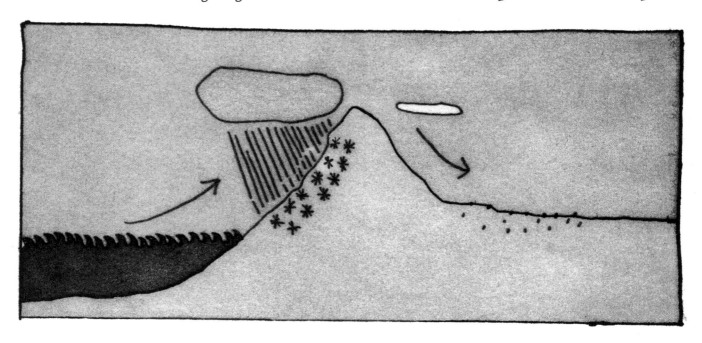

Scientists refer to the core of the Atacama as "absolute desert." It is a rocky, barren region of bleak beauty. Through the day, in the shifting light, the Atacama's sands turn gold, orange, and scarlet. In the shadows, the landscape is blue, green, violet. Treeless, plantless expanses of stark grandeur roll out like a Martian landscape. In fact, NASA uses the Atacama as a proxy for the red planet, studying the extreme conditions of the desert to aid the search for life on Mars and other extraterrestrial bodies. Robotic rovers designed by NASA scientists and researchers from Carnegie Mellon University wheel over the Atacama floor, probing for microorganisms and bacteria.

The Atacama Desert lies between two mountain ranges, a coastal range to the west and the Andes to the east. A so-called "rain shadow" effect blocks moisture coming from the Amazon Basin from reaching the central part of the Atacama: warm, humid air is trapped against the east slope of the Andes, cooling and condensing before it is able to cross the mountains. Moisture from the west is also largely prevented from precipitating out over the Atacama; the Pacific's cold Humboldt Current creates an "inversion layer," where temperature increases with height. The inversion layer restricts moisture from crossing over the coastal mountains into the desert.

According to Julio Betancourt, professor of geosciences at University of Arizona,

"You're in the sweet spot between where the winter rains can't reach and the summer rains can't reach."

The dryness creeps into your throat and sucks the moisture from your lips and skin.

Weather patterns in the desert can shift during El Niño and La Niña years. El Niño, or the warm phase of the El Niño-Southern Oscillation (ENSO), brings its warmer temperatures to the equatorial Pacific Ocean at irregular intervals every few years and, in the interaction of ocean and atmosphere, a cascade of weather effects follow around the globe. ENSO's cool phase, known as La Niña, comes with cooler waters and causes its own weather changes. For the Atacama Desert, these shifts can mean rain.

With even a small amount of rain,

the desert—particularly the fringes that surround the Atacama's core—

springs to life.

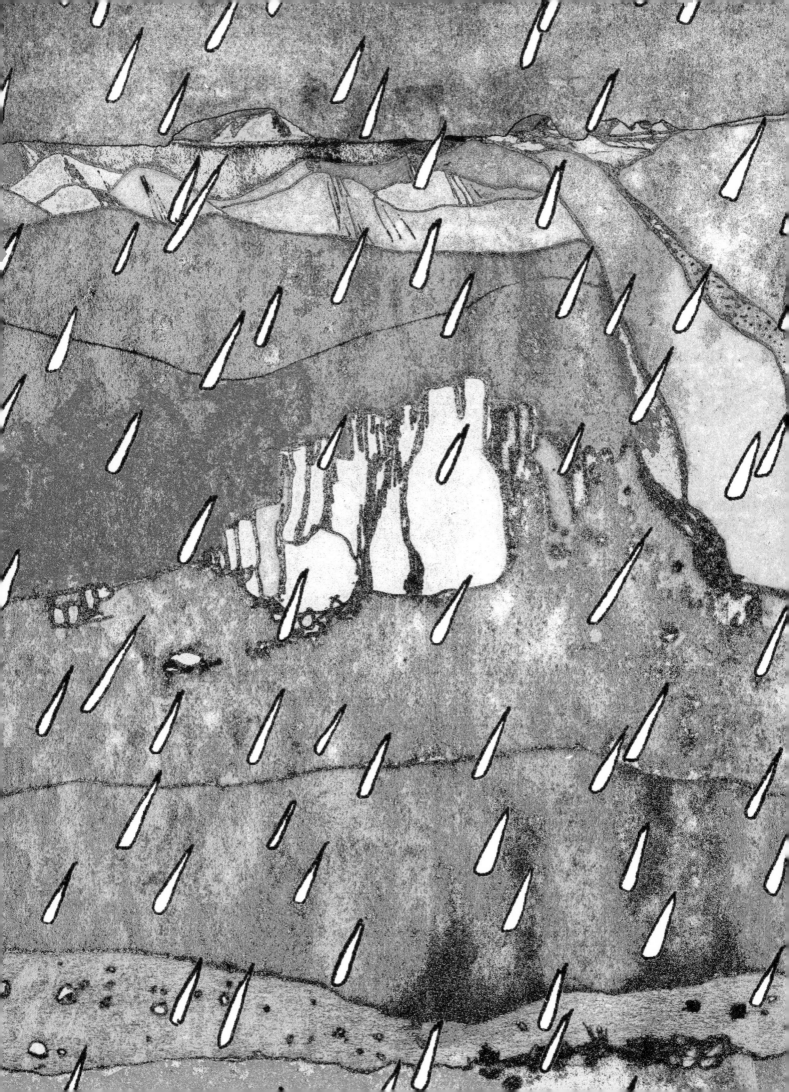

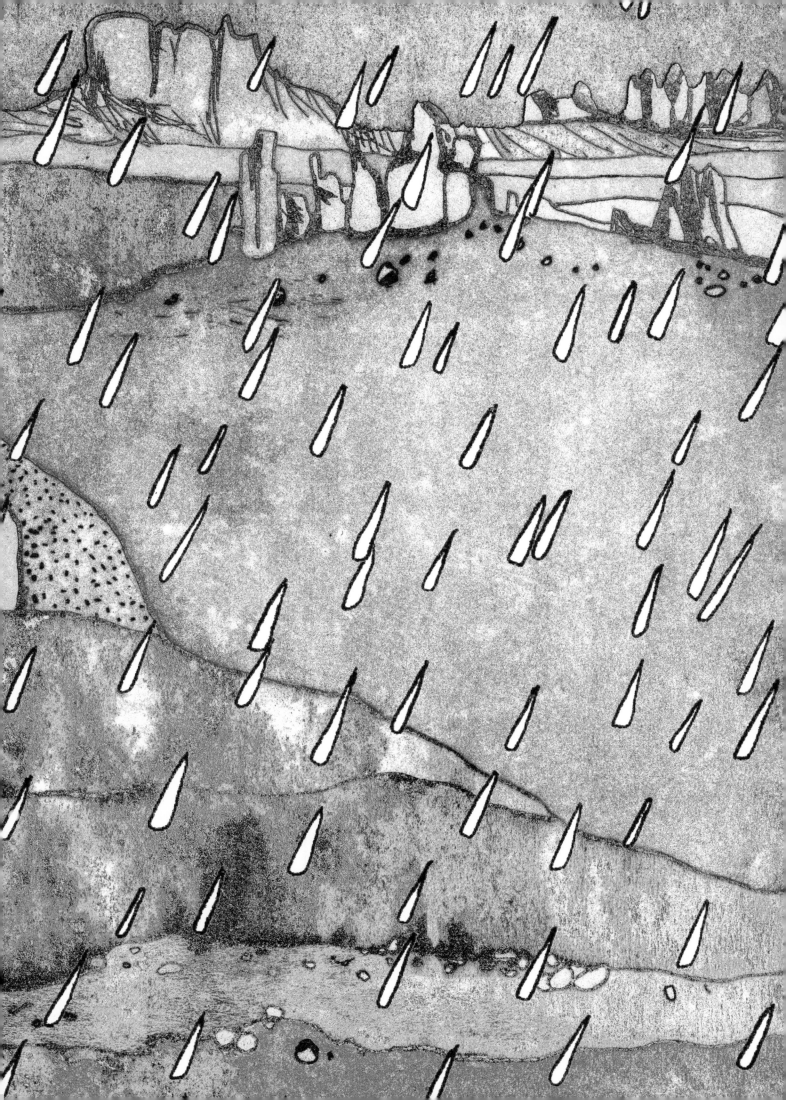

Pilar Cereceda, geographer and director of the Atacama Desert Center at Santiago's Catholic University: "It happens usually around every seven or eight years. It rains three or four or five millimeters and then we have a flowering desert. You can find the slopes of the mountains or the basins full of flowers of all colors, great biodiversity, and lots of insects and birds and animals."

The desert is poised for these moments.

Pilar Cereceda: "The seeds maintain themselves in the surface of the soil — we say *latente* — dormant. And there are bulbs — *bulbos*. Those bulbs are very deep in the soil. They don't have oxygen, they are very dry, no humidity, no water. They can wait 30 years for water."

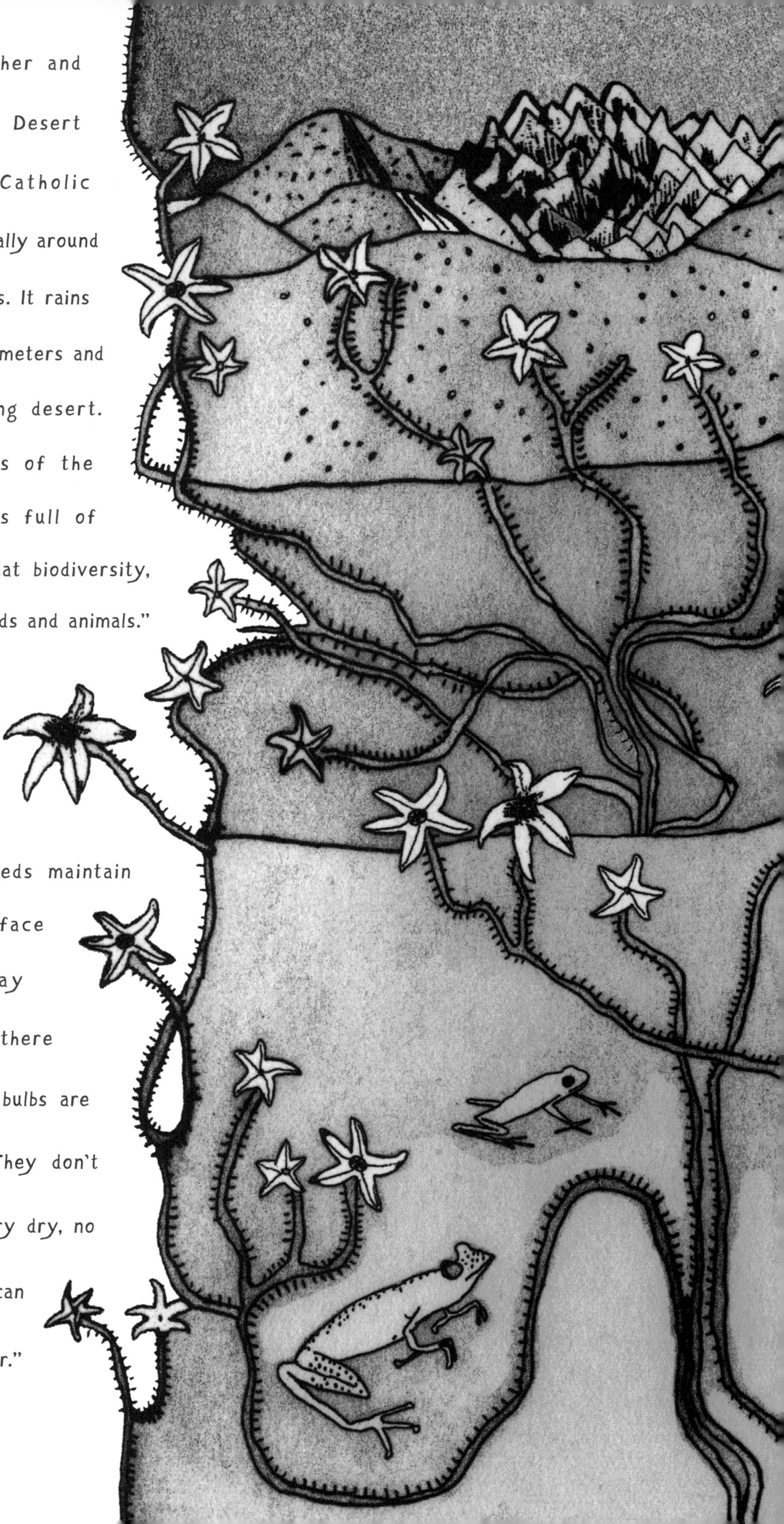

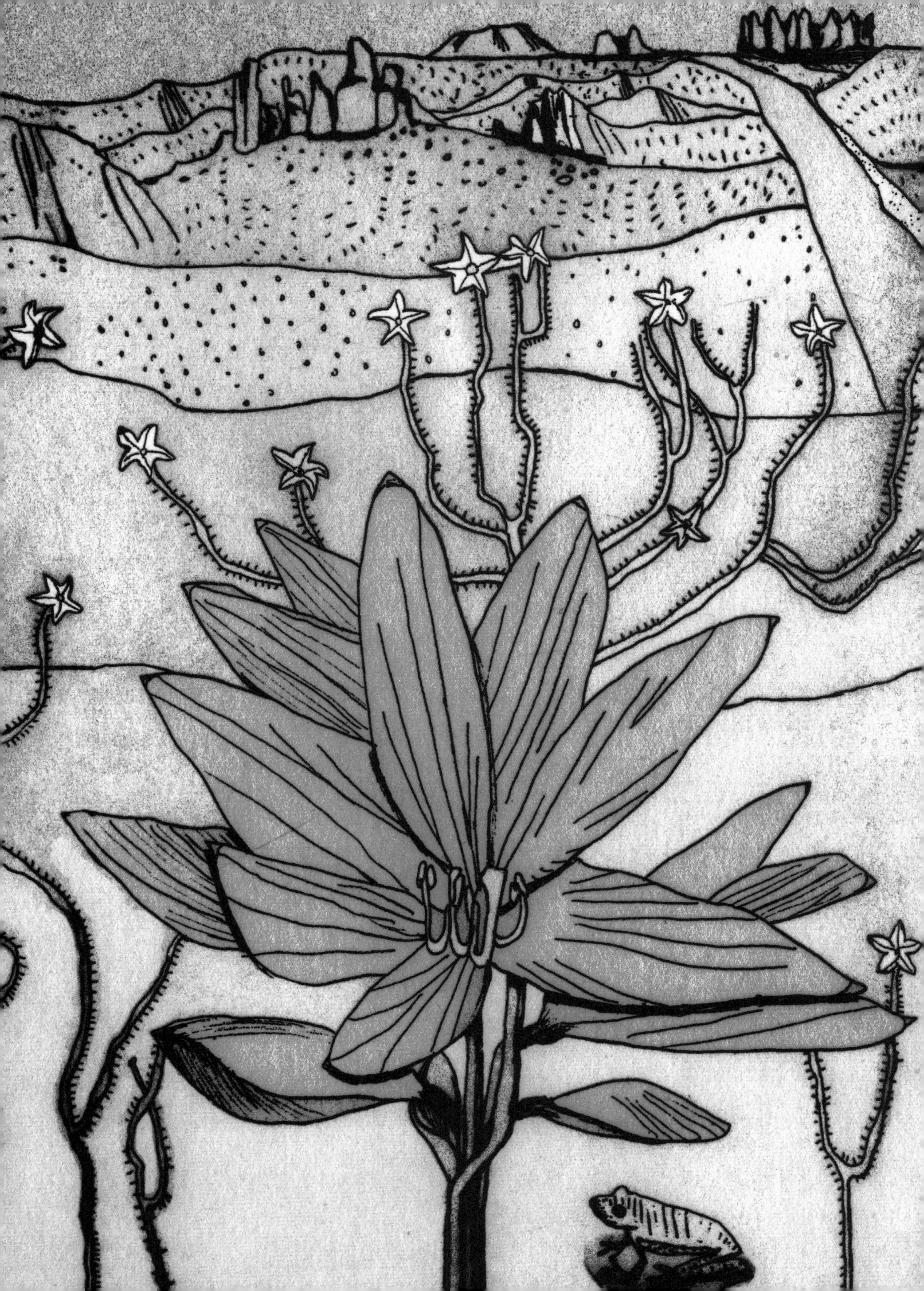

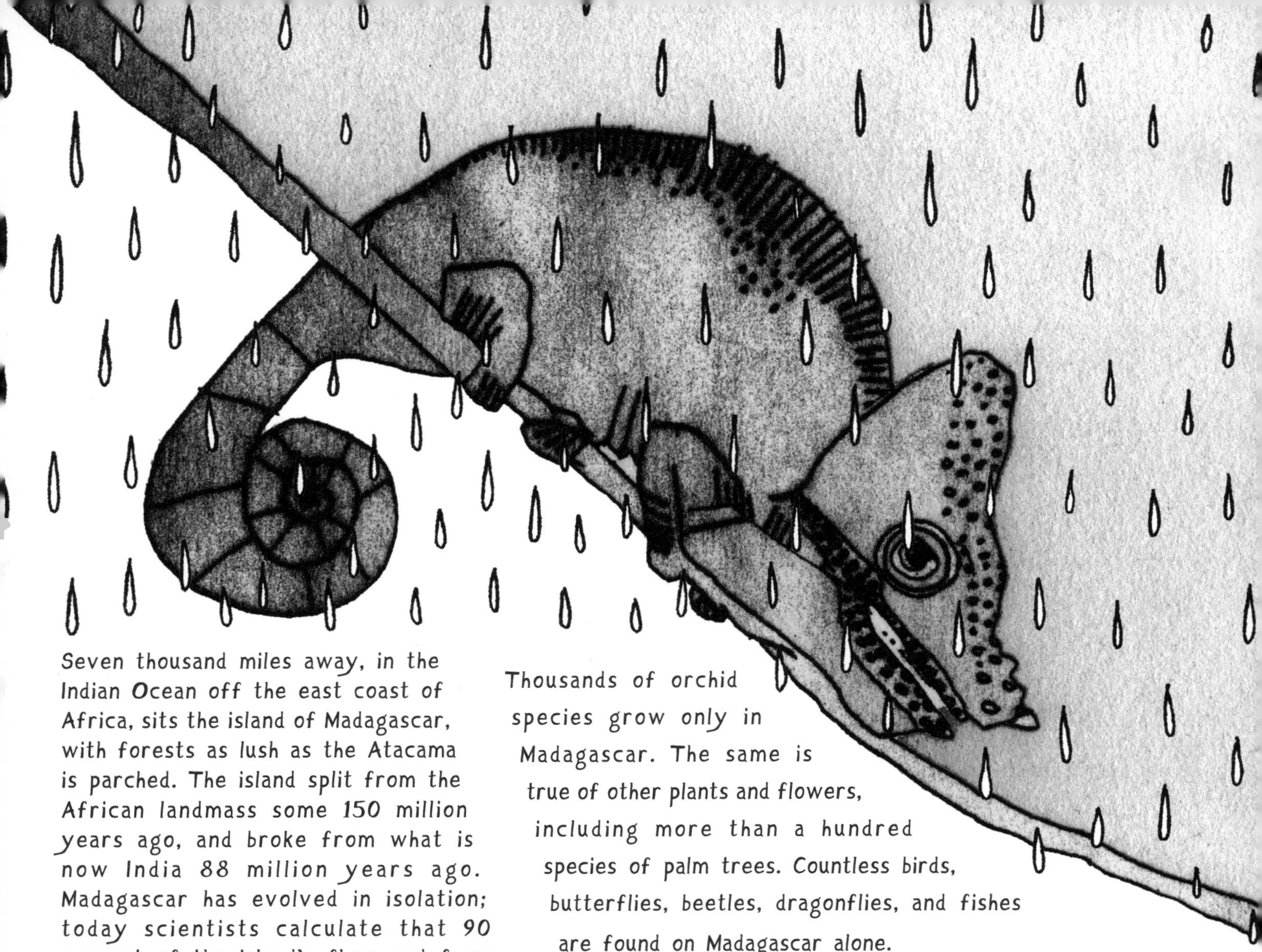

Seven thousand miles away, in the Indian Ocean off the east coast of Africa, sits the island of Madagascar, with forests as lush as the Atacama is parched. The island split from the African landmass some 150 million years ago, and broke from what is now India 88 million years ago. Madagascar has evolved in isolation; today scientists calculate that 90 percent of the island's flora and fauna exist nowhere else on earth.

Thousands of orchid species grow only in Madagascar. The same is true of other plants and flowers, including more than a hundred species of palm trees. Countless birds, butterflies, beetles, dragonflies, and fishes are found on Madagascar alone.

Herpetologist Christopher Raxworthy has been conducting research in Madagascar since 1985. He schedules his fieldwork for the rainy season — between November and April — which is also the hottest part of the year.

Christopher Raxworthy: "That combination of rain and heat is the absolute best time to look for amphibians and reptiles. This is when the animals are active."

A hundred species of lemur roam the island. The black-and-white ruffed lemur has a chin-curtain beard; it looks like C. Everett Koop. The silky sifaka lemur has a small, alien face and smooth white fur. The golden bamboo lemur gnaws on that tree's shoots. The sociable ring-tailed lemur huddles in groups for warmth and company. Some lemurs leap from tree to tree, others shimmy sideways along the ground, forearms in the air, as if making a swashbuckling advance into a fencing match.

Christopher Raxworthy: "At the start of the rainy season, the forest feels very dry. All the leaves are crisp and dried out. When you're walking around you hear little leaves crunching under your feet. If you are turning over logs, they're very dry underneath.

"Most of the animals are hiding. They're inactive. They could be in the soil, or they could be hiding under bark or in tree holes. In some cases, for instance with chameleons, they're sleeping, but high up in the trees. They're doing a thing called estivation, which is a bit like hibernation.

"Then the rains come. The first rains are usually these gentle rainfalls. And then a bit more. And then a bit more. The forest starts to get sort of soggy. It acts like a sponge, absorbing all this water.

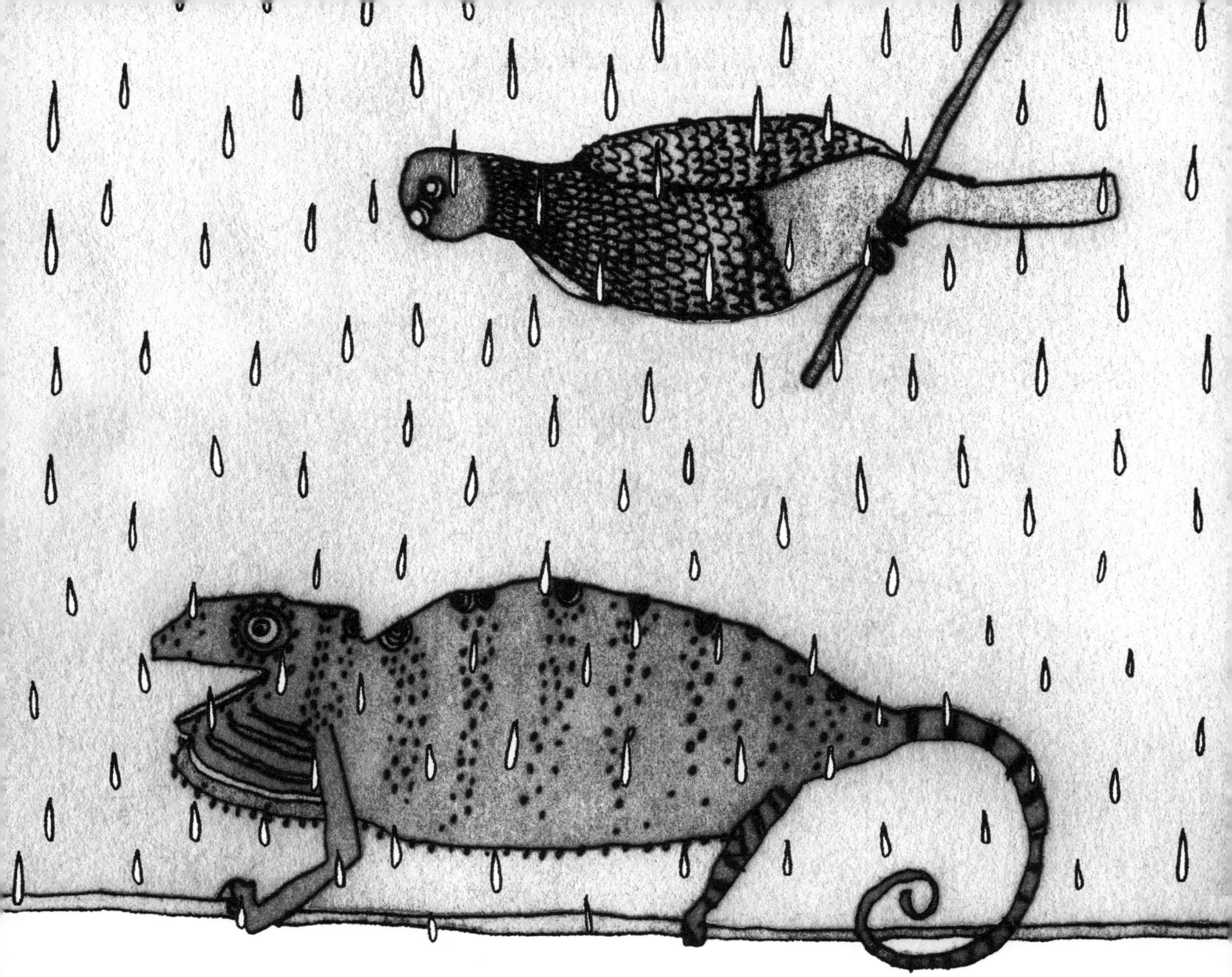

"As soon as things start shifting to this very humid environment, you get these explosive breeders. Some frogs immediately go to temporary pools and the males start calling — many varied calls — and all the females come into the area. In a period of a few days, all the breeding happens for the whole year. There are frogs everywhere.

"You have a type of frog — these long-legged things with a fair amount of webbing on their feet to help them climb, and even, in some cases, glide. Those frogs are usually green-colored, brightly marked. The males have very big vocal sacs that they blow up to vocalize with. You don't normally see them because they spend most of their time high up in the trees and the foliage. But they come down to the streams to breed during the rainy season. That's when you'll see males chorusing by the sides of the rivers and females coming in and reproducing and laying eggs in the water."

Sometimes Madagascar's wet weather convulses into violent cyclones.

Christopher Raxworthy: "You get rivers, waters rising, temporary ponds, you get soggy and wet. And you get even more intense rain if the cyclone comes through."

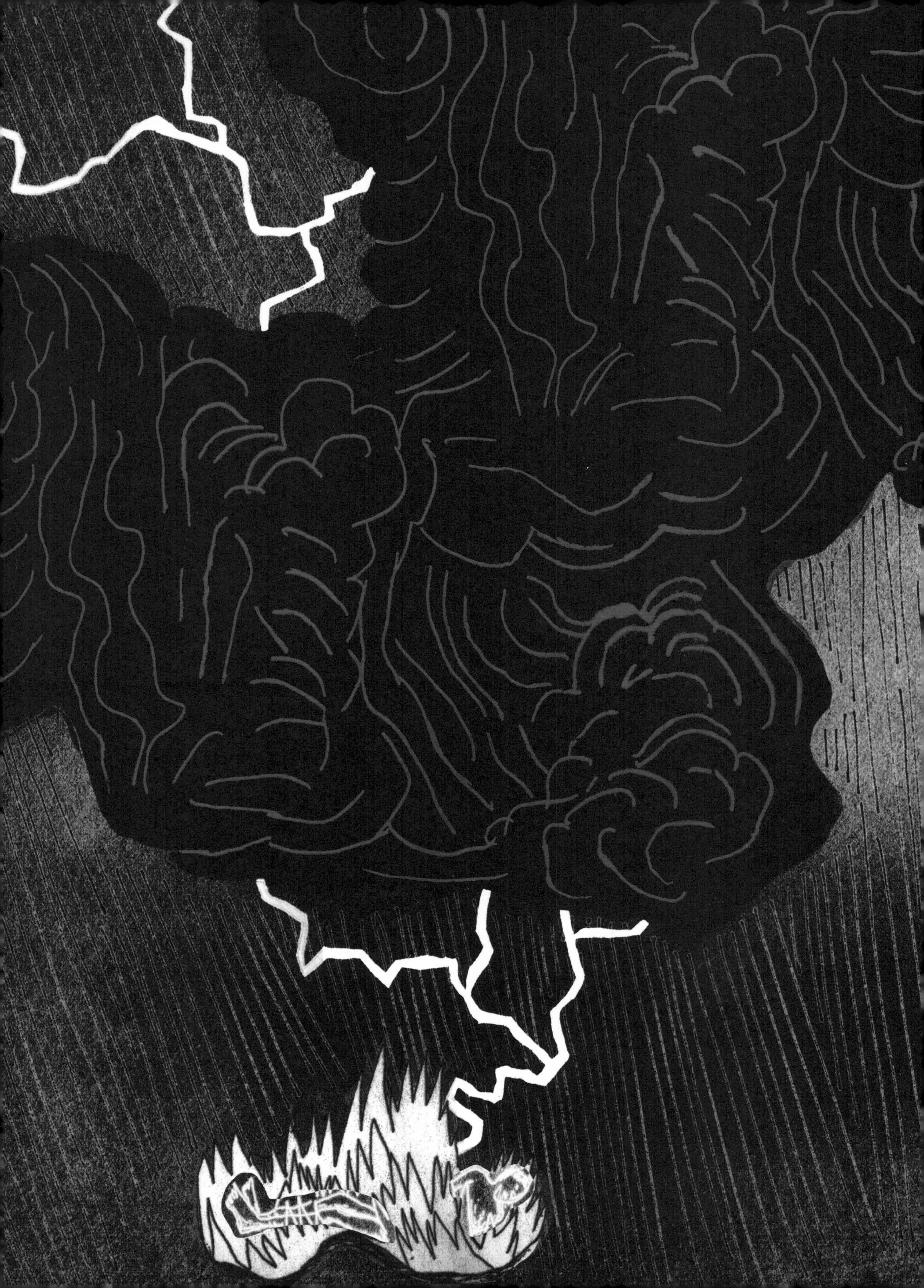

Hot air rises, we've always heard. It is that movement of warm, moist air — or updraft — that forms the massive, mushrooming cumulonimbus clouds of thunderstorms. As the warm air moves upwards, it reaches the subfreezing temperatures of the cloud top and cools, forming water droplets and ice, which then begin to fall through the cloud, merging with other droplets and thus growing in size as they are pulled to earth. The falling drops of water and ice create a downdraft; fierce air currents hurl the ice and water drops together and smash them apart. The friction builds static electricity. Turbulence redistributes charged particles within the cloud: lighter, positively charged ice crystals and water droplets rise, and heavier, negatively charged "soft hail" called graupel gathers towards the base of the cloud. As the cloud drifts, the surface of the ground below becomes positively charged. When the electrical field becomes strong enough, the tension of this charge differential is neutralized by a bolt of lightning.

Lightning — the release of atmospheric energy recognizable by its zigzag slash of light — can travel at speeds over 140,000 miles per hour and reach some 54,000°F — almost five times hotter than the surface of the sun. Some estimate that lightning can attain even higher temperatures — up to 90,000°F.

Crack, clap, rumble, roar. The sounds of thunder are shock waves of rapidly expanding heated air.

Lightning strikes with greatest frequency in tropical and subtropical areas. In the United States, Florida is most susceptible, with a so-called "lightning alley" running from Orlando to Tampa. Around the world, particularly vulnerable areas include Singapore, Malaysia, Pakistan, Nepal, Indonesia, Argentina, Colombia, Paraguay, Brazil, Rwanda, Kenya, Zambia, Nigeria, and Gabon. In 2005, NASA's Lightning Imaging Sensor documented the world's highest concentration of lightning strikes in Kifuka, a mountain village in the Democratic Republic of Congo.

In October 1998, some 300 miles southwest of Kifuka in Congo's Eastern Kasai Province, a bolt of lightning hit a soccer field during a professional match. One entire team was killed; the opposing team was untouched. The strangely lopsided nature of the tragedy caused some to interpret the lightning as suspicious — a supernatural act of sabotage.

Human beings have long invested lightning with symbolic meaning. Unlike snow or heat or fog, which blanket an area and affect the local population more or less uniformly, lightning seems to finger its targets: one person alone in a field, or, in the case of the Congolese soccer match, a select few gathered among many. The ancient Greeks believed that Zeus, king of the Gods, leveled his enemies with a thunderbolt. In Norse cosmology, hot-tempered Thor ruled the heavens, rumbling across the sky in a chariot pulled by goats that shot lightning from their hooves.

The "eccentricities of its operation" encouraged "a strong argument in favour of a diabolical origin of the thunderbolt," according to Andrew Dickson White, 19th century historian and co-founder of Cornell University. In the 13th century, Cistercian monk Caesarius of Heisterbach Abbey recounted the tale of a priest from Trèves. As a storm raged, the priest went to ring the bell of his church — a widely practiced technique for repelling lightning. Instead of foiling a strike, he was hit. According to Caesarius, the man's "sins were revealed by the course of the lightning, for it tore his clothes from him and consumed certain parts of his body, showing that the sins for which he was punished were vanity and unchastity."

Benjamin Franklin invented the lightning rod in 1752. The simple technology offered new protection, but some religious leaders of the period objected to having lightning rods installed on their church steeples, railing against the blasphemy of redirecting a striking bolt and thwarting God's will.

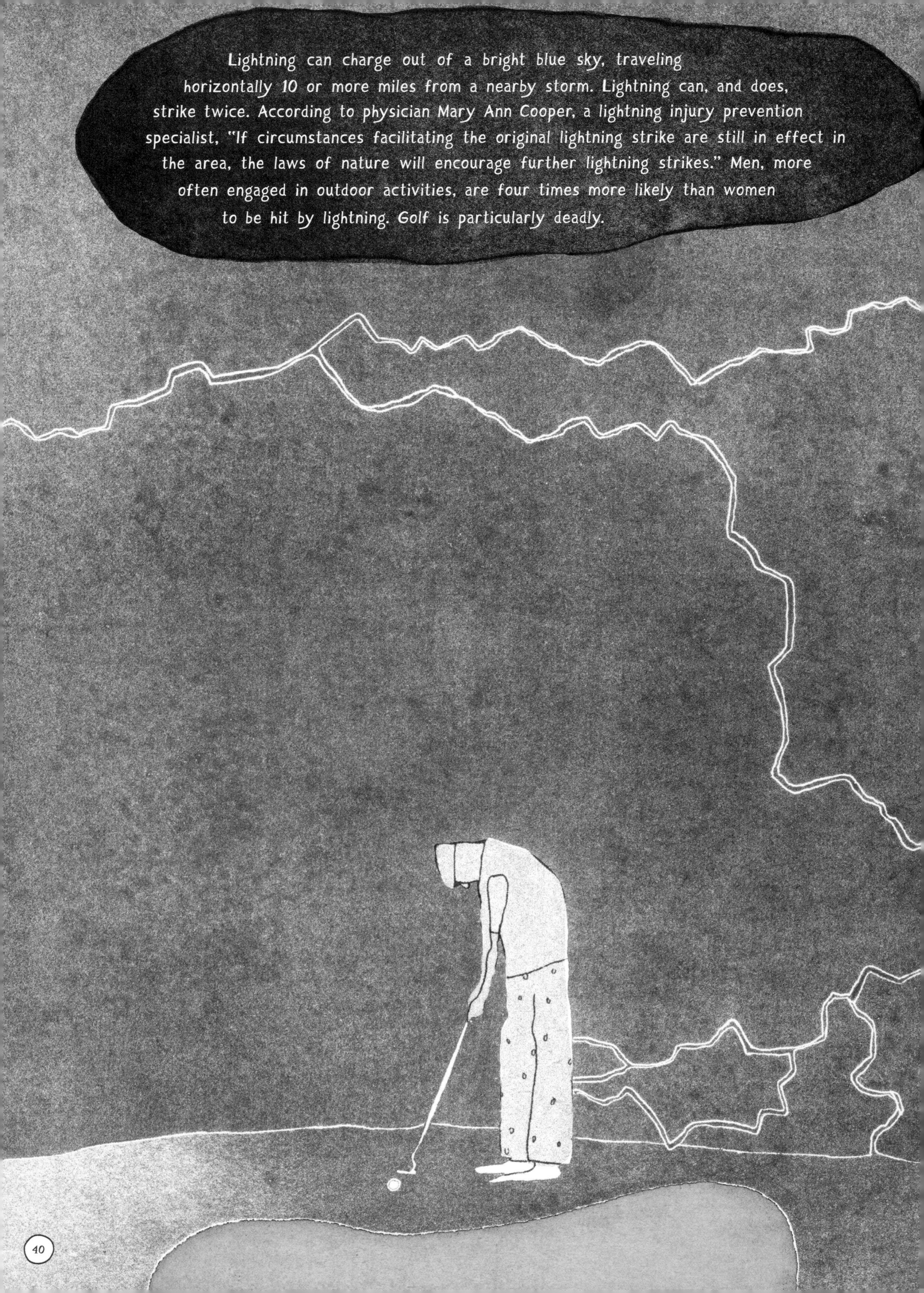
Lightning can charge out of a bright blue sky, traveling horizontally 10 or more miles from a nearby storm. Lightning can, and does, strike twice. According to physician Mary Ann Cooper, a lightning injury prevention specialist, "If circumstances facilitating the original lightning strike are still in effect in the area, the laws of nature will encourage further lightning strikes." Men, more often engaged in outdoor activities, are four times more likely than women to be hit by lightning. Golf is particularly deadly.

Steve Marshburn, with his wife, Joyce, is the founder of Lightning Strike and Electric Shock Survivors International Inc.

Steve Marshburn: "In 1969, I was struck by lightning. I was 25. It was a beautiful day. I was sitting on a teller stool inside the bank where I was employed. One bolt of lightning strayed from a storm 12 miles away and hit our drive-through window speaker.

"Back in '69 there was no direct deposit. The bank cashed your check. So we had two or three long lines flowing outside the bank. Everybody that was inside the bank saw it happen.

"The lightning went straight into my back. My feet were on the metal rungs of the stool; lightning exited one leg. I had a metal teller stamp in the other hand — I was stamping a deposit — and the lightning exited that hand.

"I thought I would never get home to my wife. I thought I would never see my child that was to be born the next month. I could hear, but I couldn't speak. I felt like the left side of my brain had blown out. I know now it had been scorched. My head was in intense pain. My back, it felt like it had been split with a machete."

A person hit by lightning may not show any visible sign of injury. In other cases, a person's skin may appear tattooed with a network of spidery branches called "Lichtenberg figures," also known as "lightning flowers": filigree bruises that mark the path of the electrical energy. In developed countries, 90 percent of people struck by lightning survive. The other 10 percent are likely to die, not from burns as is often assumed, but from cardiac arrest.

Mary Ann Cooper: "A person may have rainwater, or sweat, on them. What's going to happen to the water? It turns to steam. If you have enclosed shoes — you've got on your sport socks and your Nikes — and you've been in the rain and the socks are soaked, or you're running and sweating and whatever, you've got a lot of vapor there. You've got what's called a vapor explosion. Water expands 500 times the size when it turns to steam, it blows shoes off, explodes them off from the inside. You'll see socks melted to the inside of the shoe.

"If you take, say, an undershirt, and you hold it up to the light, and you kind of stretch it, you see light coming through and you'll see a lot of fuzz. When lightning hits it, all that fuzz burns off, leaving almost a skeleton of the fabric. Sometimes you'll see little cindering marks, like you had a Fourth of July sparkler and held it too close to the clothing."

In the volumes of stories published by Marshburn's Lightning Strike and Electric Shock Survivors International, survivors describe the moment they were hit. Carroll Dawson, former assistant coach of the Houston Rockets, was playing golf in 1990 when his club caught a streamer. "I felt lit up like a Christmas tree." Dawson lost his eyesight. Steven Melvin remembers "a loud sizzle like a steak on a grill. Then a bright flash." W.J. Cichanski "saw nothing, heard nothing…felt nothing" when he and a fellow non-commissioned officer were struck by lightning in 1945 in Mineral Wells, Texas. But a witness reported seeing a "halo of fire" around their heads and their clothes bursting into flames. Cichanski spent four months in the hospital and is unsure which of his subsequent ailments (hearing loss, arthritis, insomnia) might be due to the strike. The other officer was killed.

Electricity and heat-shock find idiosyncratic pathways through each body; the brain, heart, and other organs are all vulnerable. Victims may experience seizures, deafness, blindness, chest pains, nausea, headaches, confusion, or amnesia. Some patients have described dramatic weight loss, tingling hands, muscle twitches, loss of temperature sensation, extreme thirst, temporary paralysis, and moments of clinical death.

Steve Marshburn: "Usually it takes a while for certain problems to manifest."

Forensic pathologist Dr. Ryan Blumenthal of South Africa's University of Pretoria compares some lightning injuries to those caused by the shock wave of a bomb detonation: broken bones, ruptured eardrums, torn and tattered clothing, metal jewelry or clothing fasteners melting into flesh. Victims may also be impaled by shrapnel-like debris scattered by the strike's impact.

Steve Marshburn: "Why I was spared I don't know. There was a purpose. Maybe mine was to begin the organization. I don't know. I do not get nervous during a storm. I enjoy watching lightning. I really do. Some of our members — it takes them into a seizure. Others it petrifies. And many of us enjoy watching the lightning. It's beautiful to me. I'll tell you what I think: it is the marvelous handiwork of the Lord."

The relative rarity of getting hit by lightning means that survival can be accompanied by a peculiar sense of chosenness. Some victims describe becoming a kind of celebrity — or a sideshow attraction. After he was struck in 1994, Garry Joseph Shaw, an employee of the Ford Motor Company, spent twenty-four days in the burn unit at the Metropolitan Hospital in Detroit. "They sent heart specialists, psychiatrists, neurologists, and plastic surgeons in to see me every day. It's a wonder they did not dissect me to look any closer....I had become an exhibit at the zoo."

Before being hit, Laurie Procter-Williams had struggled with drug addiction and other problems. She believes lightning turned her life around. "Many friends and family feel that another tragic obstacle had entered my life...[but] touching the face of death enlivened me."

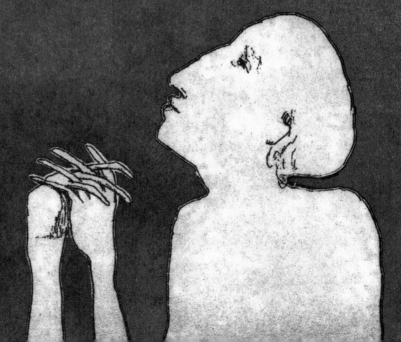

Christopher Raxworthy: "Towards the end of the rainy season, you can see some signs that the forest is starting to dry out. Certain species will start to disappear. The leaf chameleons, the dwarf chameleons, by this time of the year there are some sites where it's already too late, you've missed them. You're going to have to wait till the next year. They've retreated back underground again. As things dry out, more and more species become inactive again.

"The females are no longer reproducing. You start coming back to the ponds, you only see one or two frogs still hanging around. It's just as though they've vanished. And then for the rest of the year, if you walk around the forest, during the night or during the day, some species can be very difficult to spot. They disperse back out into the forest, and some of them are living in the ground, or in the leaf litter.

"You don't get a situation where there's nothing going on, but it's usually restricted to a few things; ground-dwelling lizards, and skinks, some of the more common species tend to be active throughout the year. The whole forest goes into this state of sleepiness till the next rainy season."

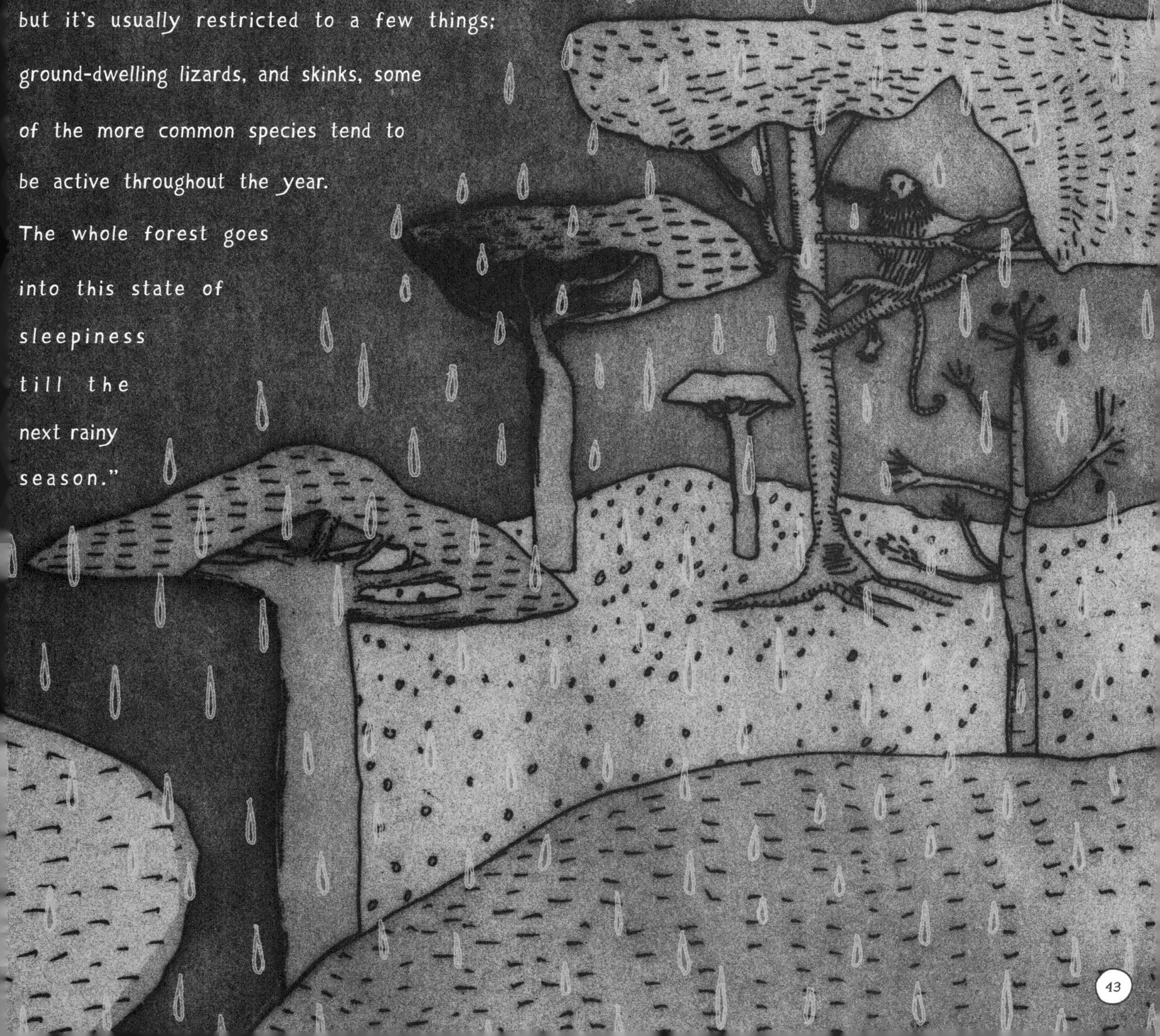

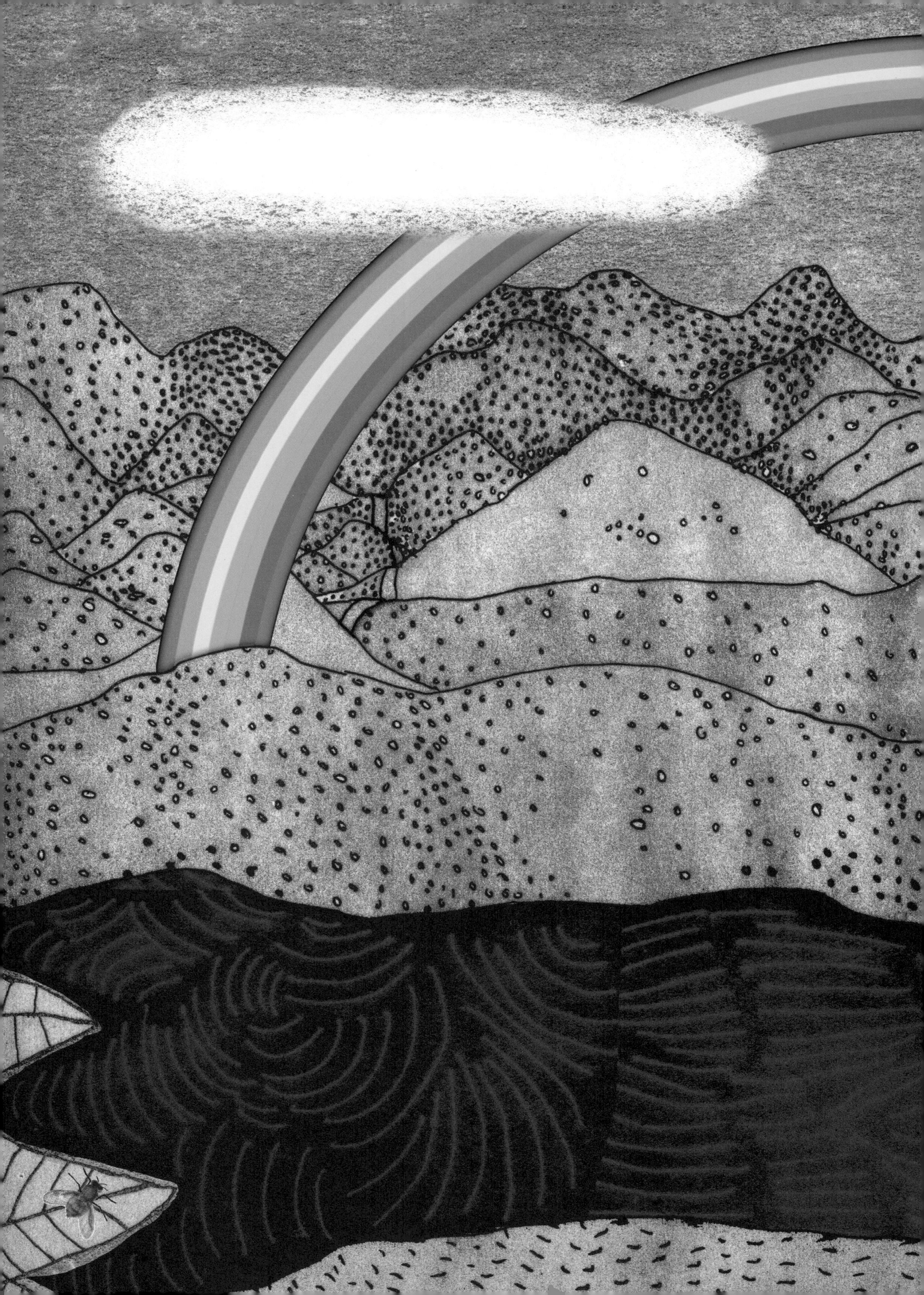

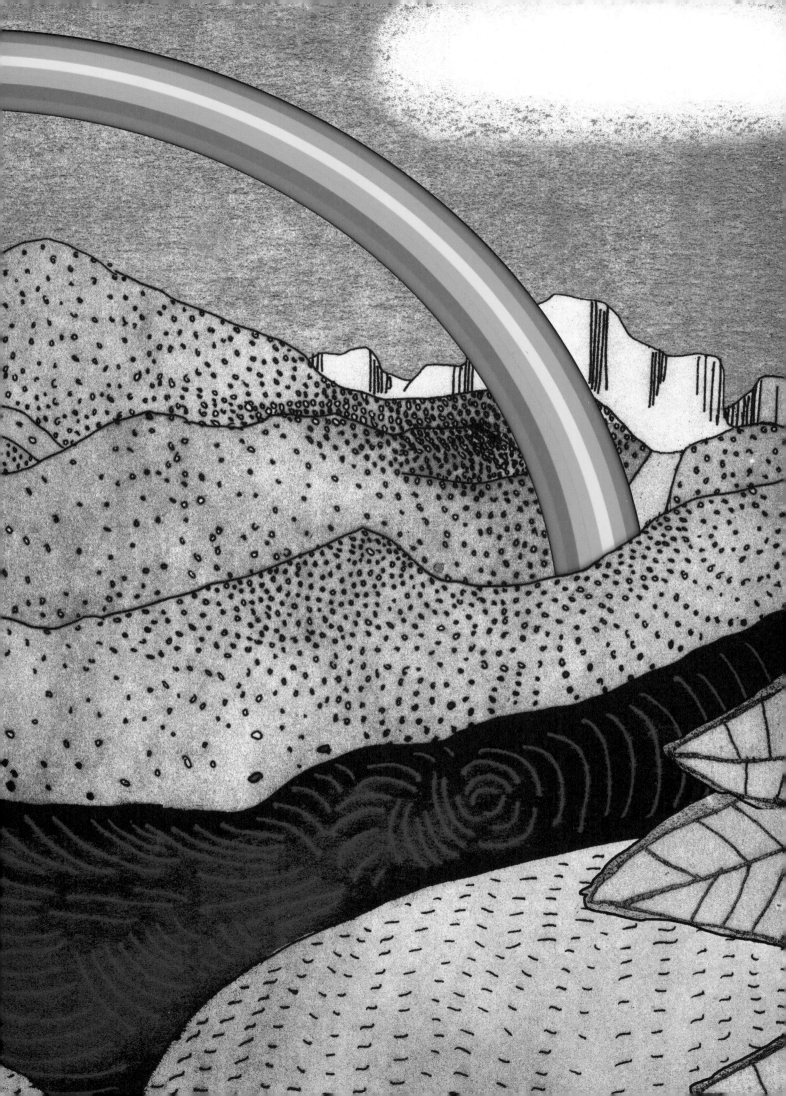

CHAPTER 4:

FOG

Twenty-one years old and married just shy of two years, Diana, Princess of Wales, accompanied her husband, Charles, on an official trip to Canada. On the 11th day of their trip, Charles and Diana visited Newfoundland to mark the 400th anniversary of the island becoming a British colony. Diana wore a turquoise feathered hat and matching suit, which hung loosely over her body from broad shoulder pads. It was 1983.

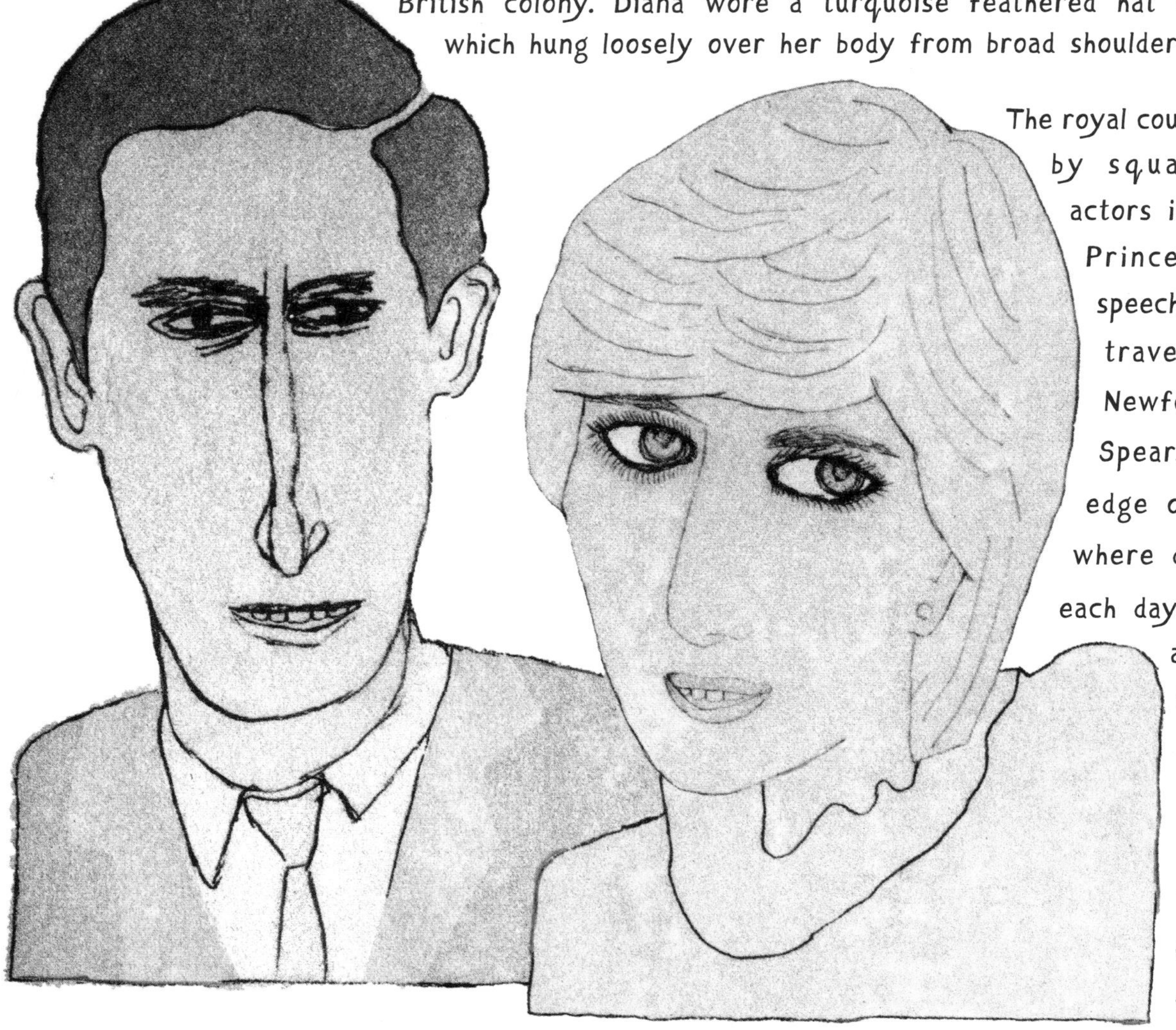

The royal couple was entertained by square dancers and actors in Elizabethan garb. Prince Charles made a speech. Charles and Diana traveled to the tip of Newfoundland—to Cape Spear, the easternmost edge of North America, where dawn arrives first each day on the continent, and where there is a lighthouse. Television news teams were on hand to capture the scene, but the cameras caught few clear pictures.

Gerry Cantwell was the lighthouse keeper at the time, the sixth generation in his family to occupy the position. "By the time Charles and Diana got there, at about 11 o'clock in the morning, you actually couldn't see your hand in front of your face with the fog."

It was not the first time a prince had gotten lost in the fog at Cape Spear.

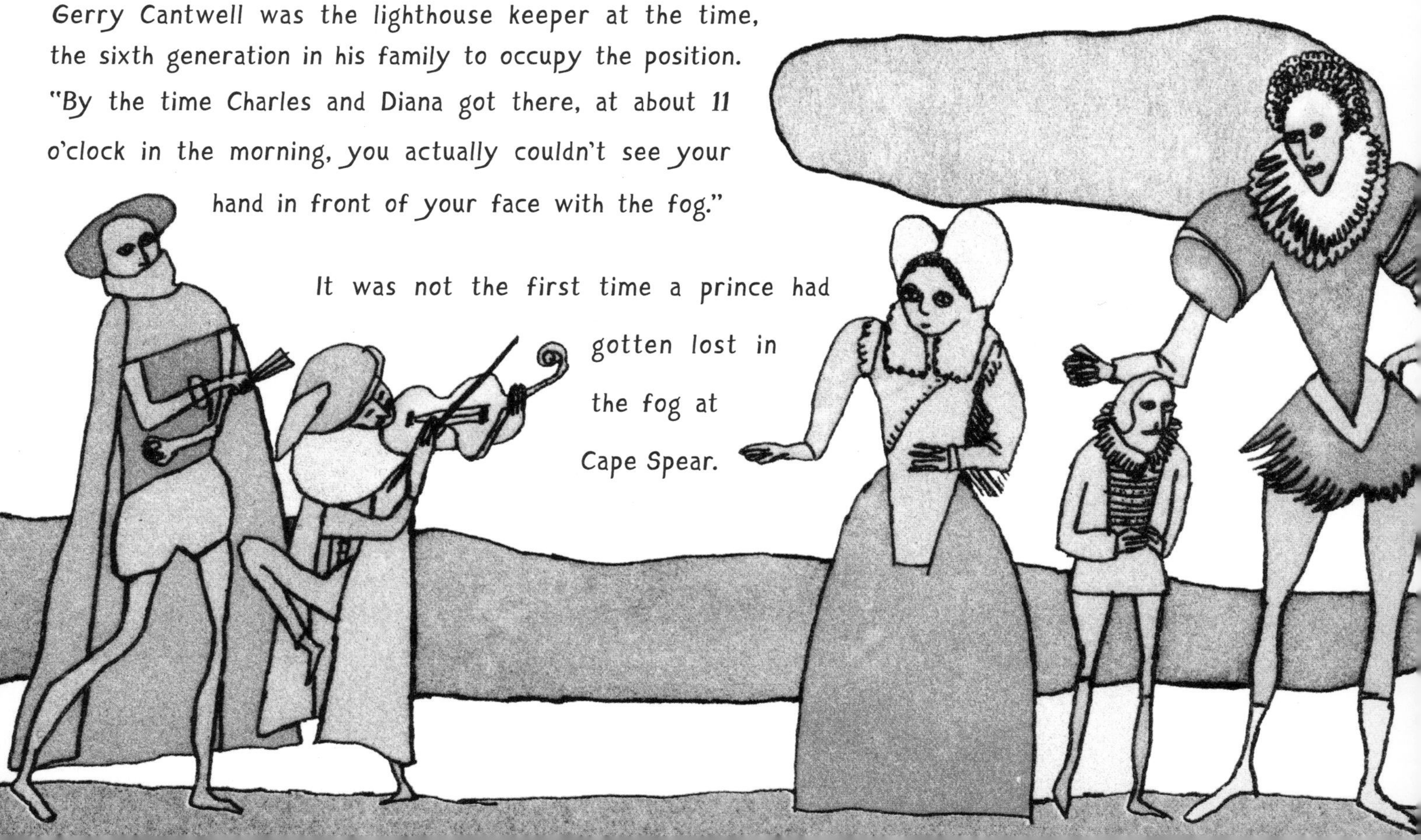

Gerry Cantwell: "It happened in 1845. The prince of the Netherlands was coming over here on a state visit, it was called. Of course the only way to Newfoundland was by boat. Everyone was dressed up in their finery, standing around. No prince. There was thick fog, of course, off Cape Spear at the time—and the captain of the prince's boat couldn't find where in the name of goodness St. John's Harbor was in order to get into it. So they sent out all the harbor pilots to see if they could find him, because they knew he was out there.

"My great-great-great-grandfather, James Cantwell, was a master mariner and a harbor pilot. He had a longboat, and a crew rowing it. These harbor pilots were so used to coming in and out of that fog and finding the hole in the cliff where the entrance to St. John's Harbor is. When a boat was lost, they would row out until they found the boat, and the person who would bring it in would get paid by the ship line that he brought it in for. That's how they made their money, you see?

"That's what James Cantwell did. He went out, he found the prince's boat, he boarded the boat. He took the wheel of the vessel, and he just took it in, very easily, in to St. John's Harbor in this blanket of fog. You couldn't see the bow of the boat because the fog was so heavy.

"Coming through St. John's Harbor it was a beautiful, sunshiny day, which happens all the time—out off the water it's cold and foggy, and inside the city, it's warm and the sun is shining.

"We were an outpost of England, and a special visitor, if you will, coming to the island, he could grant a wish to someone that he thought deserved one, you see?

"The prince asked James Cantwell what would he like 'cause he was so amazed that this person had found the mouth of the harbor, 'cause even the captain of his boat didn't know where it was. 'What would you like for this feat of heroism?'

"They were building the lighthouse at Cape Spear at that time. And James Cantwell said, 'If the job at Cape Spear ever becomes open, I'd like to be light keeper, the tender of the light.'

"This was written down. And in 1846 he got the job at Cape Spear. The wish was granted."

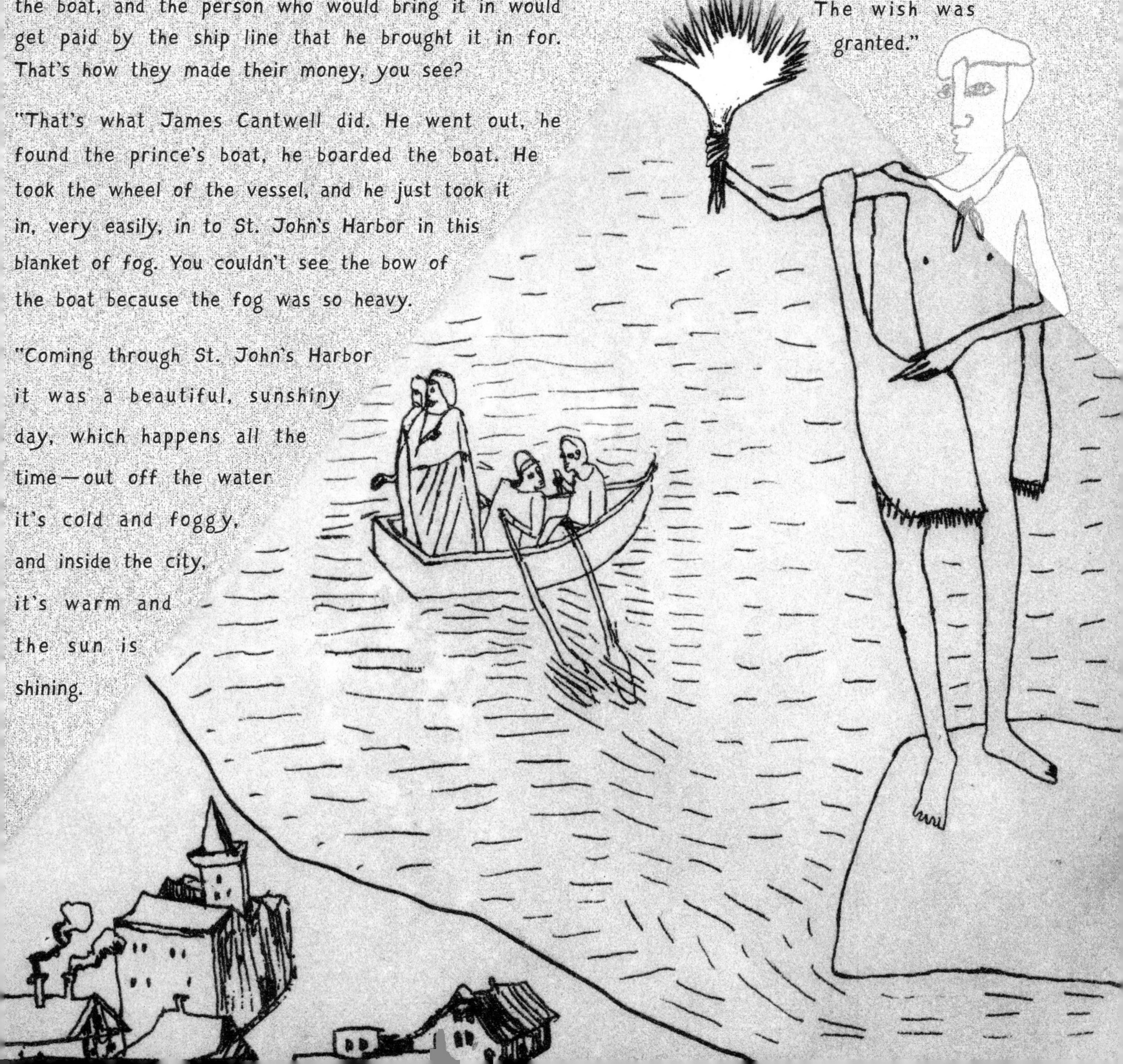

Fog is a cloud near the ground. Moisture from the air condenses into tiny water droplets, or ice crystals, that hover over earth's surface. Fog can form in a variety of circumstances. San Francisco's "advection fog" appears when warm, humid air blows across the cold ocean water. "Radiation fog" typically forms overnight: earth that has absorbed daytime sunshine radiates that heat upwards. As the ground cools, so does the still, humid air above, which condenses into fog. Wintertime fogs in California's Central Valley are radiation fogs. Fog can form astride a mountain if moist air is blown upwards, cooling to its dew point as it rises. This is called "upslope fog," seen, for instance, on the eastern slopes of the Rocky Mountains.

During London's infamous "pea soupers," soot pumped out by domestic coal fires and factories combined with moist air from prevailing eastward winds to form dense smog that could hover for days. An 1889 *New York Times* article described the effects: "When one of the thick, yellowish compounds known as a 'pea-soup' fog falls on London it makes day darker than night; it arrests all traffic, obliterates all landmarks, and as Mrs. Browning says, it looks 'as if a sponge has wiped out London.' The city is transformed into ghostland. Individuals move about like huge phantoms, and all sound is deadened to a muffled, ghostlike tone....You can see it—see little else, in fact; you can feel its wet, clammy touch; you can smell its unhallowed flavor, and it is thick enough for you to taste. There is no keeping it out, however much you may wrap and muffle yourself up. It will creep down your neck."

When a pea souper descended, crime surged. "The fog helped thieves to make a good haul from a London department store," wrote the *Times* in 1959. "With visibility near zero late last night, they blew open two safes and escaped with an estimated £20,000 ($56,000)." In the obscurity, trucks were known to veer into the Thames. Trains collided with people, automobiles, and other trains. Once, an airplane overshot the runway and exploded. ("In the interval between the crash and the fire, rescue crews could not locate the wreckage.") On foggy days ambulances had to be accompanied by guides on foot. Schools closed. At least one funeral procession was scattered and lost. Even indoors, visibility could diminish to mere feet. Theatergoers were unable to see actors onstage.

When an area of high pressure moved over London during a cold snap in early December 1952, the warm air trapped the cooler air beneath it. Fighting the chill, people lit more coal fires, further polluting the atmosphere. In nearly windless conditions, the smog thickened and settled onto the low-lying city. Thousands died, mostly young children and the elderly who succumbed to respiratory illness aggravated by the foul air.

Four years later, the British Parliament approved the Clean Air Act of 1956. A second antipollution law was ratified in 1968. Pea soupers passed into London history and lore. Today it is in other cities — Beijing, Shanghai, Tehran, New Delhi, Los Angeles — that weather, topography, and pollution combine to drape the landscape in thick and grimy air.

Newfoundland fog feels as pure and fresh as city smog is filthy. It is formed by the mixing of two ocean currents: cold air off the Labrador Current cools the warm, humid air of the Gulf Stream, condensing it into tiny droplets of fog.

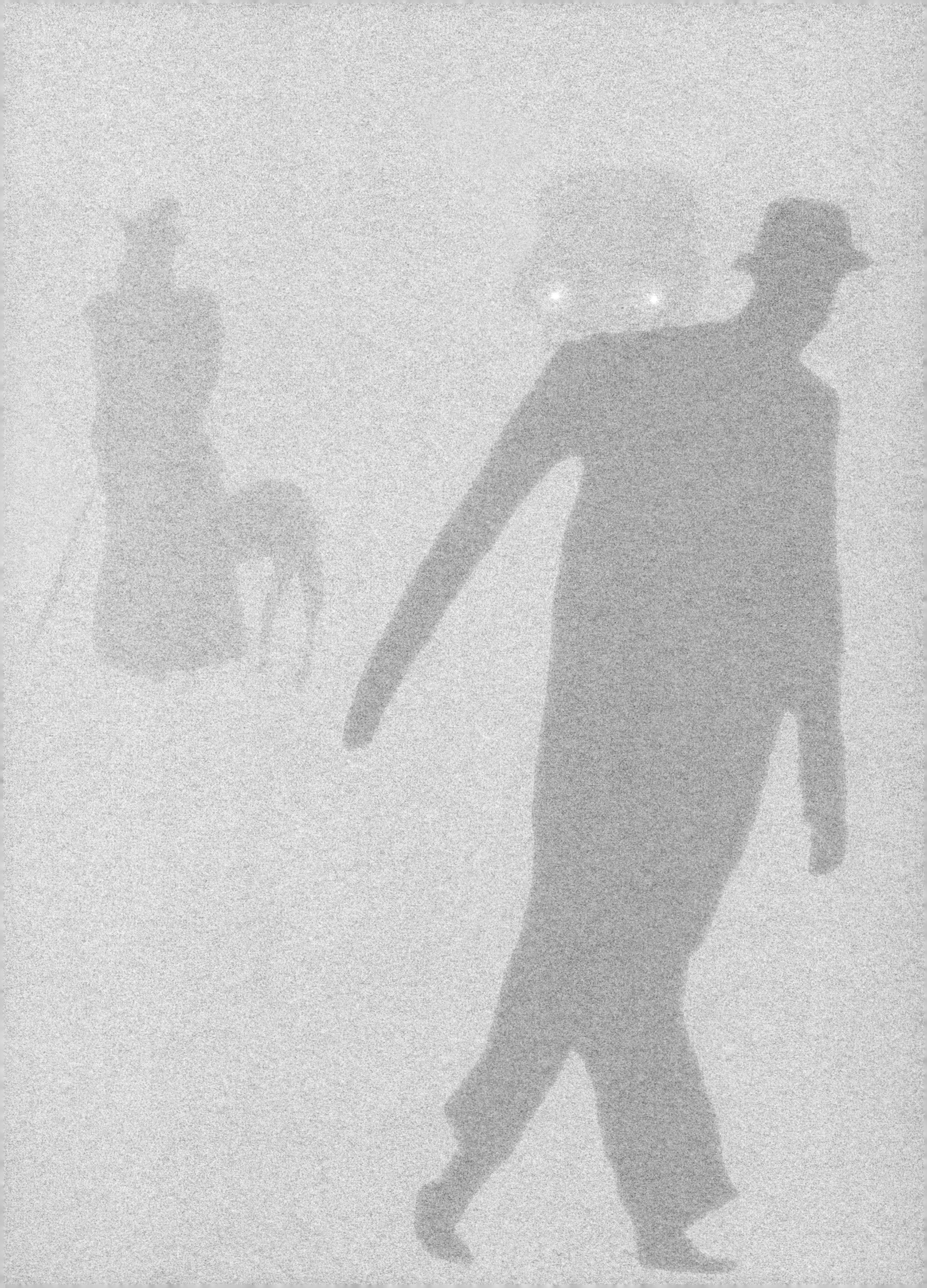

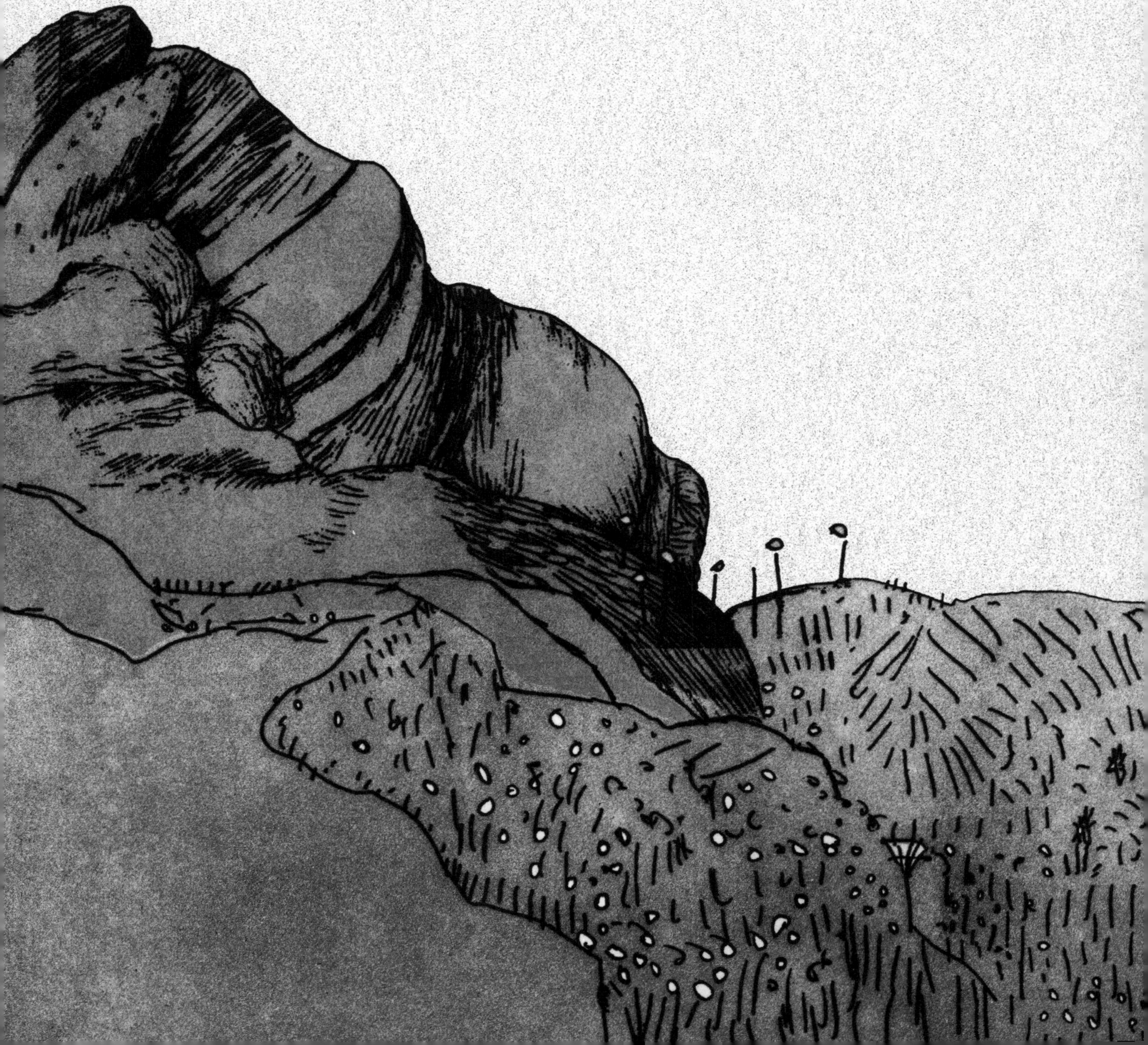

Gerry Cantwell: "It don't drop like a blind, but it's pretty close."

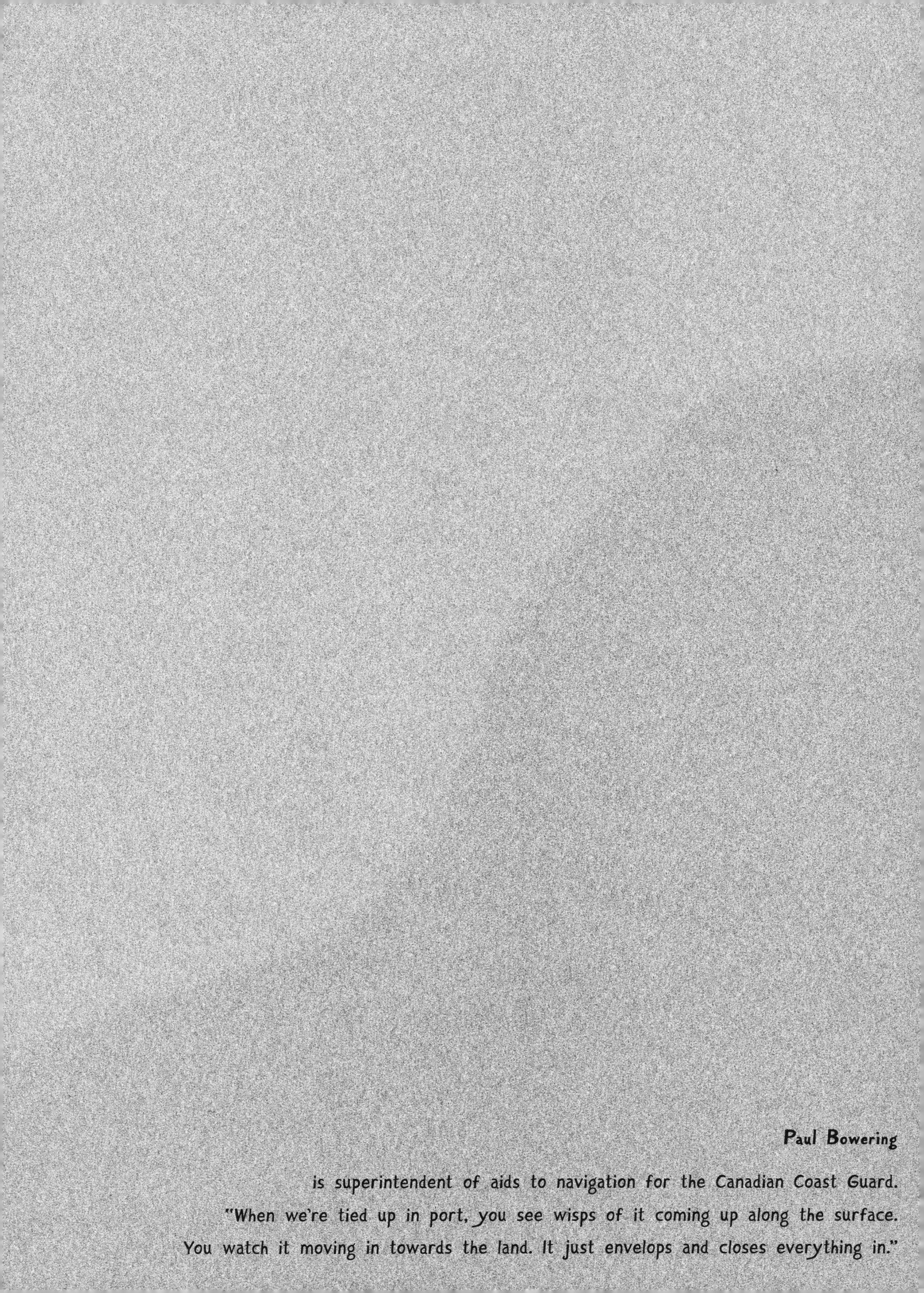

Paul Bowering

is superintendent of aids to navigation for the Canadian Coast Guard. "When we're tied up in port, you see wisps of it coming up along the surface. You watch it moving in towards the land. It just envelops and closes everything in."

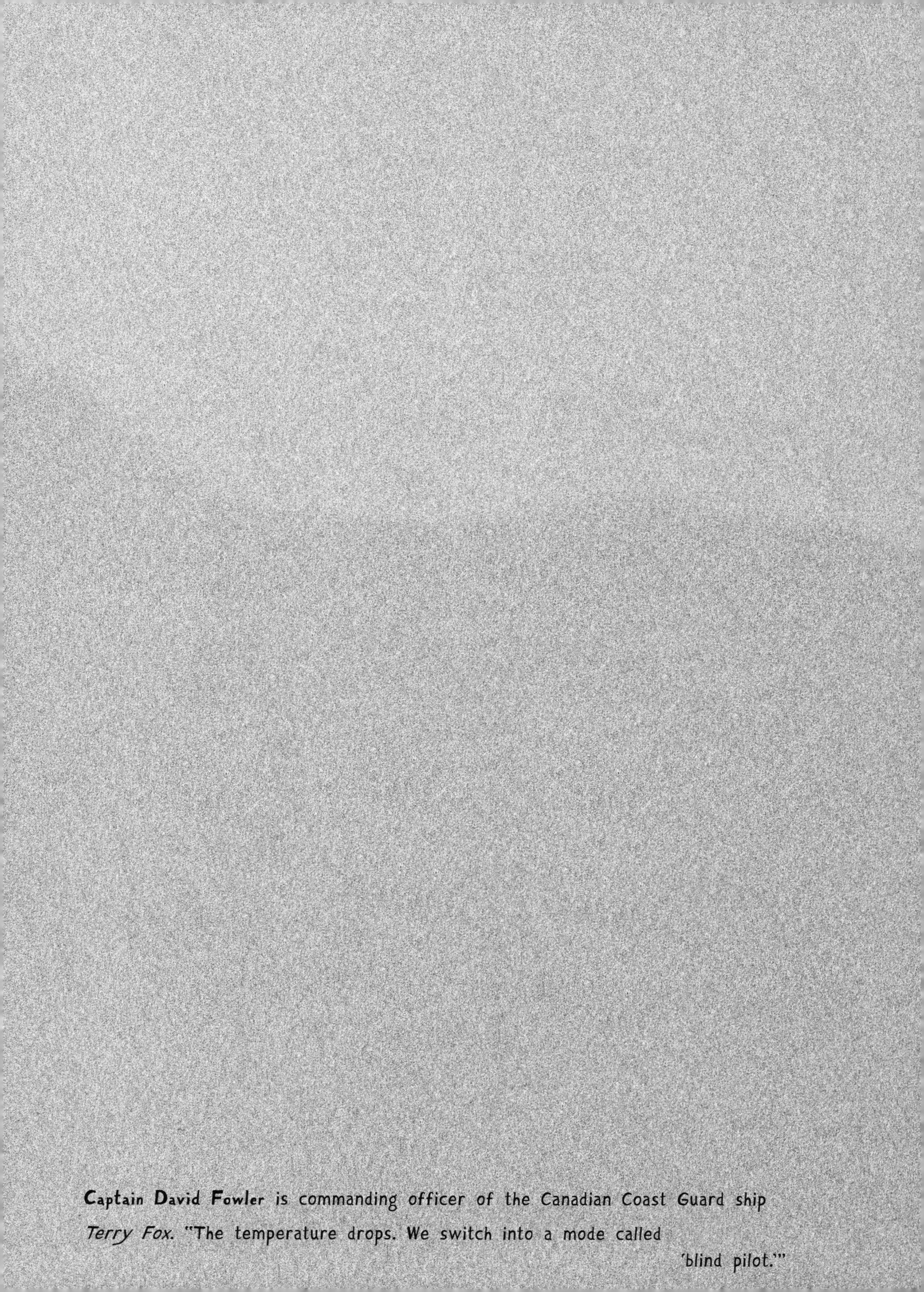

Captain David Fowler is commanding officer of the Canadian Coast Guard ship *Terry Fox*. "The temperature drops. We switch into a mode called 'blind pilot.'"

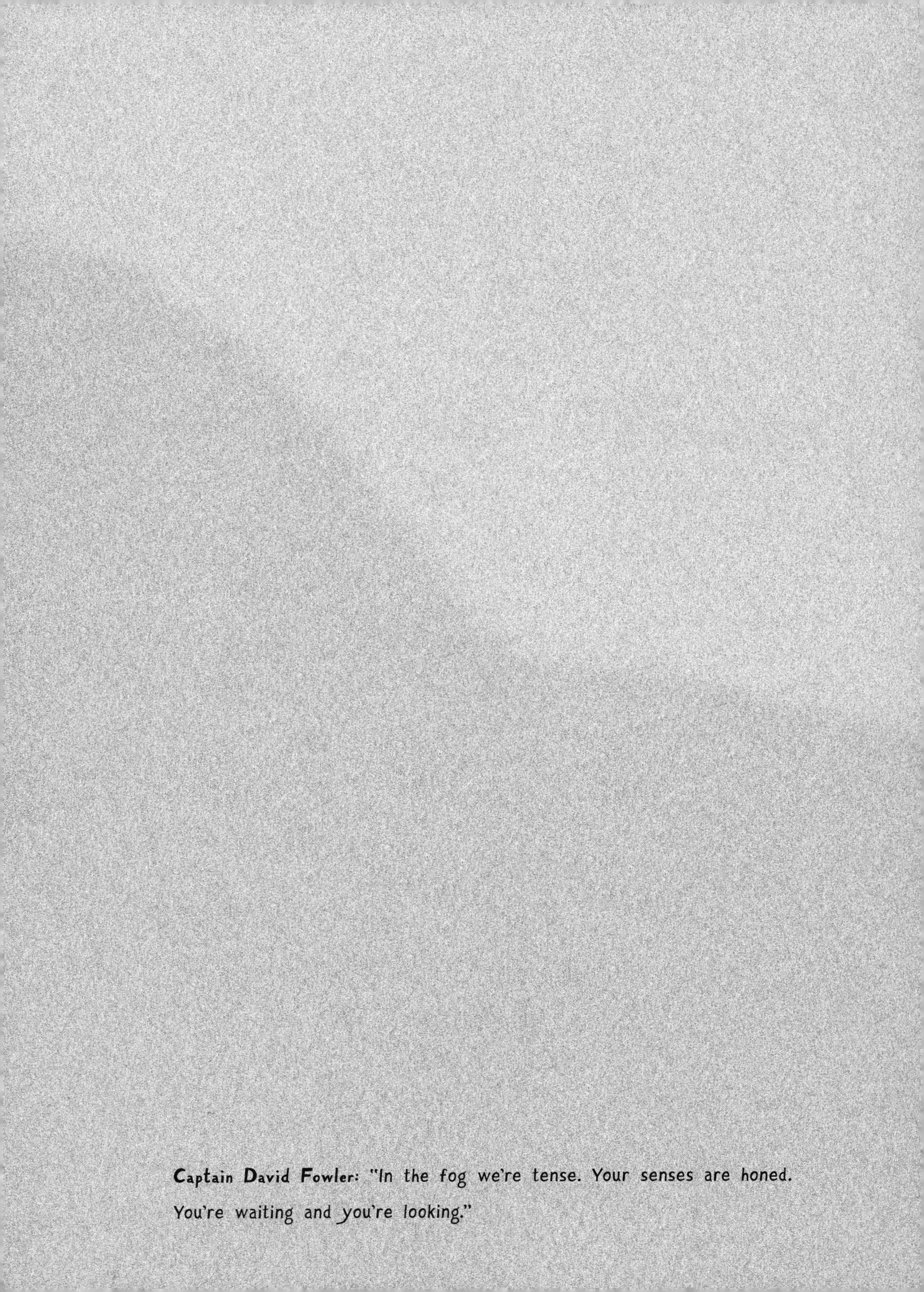

Captain David Fowler: "In the fog we're tense. Your senses are honed. You're waiting and you're looking."

Paul Bowering: "You get

pretty itchy after a while,

and you start

doubting

yourself."

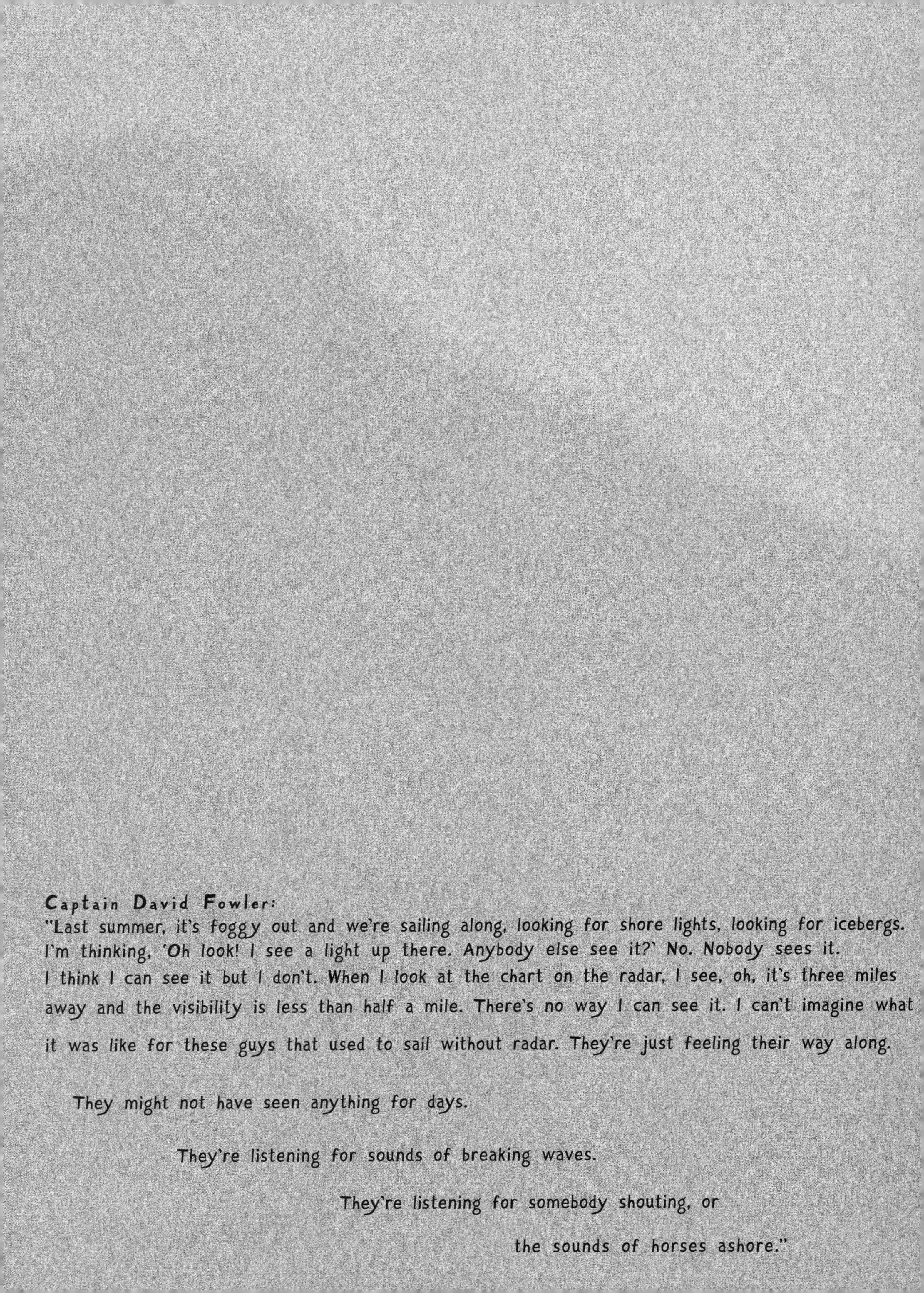

Captain David Fowler:

"Last summer, it's foggy out and we're sailing along, looking for shore lights, looking for icebergs. I'm thinking, 'Oh look! I see a light up there. Anybody else see it?' No. Nobody sees it. I think I can see it but I don't. When I look at the chart on the radar, I see, oh, it's three miles away and the visibility is less than half a mile. There's no way I can see it. I can't imagine what it was like for these guys that used to sail without radar. They're just feeling their way along.

They might not have seen anything for days.

They're listening for sounds of breaking waves.

They're listening for somebody shouting, or

the sounds of horses ashore."

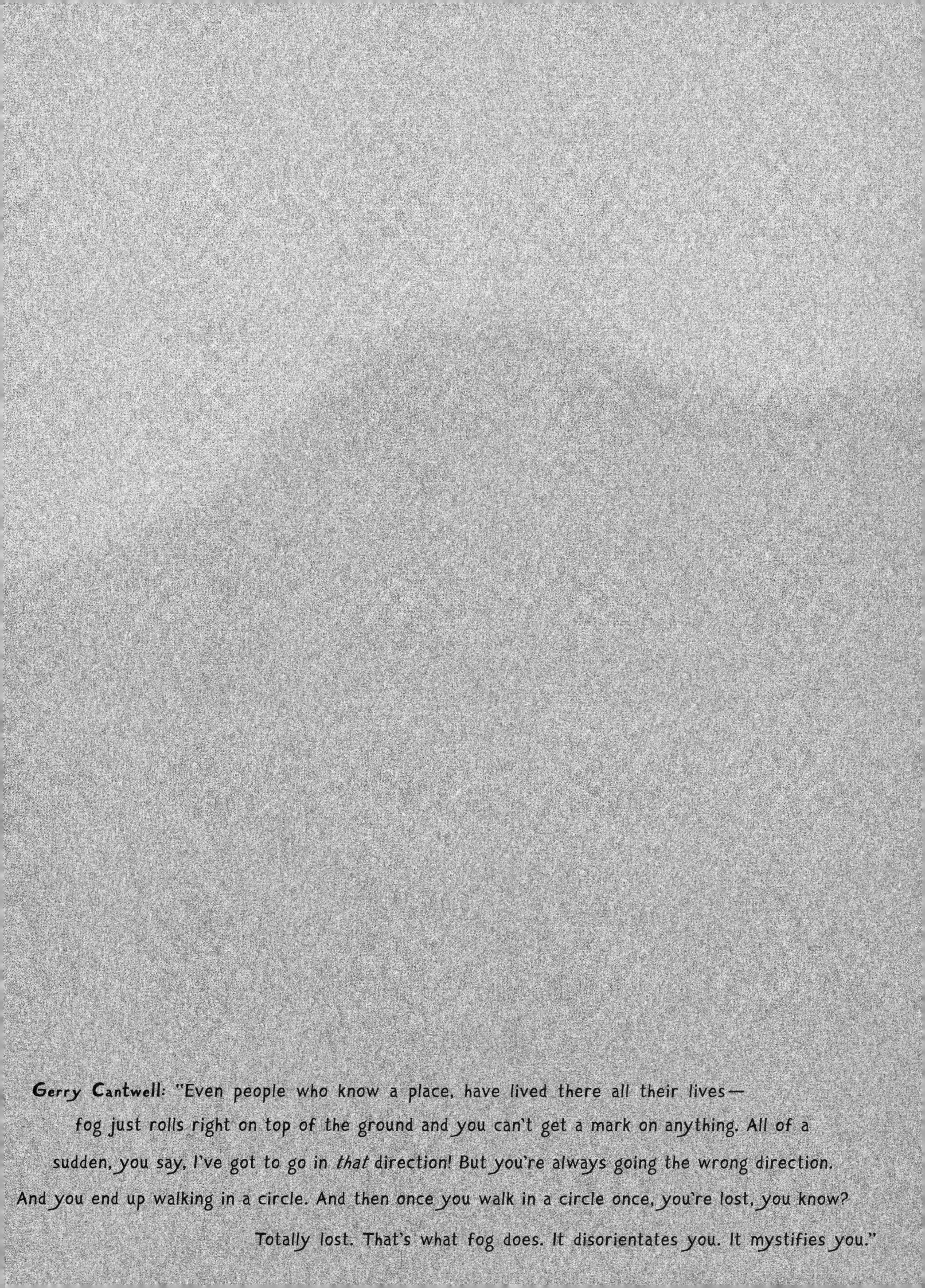

Gerry Cantwell: "Even people who know a place, have lived there all their lives—fog just rolls right on top of the ground and you can't get a mark on anything. All of a sudden, you say, I've got to go in *that* direction! But you're always going the wrong direction. And you end up walking in a circle. And then once you walk in a circle once, you're lost, you know? Totally lost. That's what fog does. It disorientates you. It mystifies you."

When the steamship *Arctic* set sail for the first time in 1850, an article appeared in *The New York Times*. "It is almost too late in the day to go into ecstasies of admiration over any new wonder of the ocean....The *Arctic* [is a] 'floating palace' and all that, but then everyone knows that already. The meanest man that floats now-a-days must be cradled in a palace as a matter of course, else he grumbles." Still, the *Times* noted that the *Arctic* had the latest in engine technology, and an elaborate system of braces and rivets for strength. American shipping magnate Edward Knight Collins had produced "the strongest ships ever built," and of them, the *Arctic* was considered the finest. She was also said to be the "fastest steamer afloat." Sixty years before the *Titanic*, the *Arctic*'s luxuries were impressive: steam heating throughout the ship, an elegant dining hall, salons for the ladies and smoking rooms for the men, all bedecked in marble, mirrors, and gold leaf. There was no steerage class on the *Arctic*.

On September 20, 1854, the *Arctic* left Liverpool, bound for New York. The plan was to complete the journey in a swift nine days. A week later, on Wednesday the 27th, the *Arctic* was sailing in the Grand Banks off Newfoundland, with Cape Spear to the north and Cape Race some 60 miles southwest, when fog blanketed the ship. It was noon. Soon a gong would sound announcing lunch. Peter McCabe was a 24-year-old waiter from Dublin preparing for the midday meal. "I was coming up out of the second cabin, bringing glasses for the table....The collision occurred as I was ascending the stairs."

Francis Dorian was third officer of the *Arctic*. "The first I heard was the cry, 'Hard a Starboard.' Then I understood that there was something wrong."

The *Arctic* had collided with another ship, a French iron propeller called the *SS Vesta*. In the fog, each had been invisible to the other until too late.

Francis Dorian: "I went on deck and found the vessels about seven yards apart. I stood watching the *Arctic*, in full expectation that she should yield to her helm. The other vessel struck her abreast the catheads."

James Smith was a first-class passenger. "I stepped out of my stateroom."

Peter McCabe: "The side of the ship was ripped open, so that water came in freely and covered the bed-plates of the engines."

James Smith: "I saw Captain Luce on the paddle-box, giving orders in one way and another, and most of the officers and men running here and there on the deck, getting into an evident state of alarm, without seeming to know what was to be done, or applying their energies to any one thing in particular."

Francis Dorian: "A perfect mania seemed to seize all on board."

There were six lifeboats, enough for about half of the number aboard the *Arctic*. The first was lowered with crew sent out to survey the damage. It disappeared into the fog.

James Smith: "Ladies and children began to collect on deck with anxious and inquiring looks, receiving no hope or consolation. Wife and husband, father and daughter, brother and sister, would weep in each other's embrace, or kneel together imploring Almighty God for help."

Peter McCabe: "I...saw four men fall from the propeller under the paddle-wheels of the *Arctic*. They could not be rescued and were never heard of afterwards."

After ship engineers and most of the officers slipped away on four of the *Arctic*'s five lifeboats, people scrambled to get onto the remaining vessel. Others saw the effort was futile and began tearing doors and planks from the ship to jury-rig rafts. Passengers seized their last moments to choke down booze and lunge at women. Some jumped, some fell into the ocean.

James Smith: "The ship began to disappear, stern foremost....I heard the gargling and rushing sound of the water filling her cabins from stern to stem as she went under, taking, I should think, from 30 seconds to a minute in disappearing, with a large number of people still upon her deck."

George H. Burns was the express messenger for Adams Express freight and cargo transport company. "I heard one wild yell, still ringing in my ears, and saw the *Arctic* and the struggling mass rapidly engulfed."

James Carnegan, brother of one of the *Arctic's* furnace stokers: "Ten minutes after the ship went down the [life] boat pulled to the place where the ill-fated vessel had gone down; nothing, however, was encountered but the bodies of several ladies with life-preservers on."

Thomas Stinson, officers' steward: "The head stewardess was particularly recognized, by her dress."

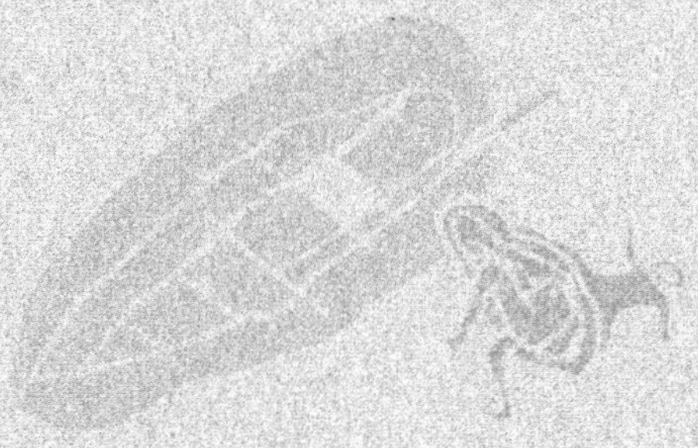

Of the 408 people aboard the *Arctic*, just 86 survived.

One recent July day, Cape Spear was veiled in fog. Every 56 seconds the foghorn bleated long and low, the sound hovering in the air for a moment after its call. If you didn't already know the place, you would have had no idea where the jagged cliffs fall away into the Atlantic, where the road you arrived on had gone. There was no view or landscape to see—just undifferentiated soft and humid whiteness. The fog's density would ebb for a moment, letting blurry shapes emerge, then close in again. When they were visible at all, the buildings at Cape Spear—the 19th century lighthouse and the 1950s

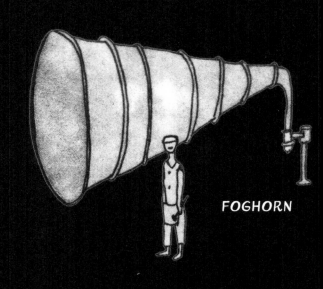

FOGHORN

light tower—were erased to the faintest ghost silhouettes. Only the ground directly underfoot was clear. There, long green grasses were combed this way and that, and in the dewy stems of wildflowers—clover, irises, buttercups, the seed-bearing parachutes of fading dandelions—spitbugs nested in foam.

ARGAND HOLLOW WICK LAMP

On clear days at Cape Spear you can look out over the ocean and crane your neck north towards the nearest landfall: Uummannarsuaq, Greenland. You can watch whales blow, surface, and then plunge back into the water in smooth, languid arcs. To the northwest you can see St. John's harbor nestled behind the V of two sloping hills. The easternmost point in North America used to be marked by a sign, but it blew away and hasn't been replaced.

Aids to navigation—buoys, foghorns, light stations—have long signaled hazards, helping mariners to determine their location and to guide them on a safe course.

Cape Spear's first light was a hand-me-down. It came, in 1836, from the lighthouse on Inchkeith, a Scottish island in the Firth of Forth and a onetime quarantine for sufferers of plague and syphilis. The light consisted of seven burners: cotton wicks that floated in fuel (sperm whale or seal oil). Shiny, copper-lined parabolic disks framed the flames like halos, reflecting and focusing the light into a concentrated beam projected out into space. Mounted on a round metal fixture, the burners rotated to produce Cape Spear's signature pattern: 17 seconds of light, followed by 43 seconds of darkness. Mariners could count the flashes of light and clock the duration of the darkness, identify the light station, and thus gauge their position.

Gerry Cantwell: "This is the oldest lighthouse in Newfoundland. I mean, an actual light*house* now—that means an actual light with a house where people lived, where they were born and where they died. That's why they called it a light*house*. It's got a life all its own. It's got a heart, the light is the heart of it. That's what's beating, you see?"

Electricity came to Cape Spear in 1930. In 1955, the light was moved from the original house where the keeper lived with his family to a nearby tower, an automated mechanism no longer requiring the constant attention of a live-in attendant.

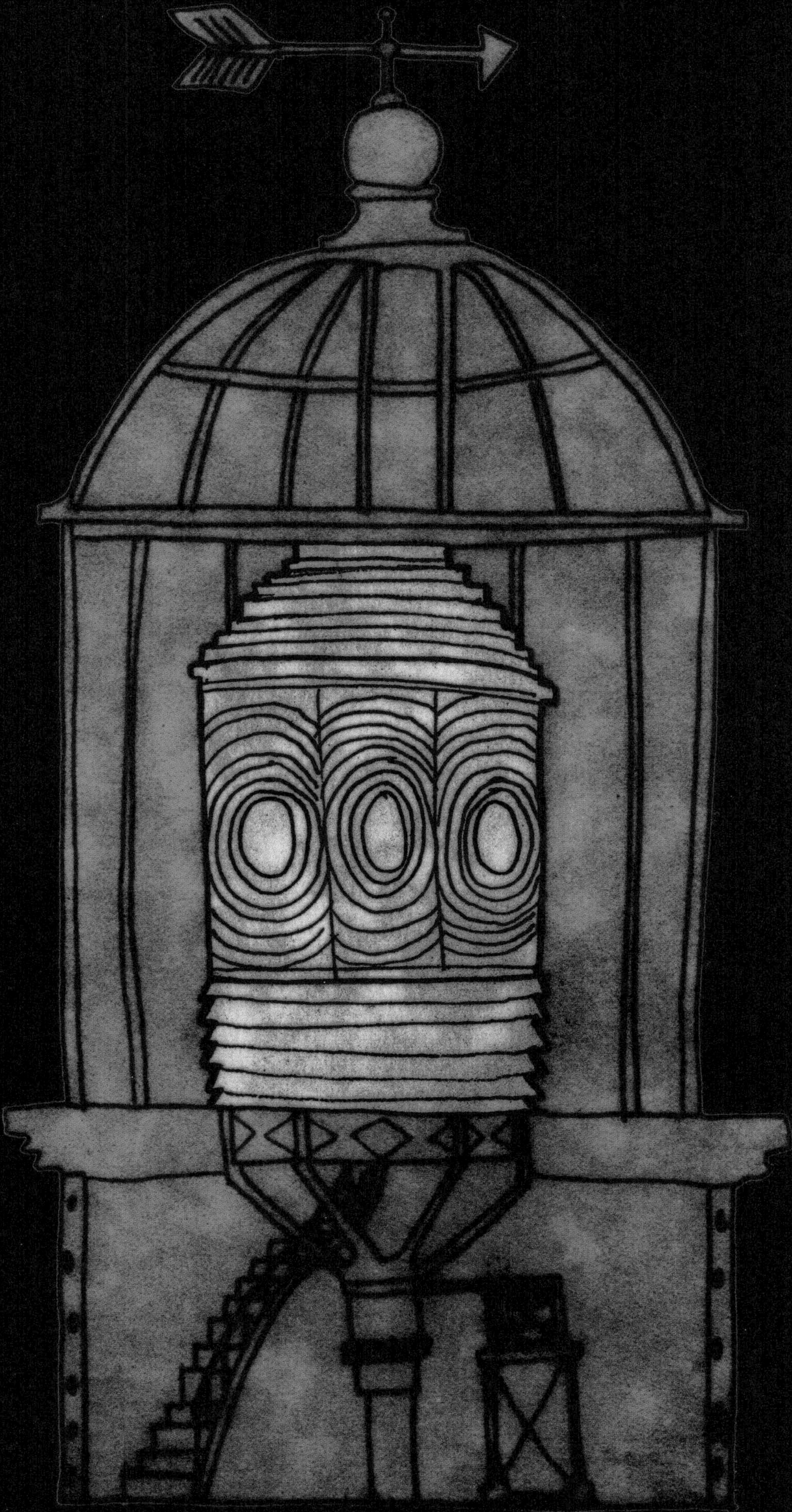

The light in the tower today is emerald green prismatic glass with three "bullseye" lenses. It is shaped like an enormous football and perches in the lantern room at the top of the light tower. Its beam is visible for up to 20 miles. The signature is now three flashes in a 15 second rotation.

One, two, three, *flash*.

One, two, three, *flash*.

One, two, three, four, five, six, seven, eight, nine, *flash*.

The original lighthouse is preserved as a museum, restored to how it looked in 1839. The master bedroom is made up with a white quilt and blue patterned wallpaper. A mustache curler rests on the side table.

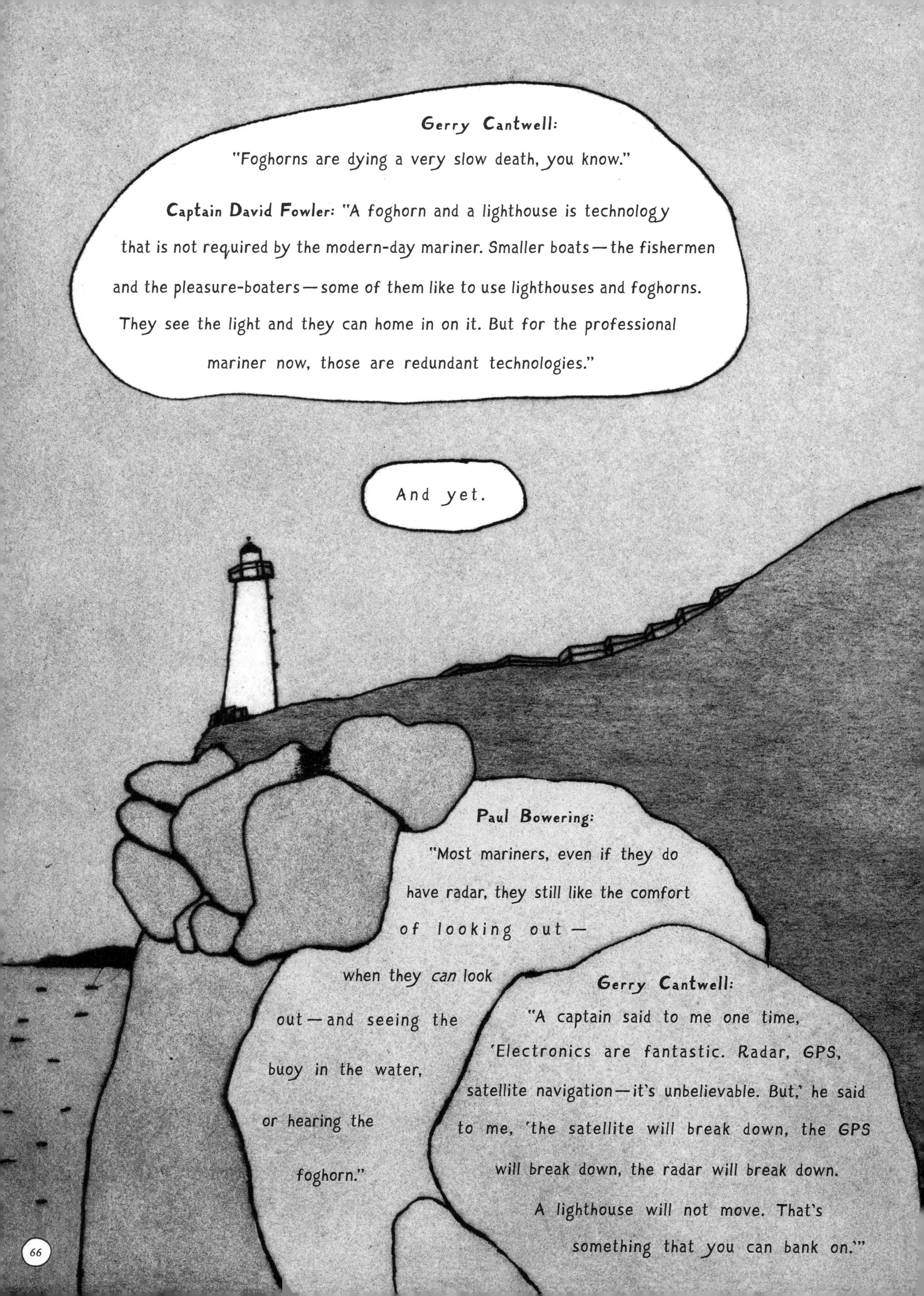
Gerry Cantwell:
"Foghorns are dying a very slow death, you know."
Captain David Fowler: "A foghorn and a lighthouse is technology that is not required by the modern-day mariner. Smaller boats—the fishermen and the pleasure-boaters—some of them like to use lighthouses and foghorns. They see the light and they can home in on it. But for the professional mariner now, those are redundant technologies."
And yet.
Paul Bowering:
"Most mariners, even if they do have radar, they still like the comfort of looking out—when they *can* look out—and seeing the buoy in the water, or hearing the foghorn."
Gerry Cantwell:
"A captain said to me one time, 'Electronics are fantastic. Radar, GPS, satellite navigation—it's unbelievable. But,' he said to me, 'the satellite will break down, the GPS will break down, the radar will break down. A lighthouse will not move. That's something that you can bank on.'"

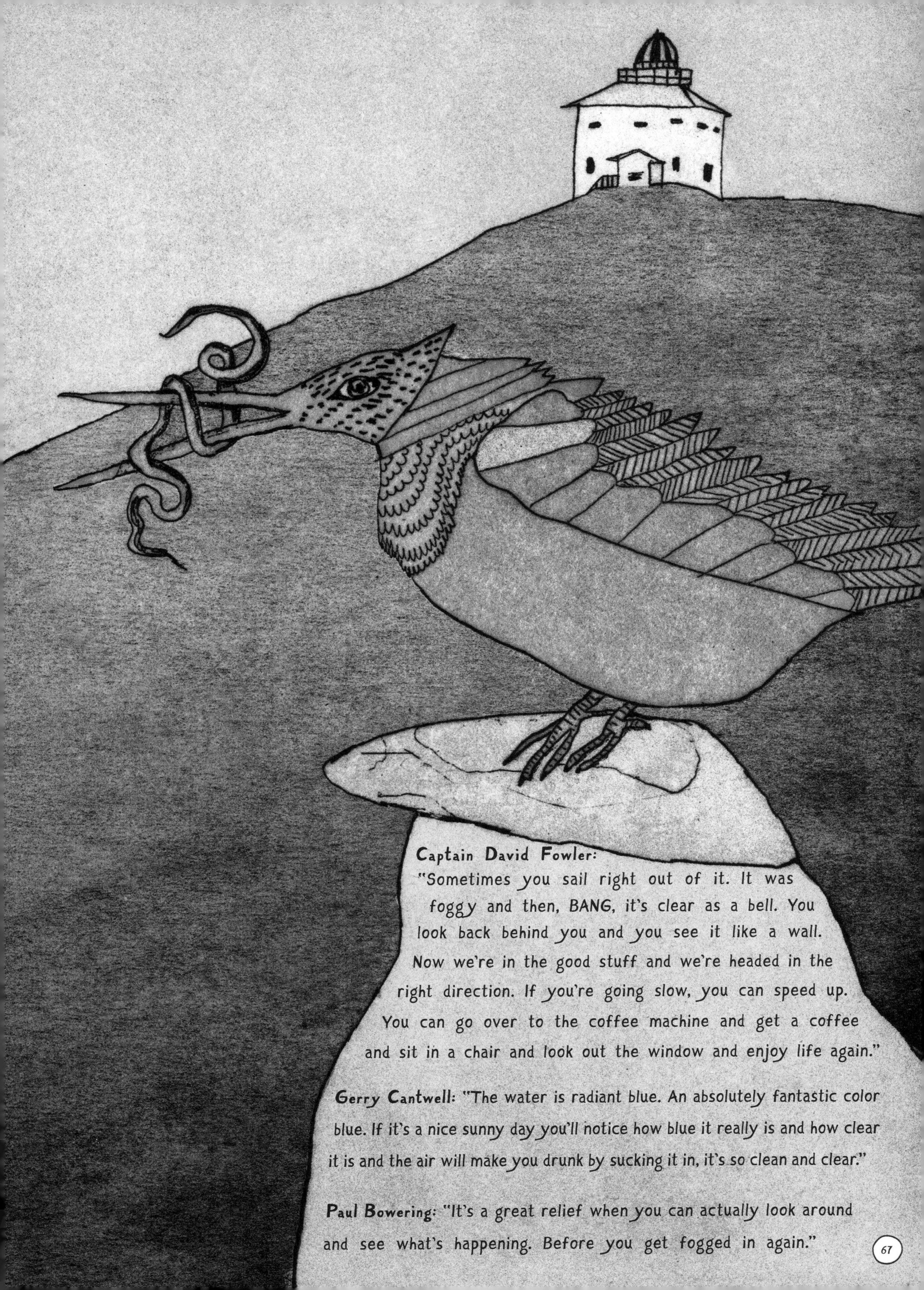

Captain David Fowler:
"Sometimes you sail right out of it. It was foggy and then, BANG, it's clear as a bell. You look back behind you and you see it like a wall. Now we're in the good stuff and we're headed in the right direction. If you're going slow, you can speed up. You can go over to the coffee machine and get a coffee and sit in a chair and look out the window and enjoy life again."

Gerry Cantwell: "The water is radiant blue. An absolutely fantastic color blue. If it's a nice sunny day you'll notice how blue it really is and how clear it is and the air will make you drunk by sucking it in, it's so clean and clear."

Paul Bowering: "It's a great relief when you can actually look around and see what's happening. Before you get fogged in again."

CHAPTER 5:

WIND

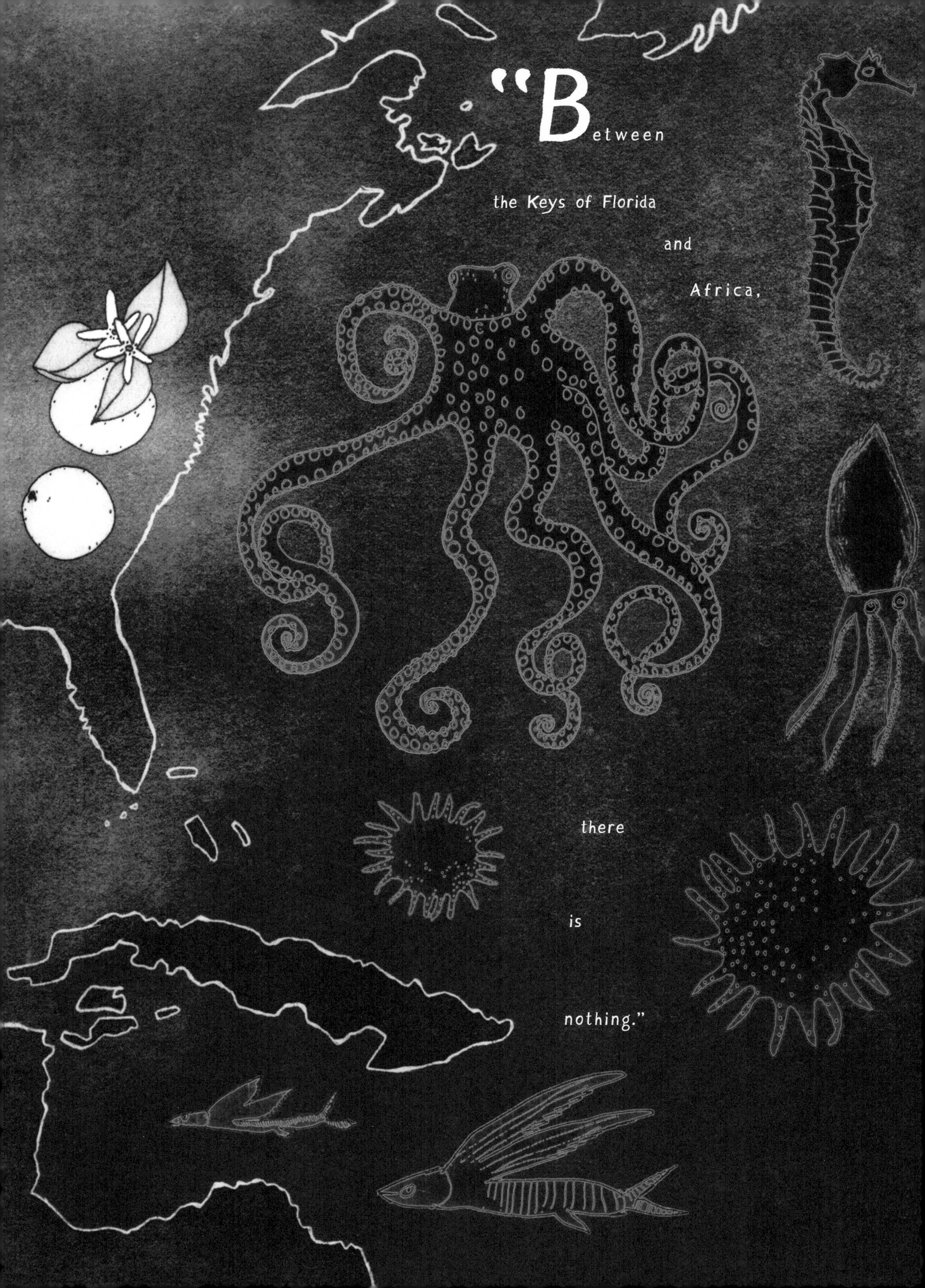
"Between
the Keys of Florida
and
Africa,
there
is
nothing."

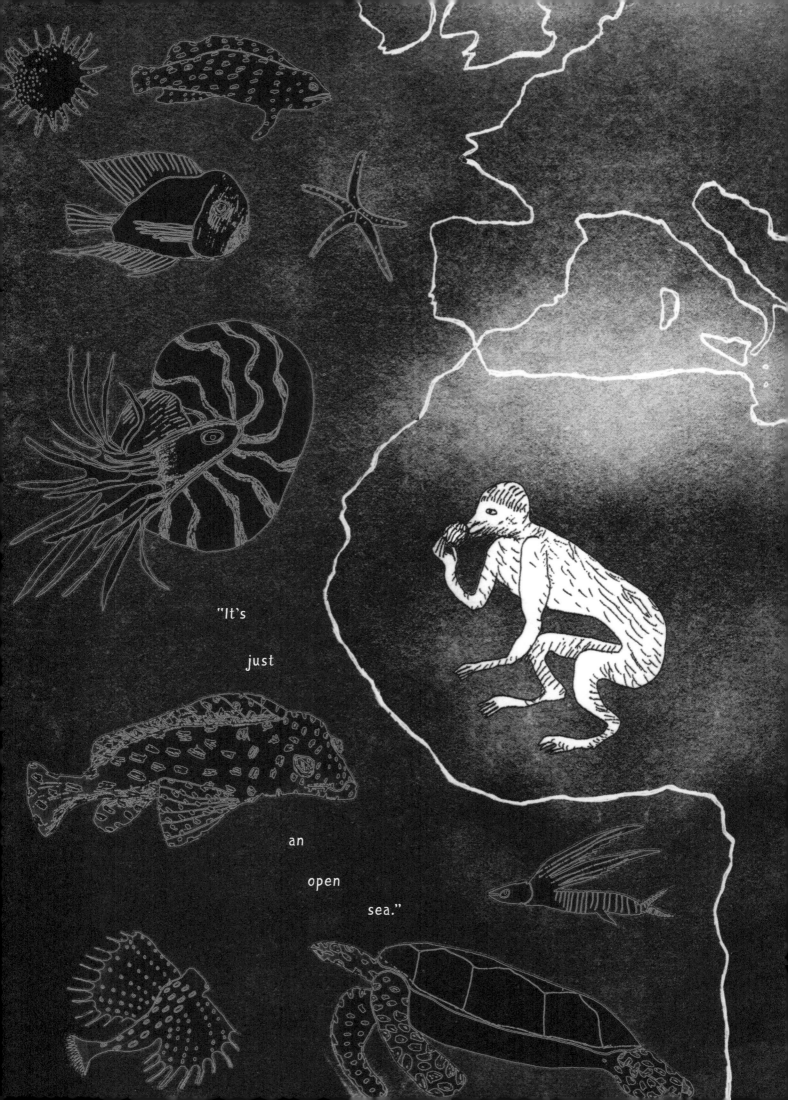

"It's

just

an

open

sea."

In September 2010, 60-year-old endurance swimmer Diana Nyad was attempting to swim from Cuba to Florida. She planned to swim for three straight days and nights without sleep. She would need to withstand nausea, asthma attacks, jellyfish stings, encounters with sharks. ("I am absolutely unafraid of pain," Nyad wrote in her 1978 autobiography.) Nyad had trained by lifting weights, biking, running, and, finally, doing 10-, 15-, and 24-hour ocean swims. She had secured the proper visas and assembled a support team of expert advisors. Now she waited for the right weather conditions.

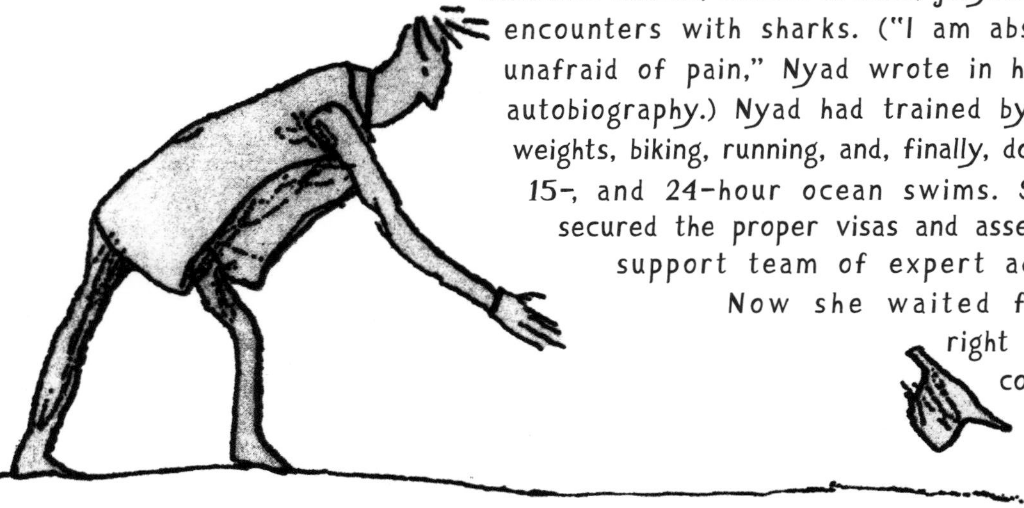

Diana Nyad: "We were so damn frustrated by the easterly winds, they just wouldn't stop. To do the Cuba swim, you either gotta get no wind—which is nice if you can get it—or southerly, or even westerly winds. But the east wind kills you. We had 90 consecutive days of—not ferocious, but very steady—east winds. One of the women there, a Whitbread round-the-world sailor, very accomplished out in the world's oceans, took us, my head trainer and I, out on the dock in Key West, on the Atlantic Ocean side, and she said, 'Put your tongue out in the air.' So the three of us are standing there with our tongues hanging out in the breeze. And she said, 'What do you sense?' We definitely sensed something grainy, crunchy, you know, in the mouth. And we said, 'Wow, it's the salt.' She said, 'No. It is the Sahara dust.' Literally grains of sand from the Sahara Desert."

In the summertime, usually in June and July, Saharan sands can blow from Africa clear across the Atlantic to Florida, causing hazy daytime skies and vivid red sunsets.

Wind is the horizontal movement of air over the earth's surface. On a global scale, the warmth of the sun and the rotation of the earth create patterns that we know as the "trade winds," the "prevailing westerlies," the "polar easterlies." These ribbons of air temper local climates and propel jet planes.

The sun's rays fall more directly on the equator than the poles, and so heat the atmosphere unevenly. The differences in temperature cause masses of air to rise and fall, push and churn. The warmer air ascends and heads towards the poles. Cooler air drives in below. All the while, the earth is spinning, wrapping the winds into invisible bands around its sphere. There are also, of course, winds that occur on smaller scales—gusts and gales, hurricanes and typhoons—as air currents are affected by seasonal temperature changes and topography: the presence of mountains and valleys, buildings and forests, the different ways bodies of water and areas of land absorb and then radiate heat.

winds can shape the personality of a place. The *scirocco*, the warm wind that comes from the south-east and can persist for three or four days," writes Peter Ackroyd, "has been blamed for the Venetian tendency towards sensuality and indolence; it has been accused of instilling passivity and even effeminacy within its citizens."

Raymond Chandler described the effects of California's wildfire-stoking winds in a short story. "It was one of those hot dry Santa Anas that come down through the mountain passes and curl your hair and make your nerves jump and your skin itch. On nights like that every booze party ends in a fight. Meek little wives feel the edge of the carving knife and study their husbands' necks."

Föhnkrankheit is the illness associated with a *Föhn* wind — a dry, down-slope wind notorious in Central Europe. The *Föhn* is said to cause headaches, insomnia, malaise, even to increase rates of suicide and murder. Judges in Switzerland have been known to consider wind as a mitigating factor in crimes committed during a *Föhn*. In his 1904 novel *Peter Camenzind*, Herman Hesse writes in the voice of his Swiss protagonist, "Day and night you could hear the Föhn howl, distant avalanches crash, and the embittered roar of torrents carrying boulders and splintered trees, hurling them on our narrow strips of land and orchards. The Föhn fever would not let me sleep. Night after night, rapt and fearful I heard the storm moan, the avalanches thunder, the raging water of the lake burst against the shore. During this period of feverish springtime battles, I was once more overcome by my old love sickness, so impetuously this time that I got up at night, leaned out the window, and bellowed words of love out into the storm to Elizabeth....It seemed to me often as if the beautiful woman were standing very close, smiling at me, yet withdrawing with each step I took in her direction....Like an infected man, I could not help scratching the itching sore. I was ashamed of myself, but this was as agonizing as it was futile.

I damned the Föhn."

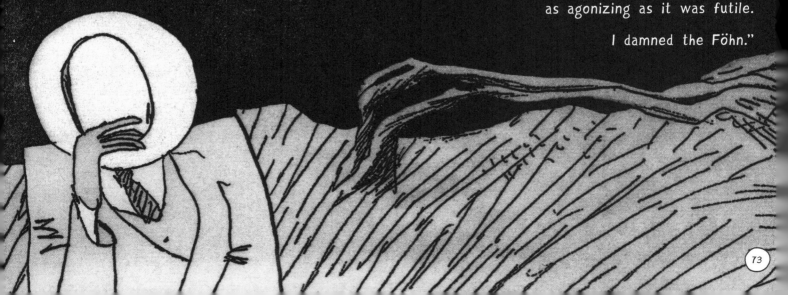

Two global wind belts—the northeast trade winds and the southeast trade winds—meet in a ring of erratic weather close to the equator. This seasonally shifting area is known as the Intertropical Convergence Zone, or, to sailors stuck in its sometimes deadly still air, the doldrums. While the hot moist air of the ITCZ rises, it doesn't always travel horizontally; the result is thunderstorms and cloud cover but, often, little wind. The term "doldrums" is used to describe windless pockets beyond the ITCZ as well. Such conditions bedevil sailors, but they are a boon to a swimmer in open waters.

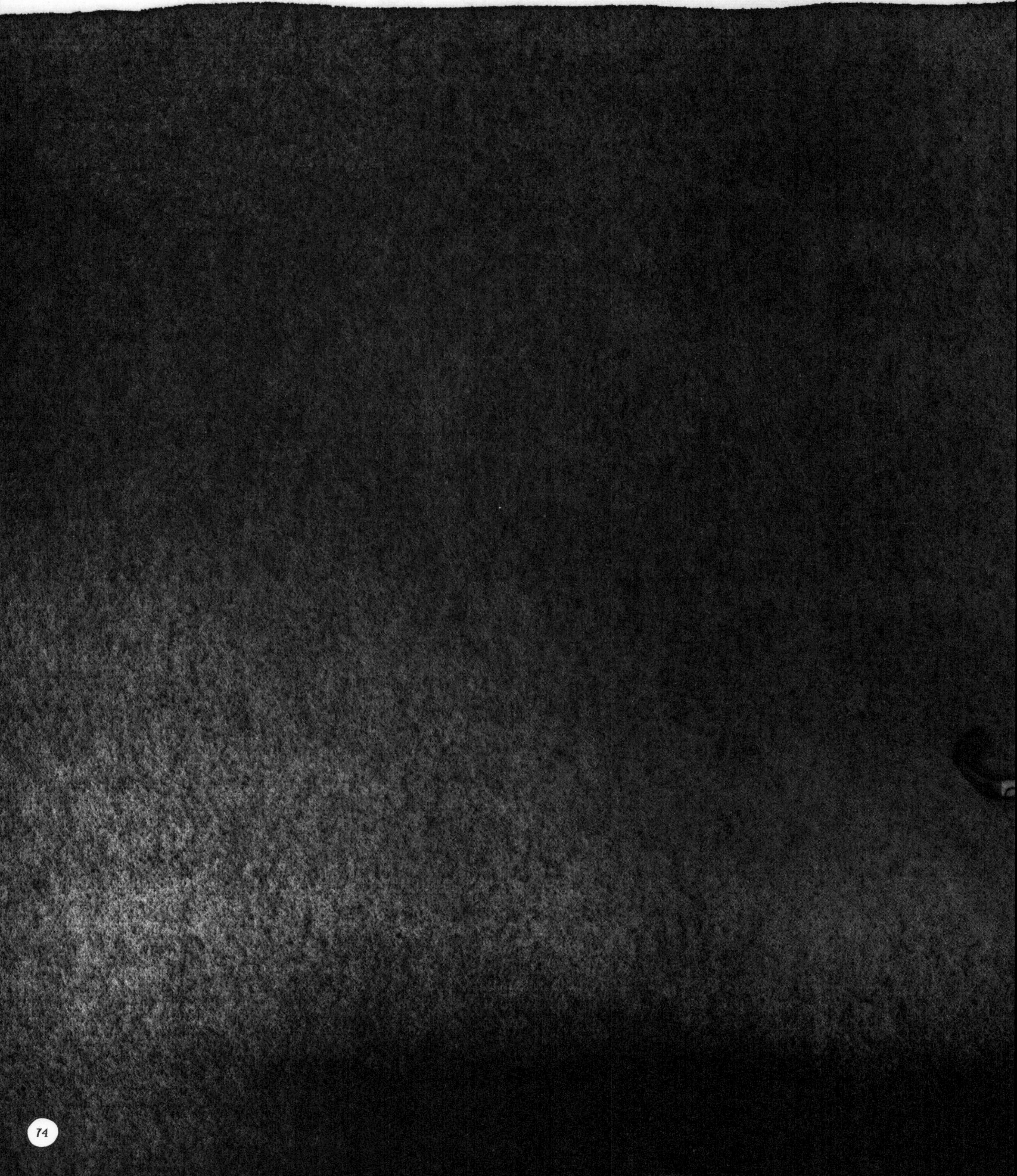

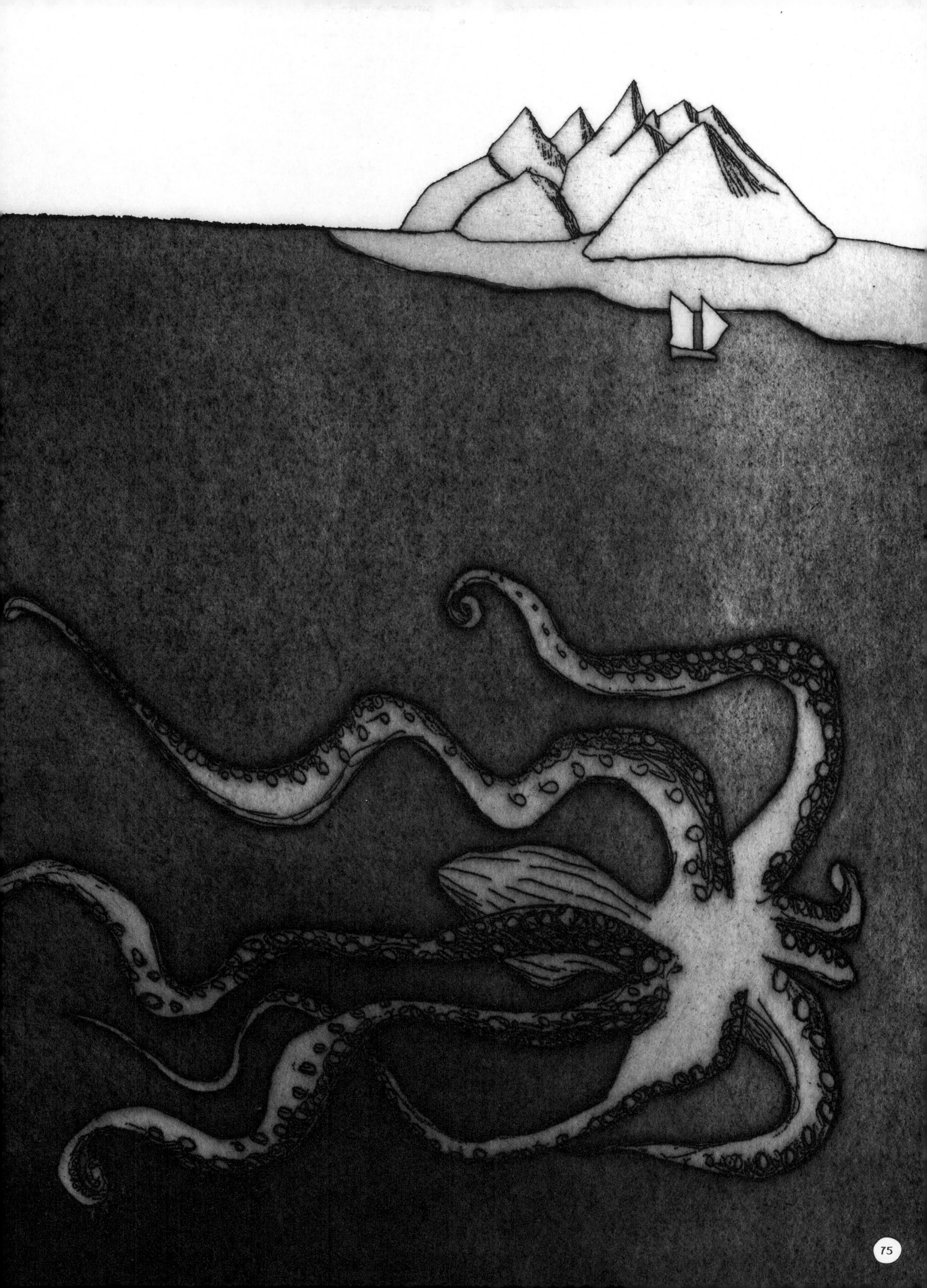

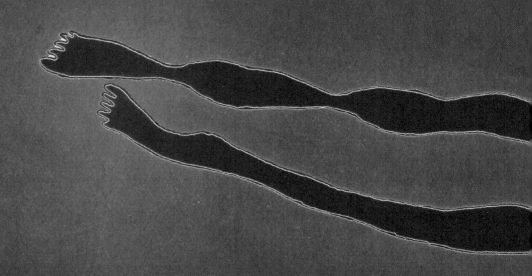

Diana Nyad: "That's what we pray for, the doldrums. Not a breath of wind. For a fisherman or a sailor, a light wind might be 15 knots, you know, no big deal. For a swimmer, picture it: your face is on the surface and you're turning your head 60 times a minute to get some oxygen. Your lips are literally at the surface, so a wave that's only six inches tall is a little pain in the neck.

"Now, a wave that's two feet high, that's considered still very low for anybody out in a boat just banging around. But two feet is a lot taller than what my arm reaches. So all of a sudden I'm having to press and push and get the head way above. Whereas, in the doldrums, it's like glass. Your hands are in, and you're just catapulting and streaming along the surface. It's as if there's a force under the water tugging down the surface and keeping it flat, holding the waves down."

When she was a child, Diana's Greek stepfather, Aristotle Nyad, a handsome and mercurial figure who made his living "gambling, lying, and stealing," read to her from *The Odyssey* and took her deep sea fishing. In 2005 she wrote about him for *Newsweek* magazine.

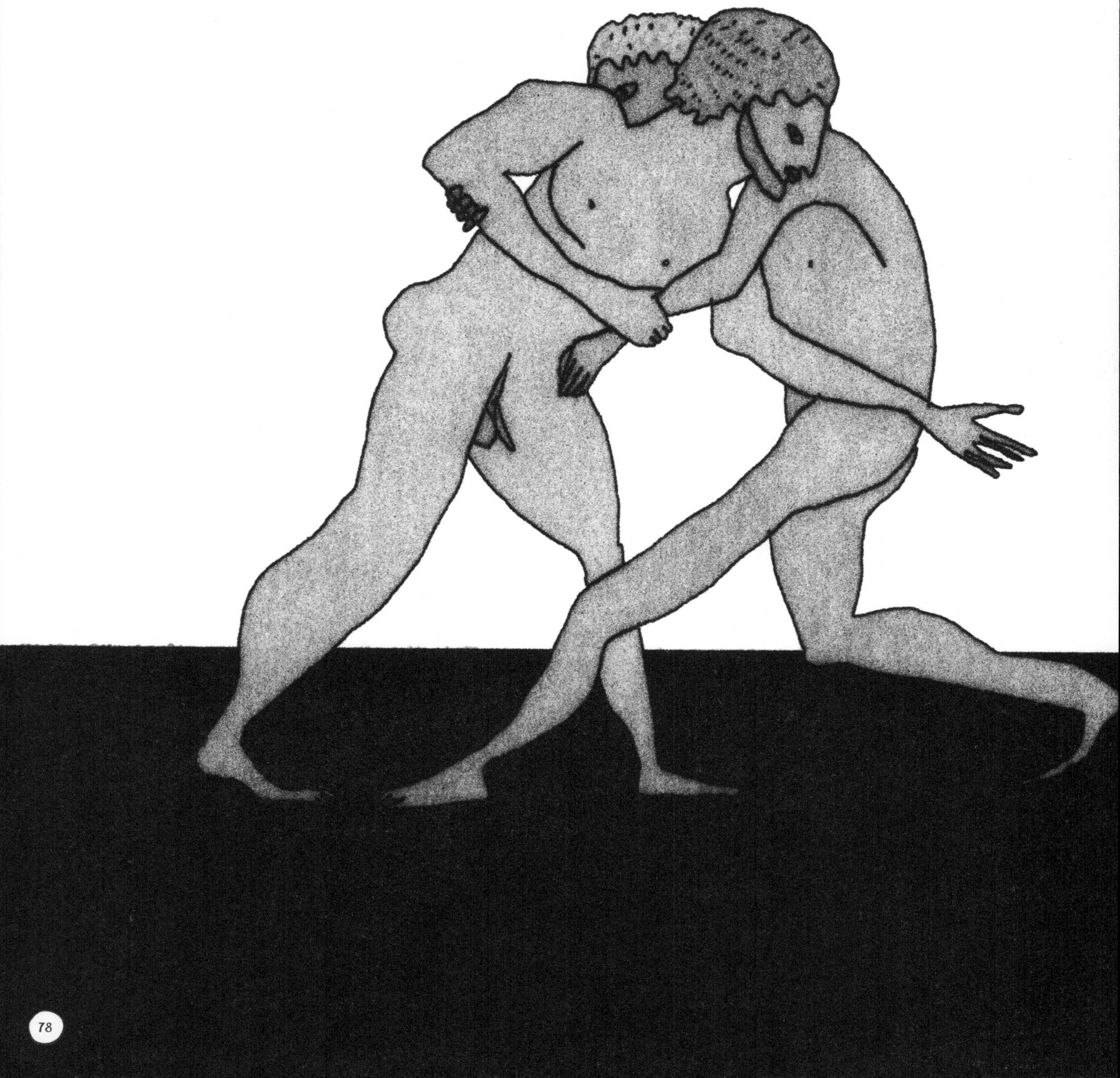

"One night, when I was five or six, Aris called me over as he opened our huge unabridged dictionary. He thumbed through the pages to the N's and pointed out the word *naiad*. There was my name (changed to Nyad by Aris's family several generations ago)....The first meaning of naiad: 'from Greek mythology, the nymphs that swam the lakes, fountains, rivers, and seas to protect them for the gods.' The second meaning: 'girl or woman champion swimmer.' Aris winked at me, and we both understood this was my destiny."

In *The Odyssey*, Aeolus, keeper of the winds, gives Odysseus, after years of trial and adventure at sea, an ox-skin bag holding "the winds that howl from every quarter" and unfurls for him a favorable breeze from the west. Aeolus's gift nearly carries Odysseus and his crew home, the shore within view. "We were so close," Odysseus says, "we could see the men tending fires."

Exhausted and relieved, Odysseus falls asleep. His restless sailors hover, eying the leather sack. A trove of gold and silver must be hidden within, they imagine, riches that will unjustly be enjoyed by their captain alone. And so, while he sleeps, they ransack the parcel. "A fatal plan...All the winds burst out."

The atmosphere is roiled into a violent storm that pushes the men back to sea, back to Aeolus's island. Odysseus begs for another fair wind, but Aeolus is unforgiving. "Away from my island — fast — most cursed man alive!...Crawling back like this — it proves the immortals hate you! Out — get out!" The blunder adds nearly a decade to Odysseus's journey.

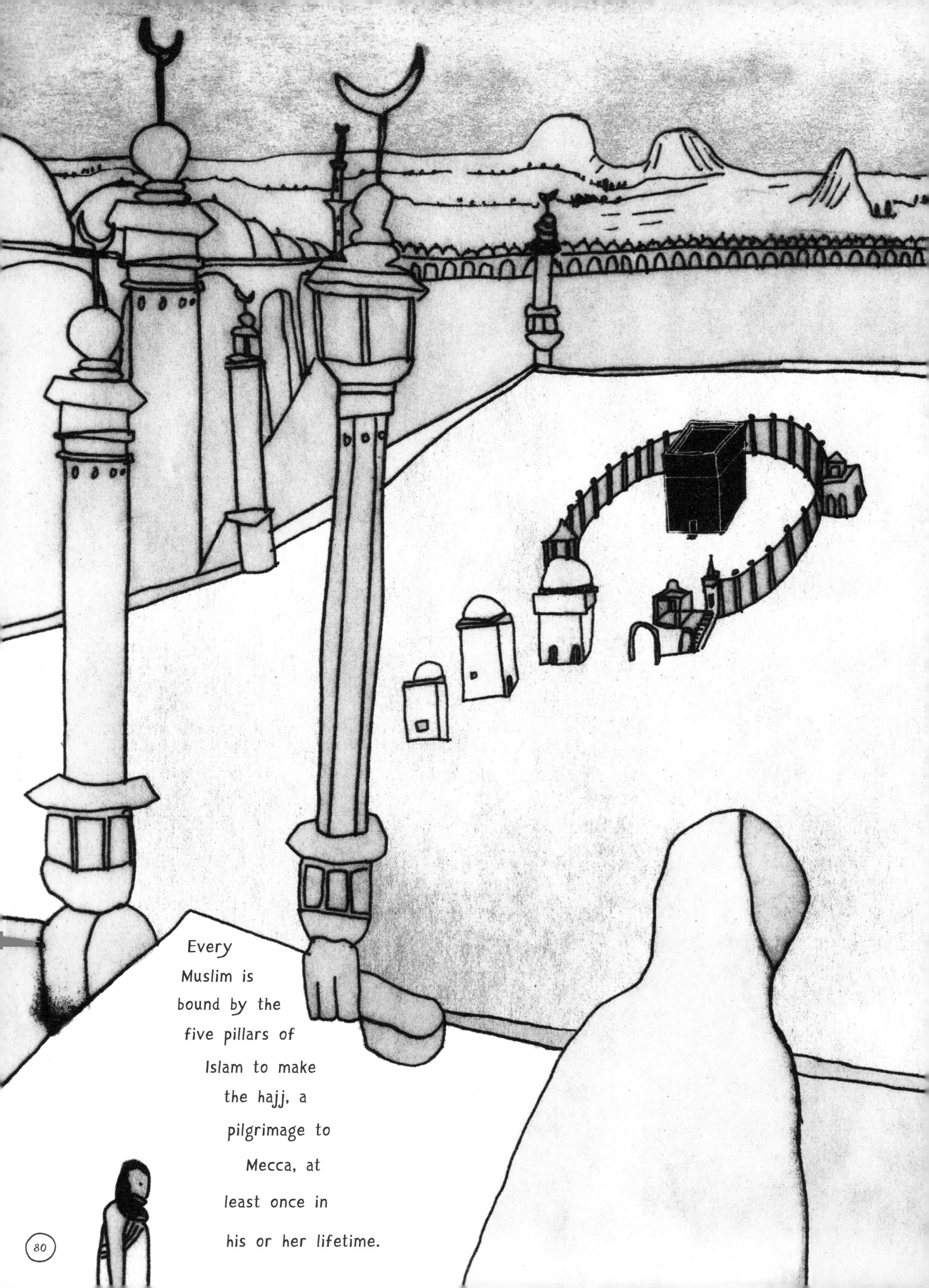
Every Muslim is bound by the five pillars of Islam to make the hajj, a pilgrimage to Mecca, at least once in his or her lifetime.

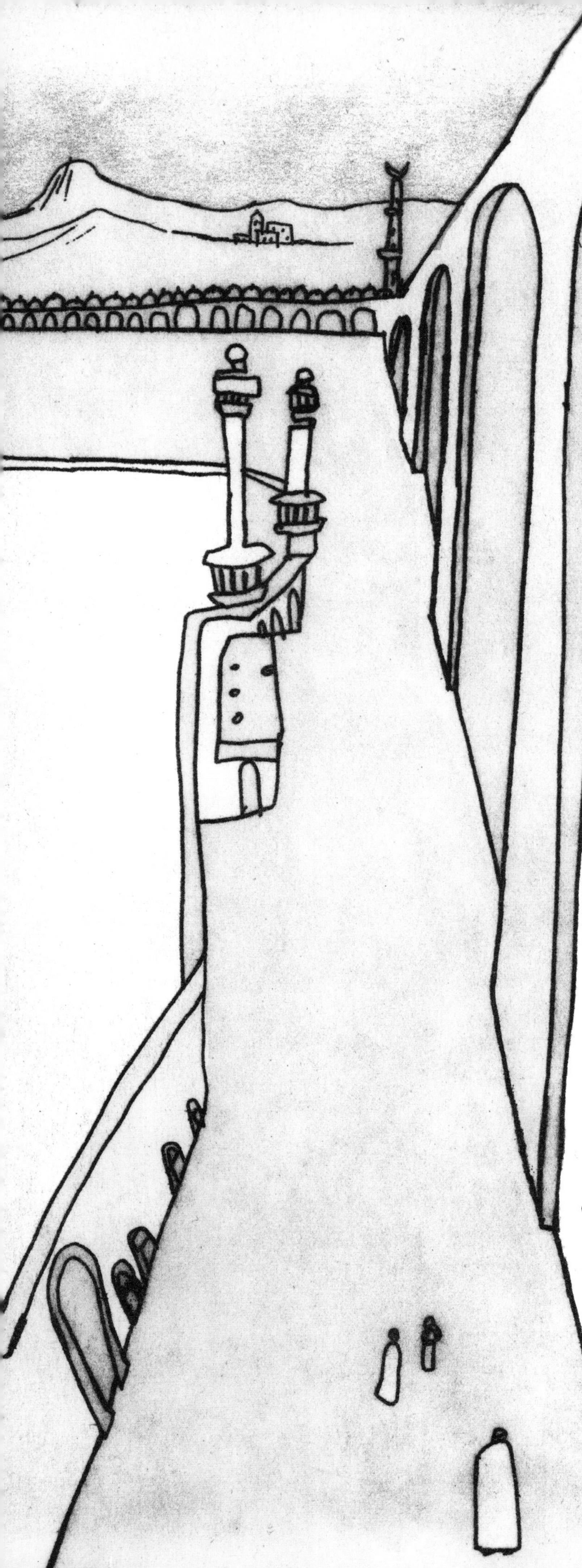

Muslims have made this journey since the 7th century, retracing steps the Prophet Mohammed is believed to have made. In the past, reaching Mecca could be arduous, taking months, even years, with some never returning home. Today, most of the more than three million annual pilgrims fly into the dedicated Hajj Terminal at Saudi Arabia's King Abdulaziz International Airport in Jeddah, about an hour's drive from the Grand Mosque at Mecca. The Grand Mosque's indoor and outdoor spaces combine to occupy nearly 90 acres surrounding a central courtyard. In the middle of that open-air plaza stands the Kaaba, Islam's most sacred site. Muslims around the world face towards the Kaaba when praying. The Kaaba is a cube-shaped building with fragments of a holy relic known as the Black Stone, said to have fallen from heaven, embedded into its eastern corner. In one of the central rituals of the hajj, called *tawaf*, pilgrims walk counterclockwise around the Kaaba seven times, each time reaching to give a symbolic kiss to the Black Stone.

When legions flock to this small spot under the Saudi Arabian sun, the courtyard at the Grand Mosque becomes its own microclimate: hot, with still air. The number of annual pilgrims has grown over the years and is expected to continue increasing. To accommodate the crowds, the Saudi government has embarked on a series of expansions and renovation projects at the Grand Mosque. Anton Davies is a wind engineer at Rowan Williams Davies and Irwin Inc., a Canadian engineering and consulting firm. RWDI was hired to study the wind dynamics at the Grand Mosque: to bring a breeze to Mecca.

Anton Davies:

"There isn't enough ventilation. There are so many people and so many buildings around this area that there is very little air movement. We're talking about the absence of wind.

"The time of the hajj follows the lunar calendar. Right now it is moving into the hotter time of the year, towards those 110, 120 degree days. You're in full sun. People are dropping from heat prostration. Three million people all crammed together in a very, very small space. Sweating. At some point, the humidity level gets to 100 percent. Moisture no longer evaporates off your skin. It just sits there as sweat, and when that happens, you start to boil. We're trying to generate wind, and this breeze we're trying to generate will actually be cool as well. The question is: how do you air condition an open space?"

RWDI looked at ways buildings being constructed on the northern side of the Kaaba could help cool the space. These buildings are five stories high and can accommodate 375,000 people. Dozens of massive umbrellas are planned to shade the area beyond the outer walls.

Anton Davies: "We ended up over-air conditioning the buildings and allowing some of the conditioned air to spill out into the courtyard. You've probably experienced it walking by a shopping center: you feel the cool air coming out of the doorways. We're talking about 200,000 tons of refrigeration. Air-conditioned air — cooled air — is much heavier than the hot air outside, so cool air will actually pour out of the building. That's enough to drive the flow. This air exchange from inside the building to outside the building creates a breeze, exchanges the air, and keeps the humidity levels fairly low.

"The cool air from the buildings is going to move underneath the umbrellas. The umbrellas act as sort of a ceiling, reducing the amount of mixing with the ambient air, which means that the cool air will stay cooler and stay closer to the pilgrims for a longer period of time. They will just feel a gentle breeze."

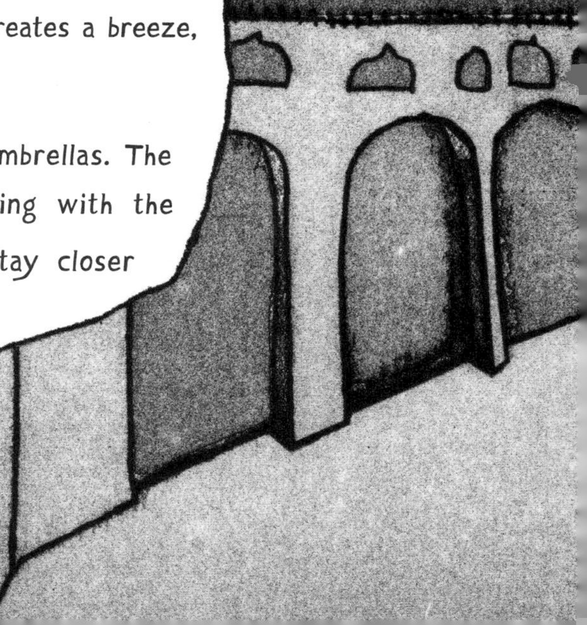

Diana Nyad:
"What am I going to do, be angry if the wind comes up?
Am I going to quit? It just doesn't do any good.
I try to swim along thinking, yeah, things seem perfect right now.
I feel great, the weather is calm as an ironing board and everything's great, but don't think this
is going to last for two and a half days. It's not. You're not going to feel great,
you're going to go through undulations, peaks and valleys, and if you're in a
valley you've gotta bear with it. When waves come up and weather and wind
comes up, whether in training or in the real swim,
I try to think to myself, it's not what's happening above
the water. I can't control if the waves are slapping around and

I'm ingesting saltwater and my
arms and shoulders are having
to bop instead of smoothly
glide along the surface, but I can
control what's happening under the water,
get my arm in, get the elbow in the right angle, and get that hand
ready to push back as efficiently as possible, to catapult me forward.

"I'm hearing my own breath. It sounds like [throaty, rhythmic]
WHOOUH-WHOOUH, WHOOUH-WHOOUH. I'm breathing right at the surface,
which magnifies the sound, because water makes sound travel. My hands are
coming over: the right one every six seconds, the left one every
six seconds. So you're hearing not only that breath but you're
hearing the TCHOO, TCHOO of the hand entering the water,
and the little kick in the back is

ba-ba-ba-

ba-ba-

ba.

"There's the metronomic sensuality of it. It
sounds almost like when you hear the
baby in the amniotic fluid
breathing."

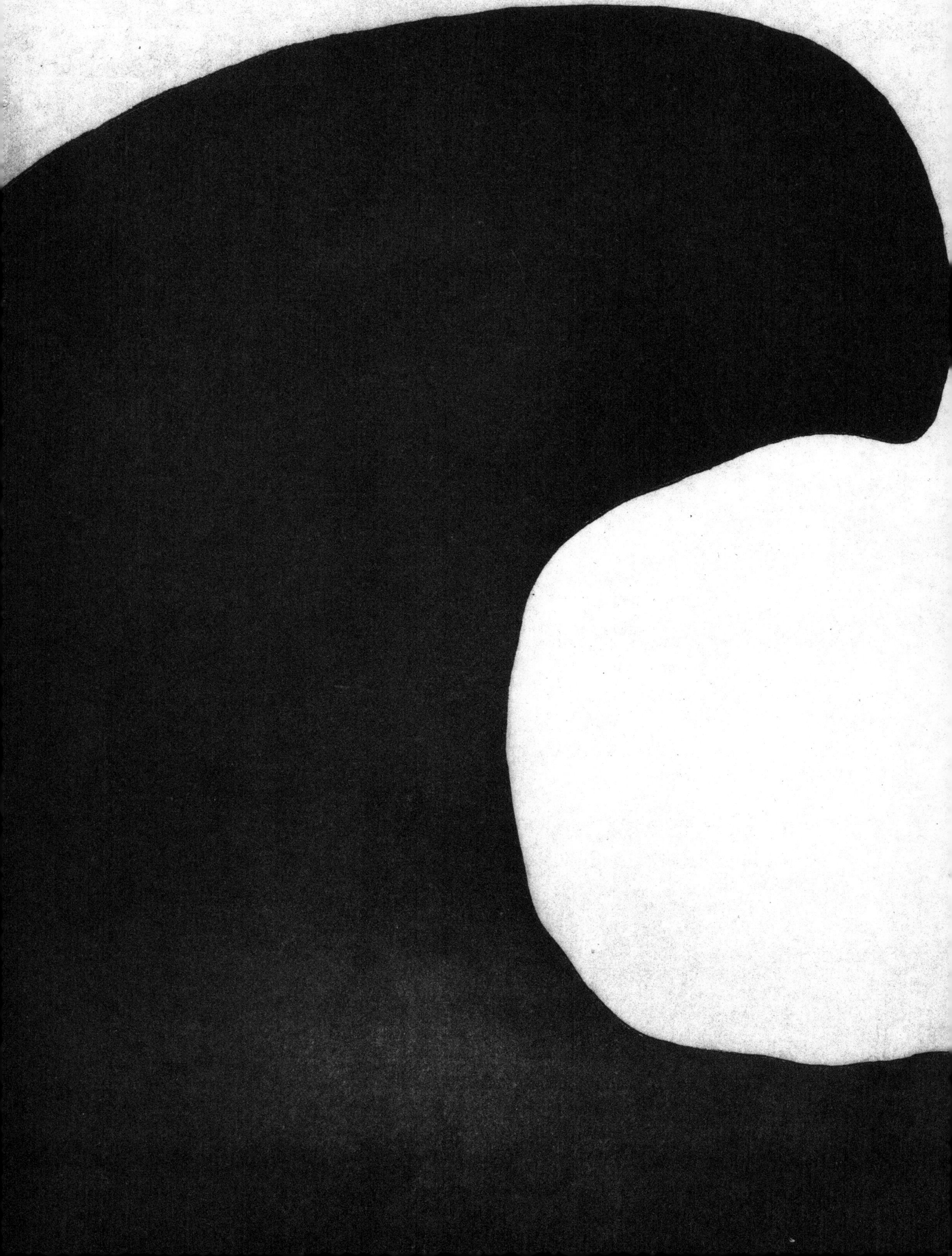

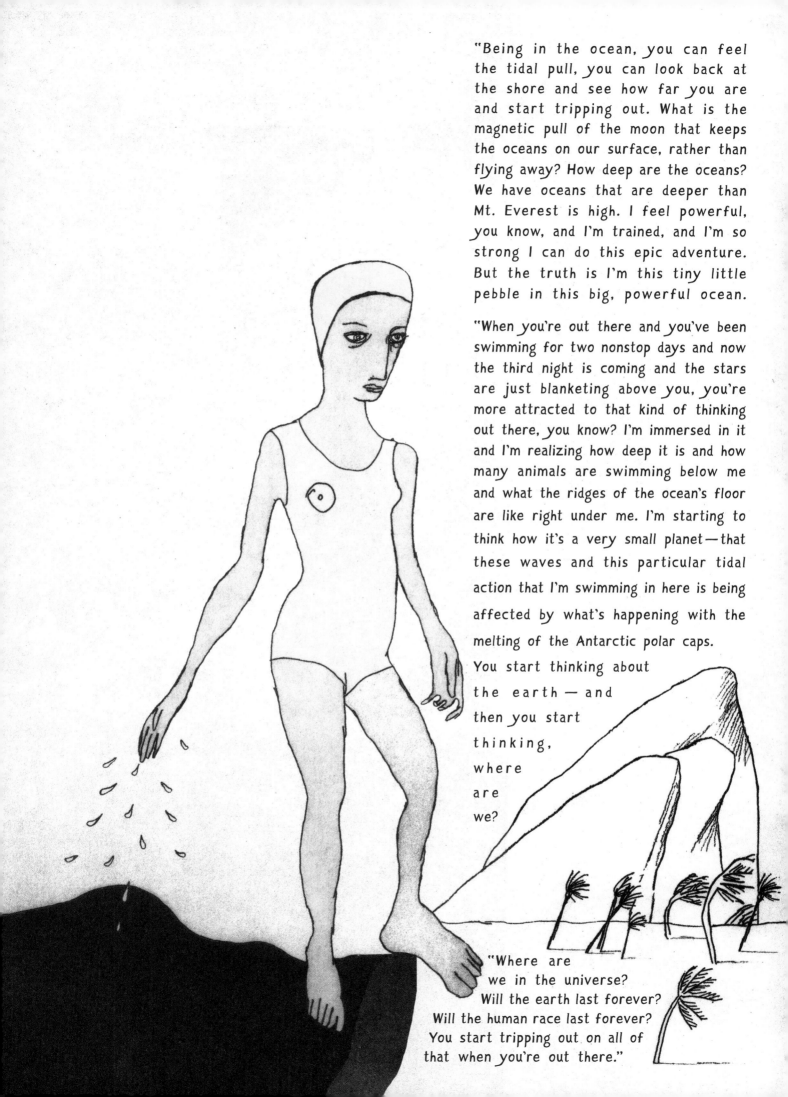

"Being in the ocean, you can feel the tidal pull, you can look back at the shore and see how far you are and start tripping out. What is the magnetic pull of the moon that keeps the oceans on our surface, rather than flying away? How deep are the oceans? We have oceans that are deeper than Mt. Everest is high. I feel powerful, you know, and I'm trained, and I'm so strong I can do this epic adventure. But the truth is I'm this tiny little pebble in this big, powerful ocean.

"When you're out there and you've been swimming for two nonstop days and now the third night is coming and the stars are just blanketing above you, you're more attracted to that kind of thinking out there, you know? I'm immersed in it and I'm realizing how deep it is and how many animals are swimming below me and what the ridges of the ocean's floor are like right under me. I'm starting to think how it's a very small planet—that these waves and this particular tidal action that I'm swimming in here is being affected by what's happening with the melting of the Antarctic polar caps. You start thinking about the earth — and then you start thinking, where are we?

"Where are we in the universe? Will the earth last forever? Will the human race last forever? You start tripping out on all of that when you're out there."

CHAPTER 6:

HEAT

Fire is a natural part of the global ecosystem. Fire can clear forest underbrush and make way for new growth. Fire can nourish soil and trigger seed germination. Humans have long relied on controlled fire for warmth, light, cooking, farming, and protection from predators.

Fire can also lay waste and destroy.

Scientists believe climate change—with prolonged droughts and record temperatures—is contributing to increasingly devastating wildfires.

"We are seeing more fire activity, more extreme fire weather, more extreme fires all over the world," says David Bowman, professor of environmental change biology at the University of Tasmania and co-author of *Fire on Earth*. "Fires are now beginning to be like a rash on this planet."

David Bowman: "We are seeing fire behaviors which are unusual. Things like fires burning through the night, fires just not going out. Fires burning for weeks, for months. Firefighters are saying that they are seeing things they have never experienced before.

"Fire is much more dramatic than a flood. It's almost instantaneous. Switching from one world to another world: a forested landscape becomes an incinerated landscape. Once you get the mechanics going, once you've lost control of these fires, there is absolutely nothing you can do. You've got to think about surviving them, not controlling them. It's like a wild animal, like a snake. It's a terrifying beauty."

Israel's Mount Carmel forest caught fire in December 2010. The forest had been desiccated by Israel's hottest year on record. Weather monitoring stations in Jerusalem and Haifa had recorded the fewest rainy days in 80 years. In the Mount Carmel fire, dozens of people were killed and more than five million trees burned. Israel sought and received help from the international community. The United States, Russia, Britain, France, Spain, Egypt, Jordan, Cyprus, Greece, Germany, Turkey, and the Palestinian Authority sent aid and fire-fighting aircraft. "This is a special type of battle," said Israeli Prime Minister Benjamin Netanyahu.

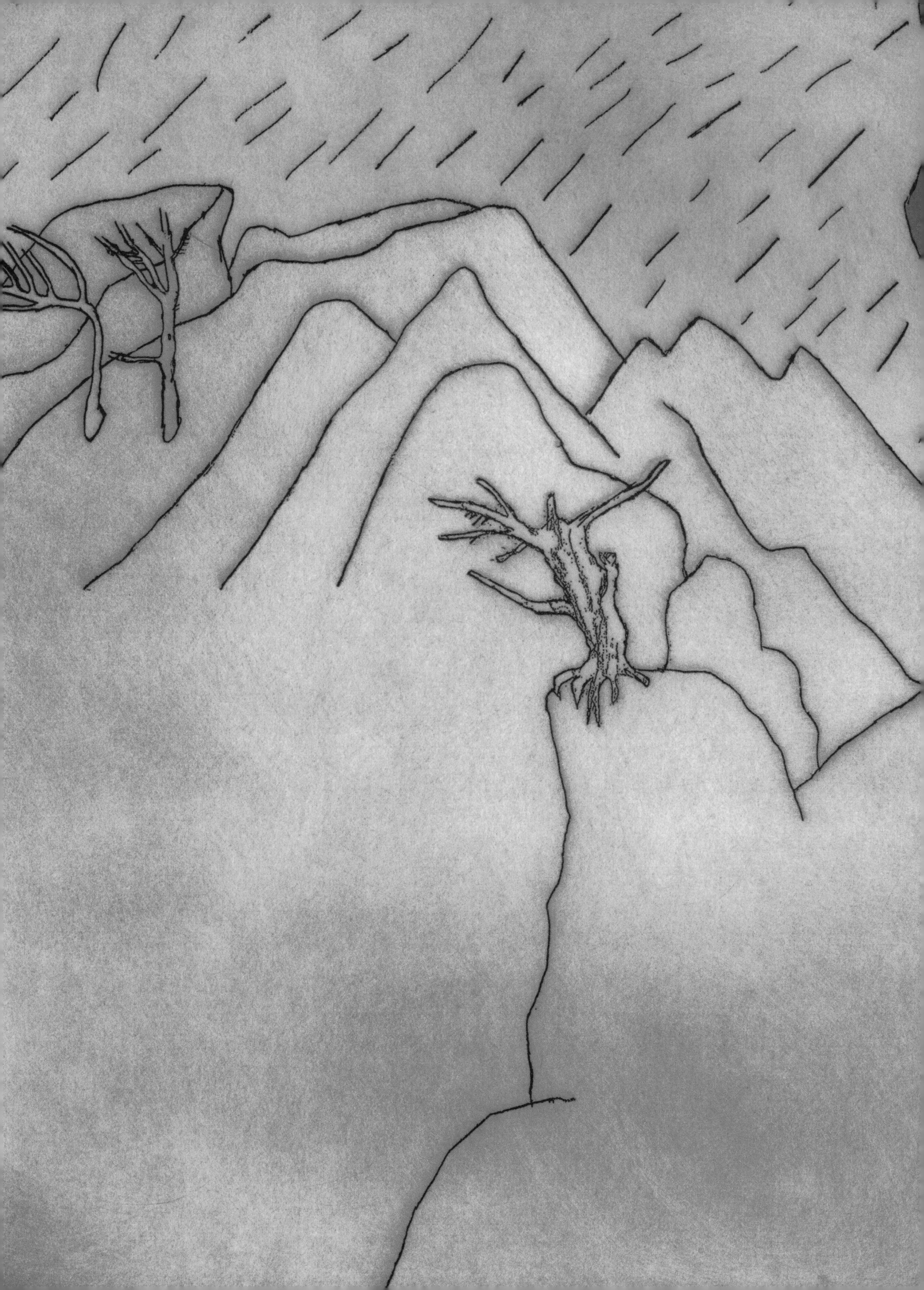

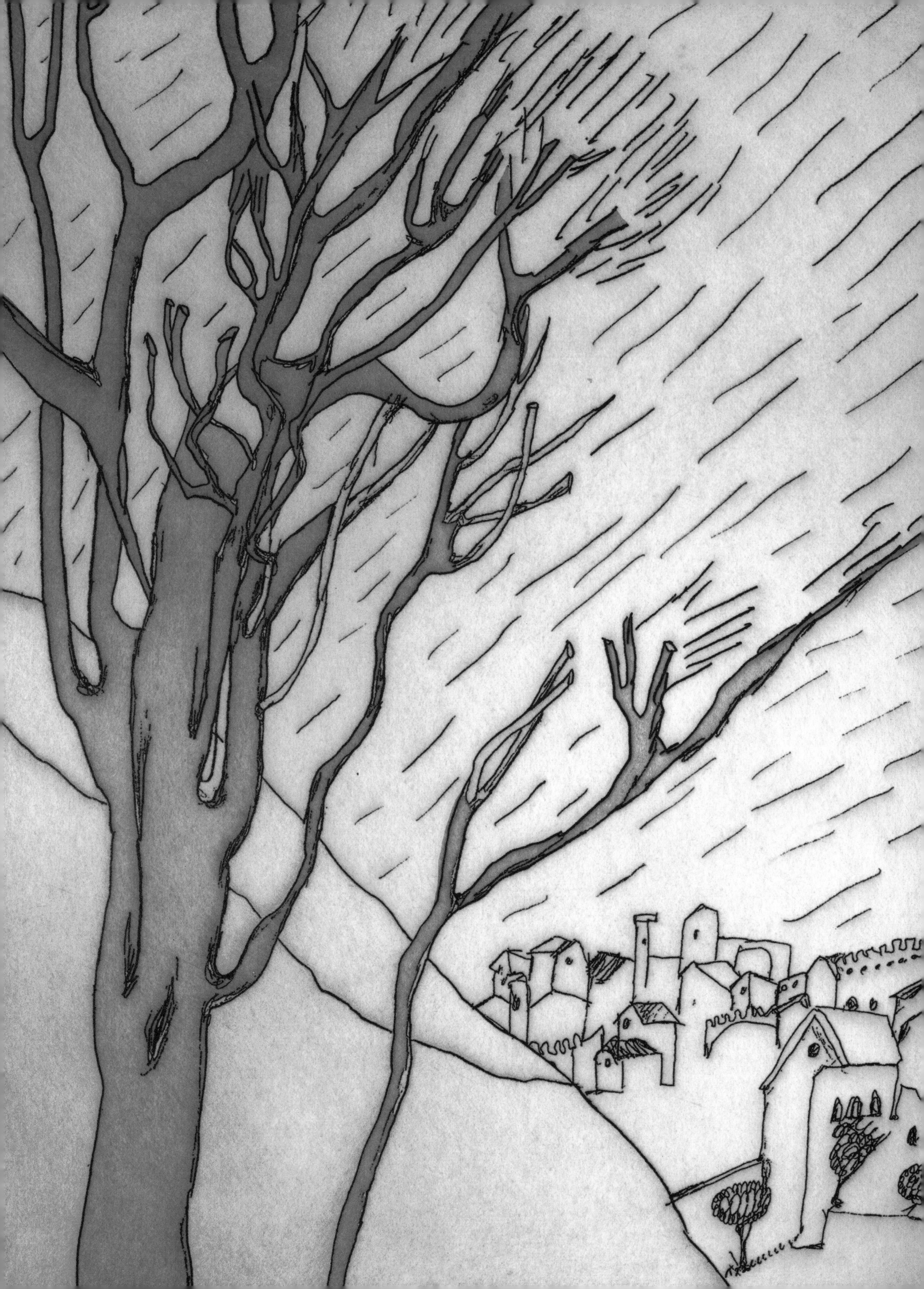

Article 5:3 of the Constitution of the Kingdom of Bhutan establishes that a minimum of 60 percent of the country's land area "shall be maintained under forest cover for all time." Bhutan is celebrated for its dramatic landscape and its biodiversity. It is home to the clouded leopard, the one-horned rhino, the red panda, the goat-antelope, the barking deer, the golden langur, tigers, black bears, wild boars, wolves, and blue sheep. Fifty species of rhododendron bloom.

According to government statistics, one in five Bhutanese live below the poverty line. Most working Bhutanese are farmers, growing crops or raising cattle. The cheapest way to clear land for cultivation is to set it aflame. Particularly during dry winter months, agricultural burning can quickly spread and become a forest fire. Fires are also attributed to "children playing with matchsticks, cow herders, lemongrass harvesters." Winds and mountainous terrain make these fires difficult to control, threatening people and wildlife.

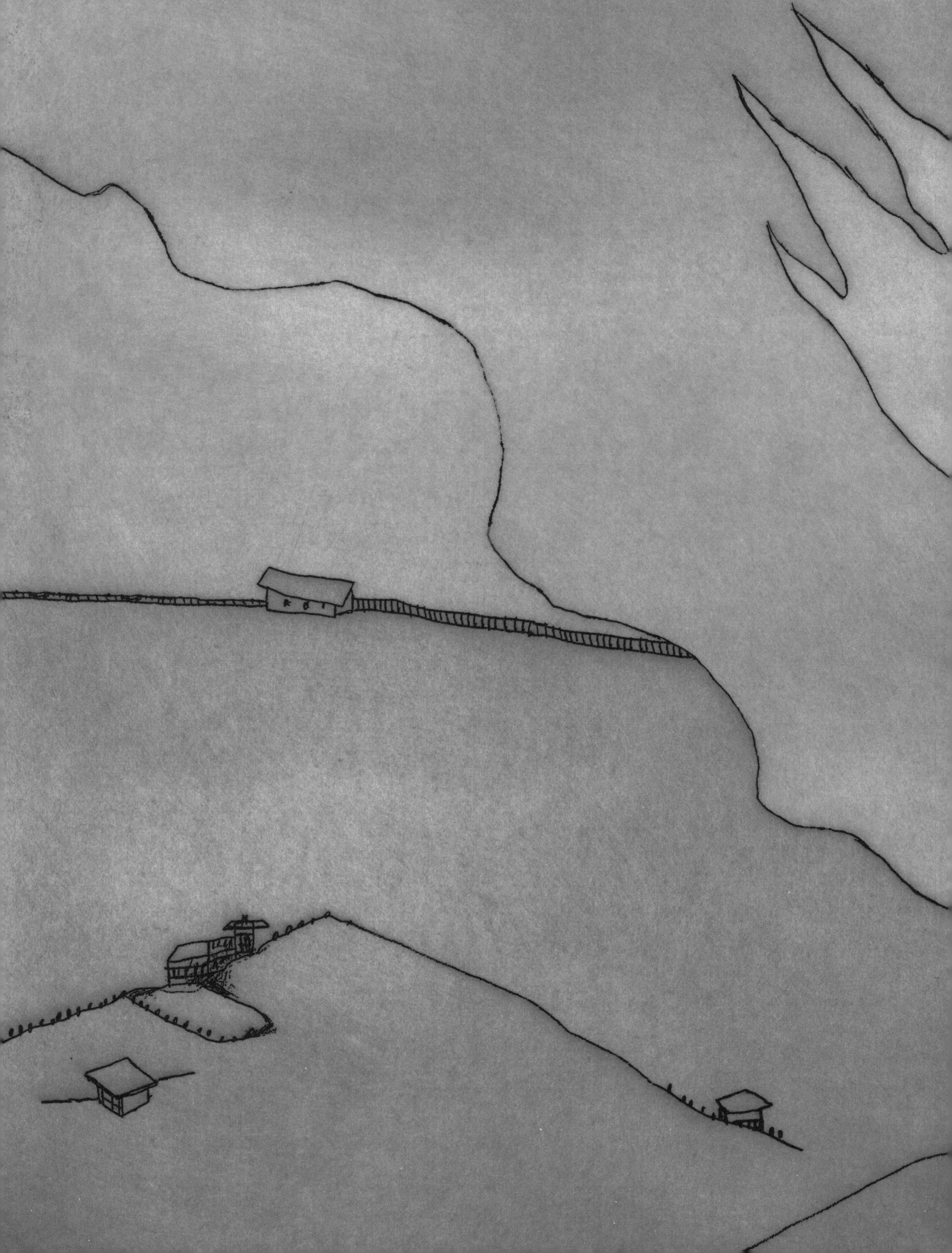

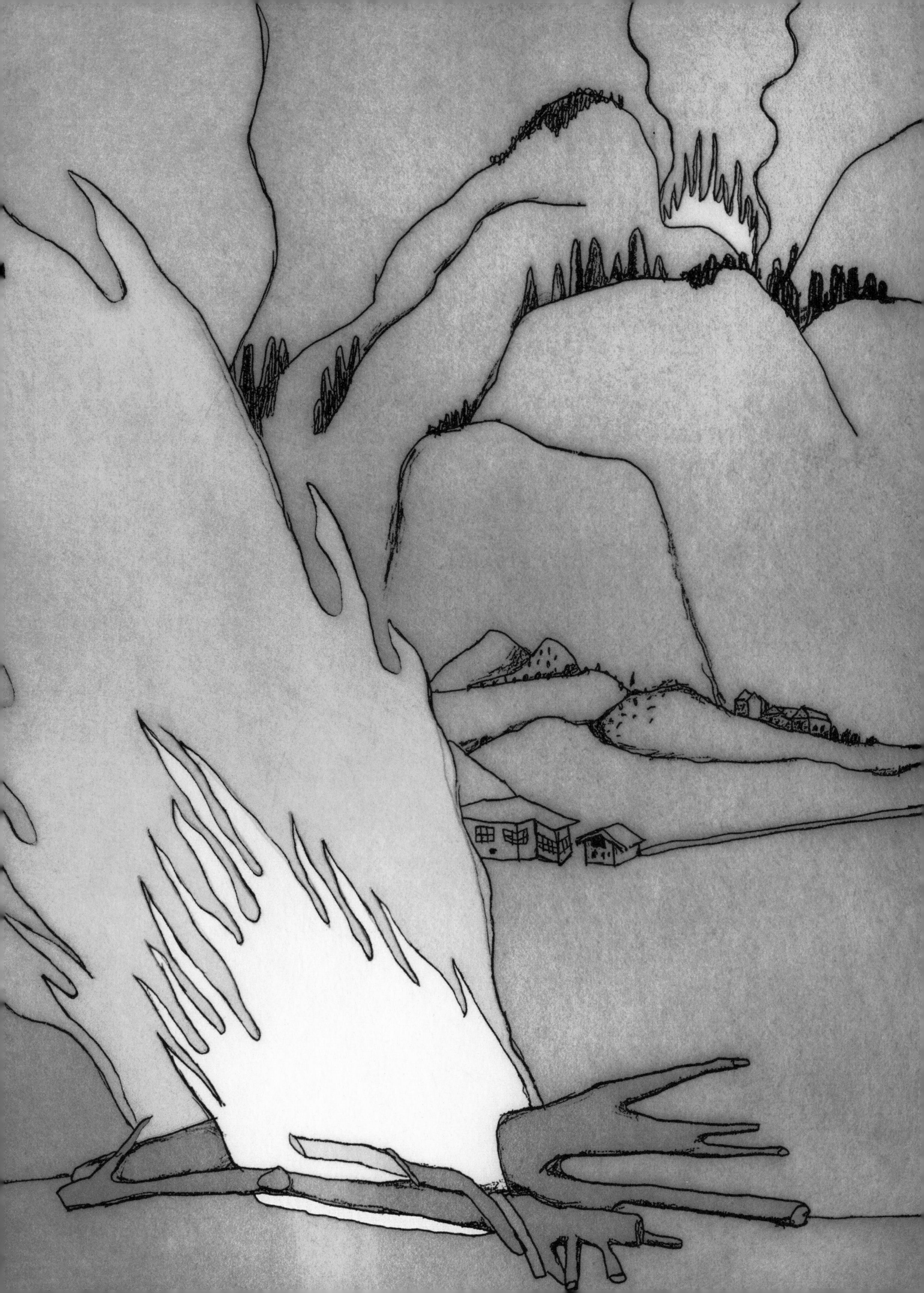

Black Kites, raptors with dark eyes and forked tails, can be seen flying in large flocks over parts of Africa, Europe, Asia, and Australia. Black Kites are drawn to burning landscapes. When other animals flee the flames, this hunter-scavenger swoops in. In *Eagles, Hawks, and Falcons of Australia*, David Hollands writes: "Fire is one element which the Black Kite does exploit with great skill....I have seen a thousand birds appear in minutes in front of a Darwin grass fire, gathering before its front to hang where the air rises fastest and then plunge in their scores through the smoke and almost into the flames, swerving and weaving through the gloom as they snatch the insects which flee before the advancing inferno."

Australia is hot, dry, and drought-prone. Its people, flora, and fauna are adapted to fire. Aborigines used fire to hunt and to fish. Eucalyptus trees, which dominate the Australian arboreal landscape, contain flammable oils; burning trees seem to explode as they reach their ignition point and burst into flame. New eucalypt seeds thrive in a fire's wake. "Kangaroos, wallabies, and wombats," writes fire historian Stephen Pyne, "need the nutritious new growth that springs up after a burn."

Australians plan for fire. Recent fires, though, have proved catastrophic for even the most prepared. The Australian state of Victoria began 2009 with a heat wave. Temperatures broke records, setting a new high of 119.8°F. It had hardly rained for months; Victoria was in its thirteenth year of drought. On Friday, February 6, Victoria's premier John Brumby told people to cancel their Saturday plans and stay home on what he predicted would be the "worst day in the history of the state." The next afternoon the temperature in Melbourne passed 115°F degrees. Relative humidity hung below 10 percent.

At 11:47 a.m., smoke was seen rising behind a hill in the town of Kilmore East. Firefighters arrived within minutes, but could not contain the blaze. Crew leader Russell Court reported seeing "two tongues" of fire, one spreading to the south and another burning eastward. High winds fanned the flames and hurled embers into the air, igniting new spot fires. By 1:19 p.m., the fire had "multiple tongues." The fires fed on abundant, parched vegetation, burning up mountains and jumping across highways. Spot fires merged and spit out more embers, starting new burns.

Throughout the day, fire after fire ignited across Victoria. A long swath of fire pushed southeastward. Around 5:30 p.m., the winds changed.

"And what happens on the wind change," University of Melbourne fire ecologist Kevin Tolhurst told the Australian Broadcasting Corporation, "is that 50 kilometer flank of the fire then suddenly becomes the head of the fire. So instead of having about a five, six kilometer wide fire, we've got a 50 kilometer wide fire." Jim Baruta watched from his hilltop home in St. Andrews as the fire approached. "It was a hurricane, but it was on fire."

On what came to be known as Black Saturday, fires in Victoria burned over a million acres and killed 173 people.

In the town of Marysville, 90 percent of the town's buildings were destroyed and 34 people killed. Every firefighter in town lost his home. Daryl Hull was working at Marysville's Crossways Inn that day. He testified before the Royal Commission that investigated Black Saturday. As trees burned and fell around him, Hull climbed into a lake. "Everything was on fire....There were fingers of orange flame creeping through the grass on the banks, as though the fire were a living thing....I moved out into the middle of the lake....There was an explosion and everything was luminous orange, and embers began to shower down on me. The embers hissed as they hit the water around me. To take cover from the embers I ducked underneath the water. From under the water I could see the embers descending, like orange lights through green glass....When I surfaced I could see the school going up in flames in front of me. At one point two cars came racing down to the lake. I heard car doors opening and the voices of two men and a child. I thought for a minute that they would get into the lake too, but they didn't. I don't know where they went." Later on the cars exploded."

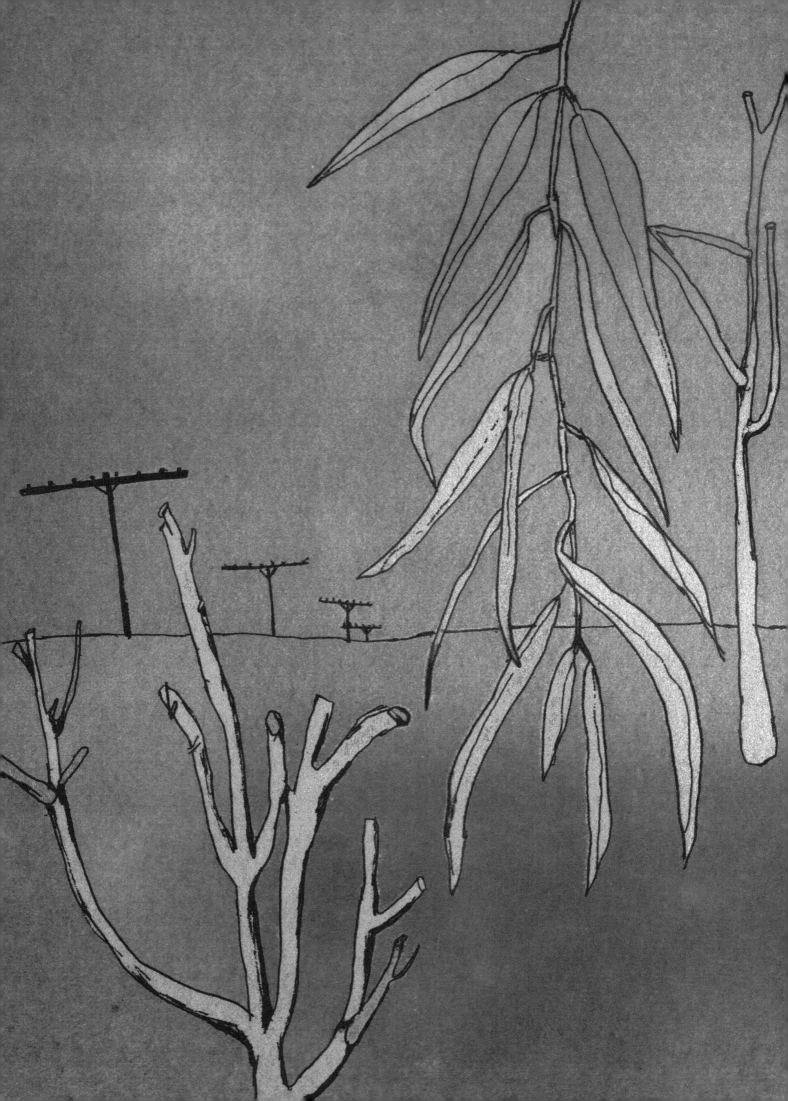

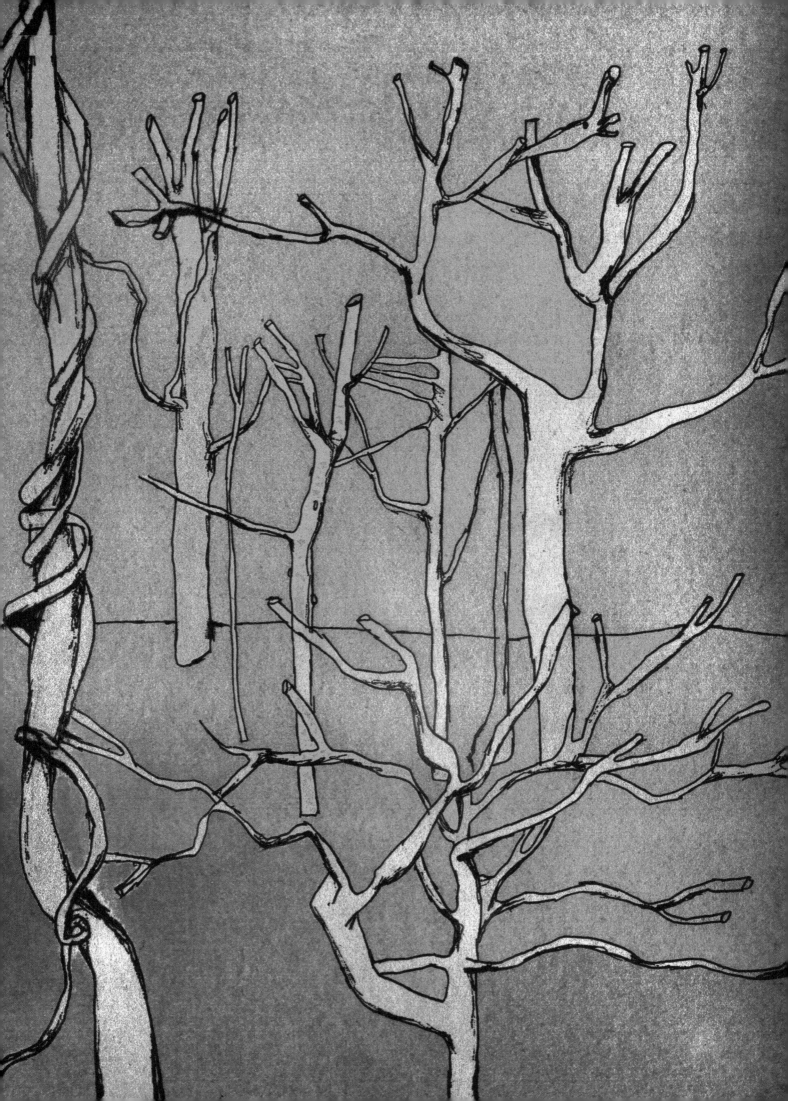

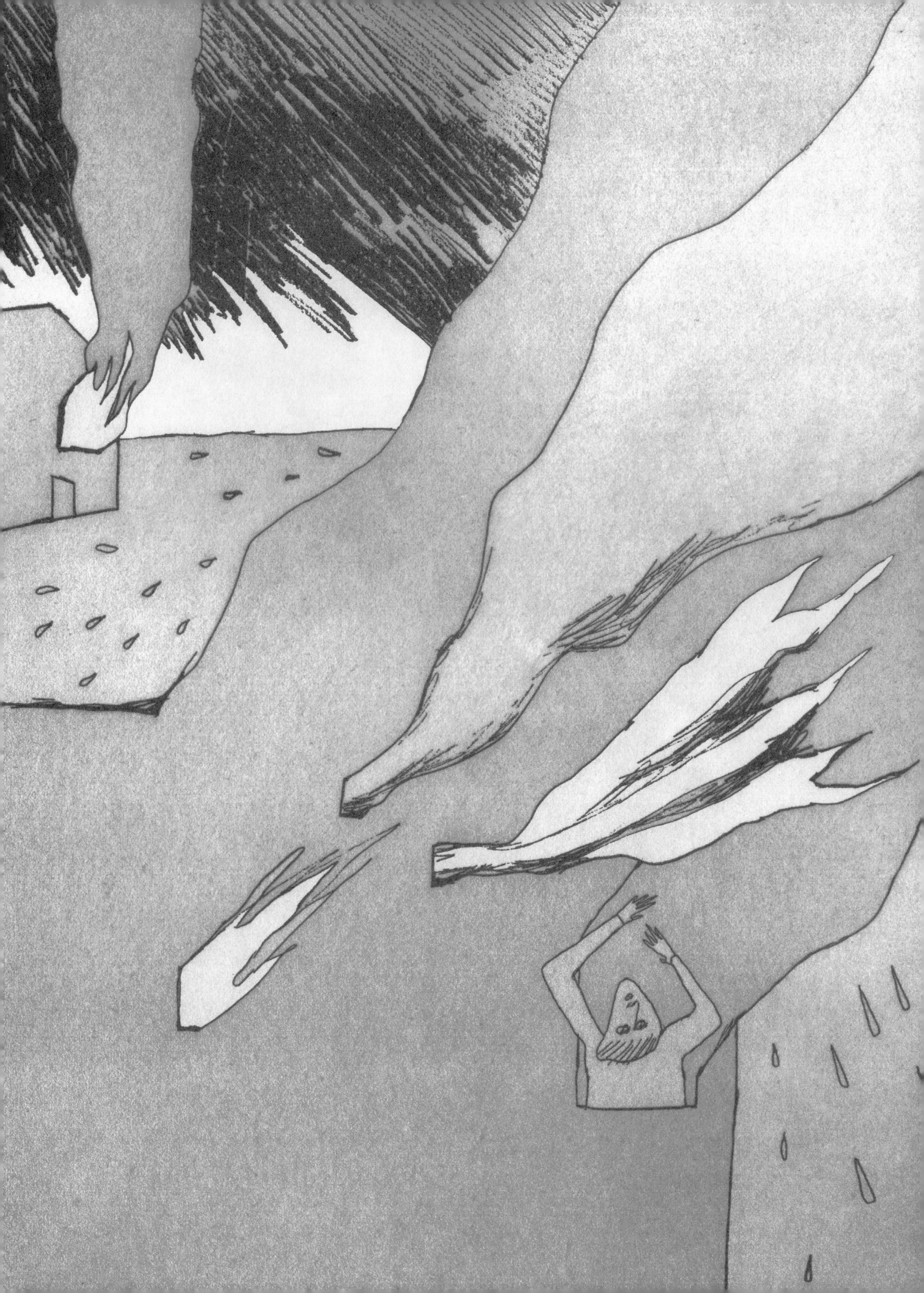

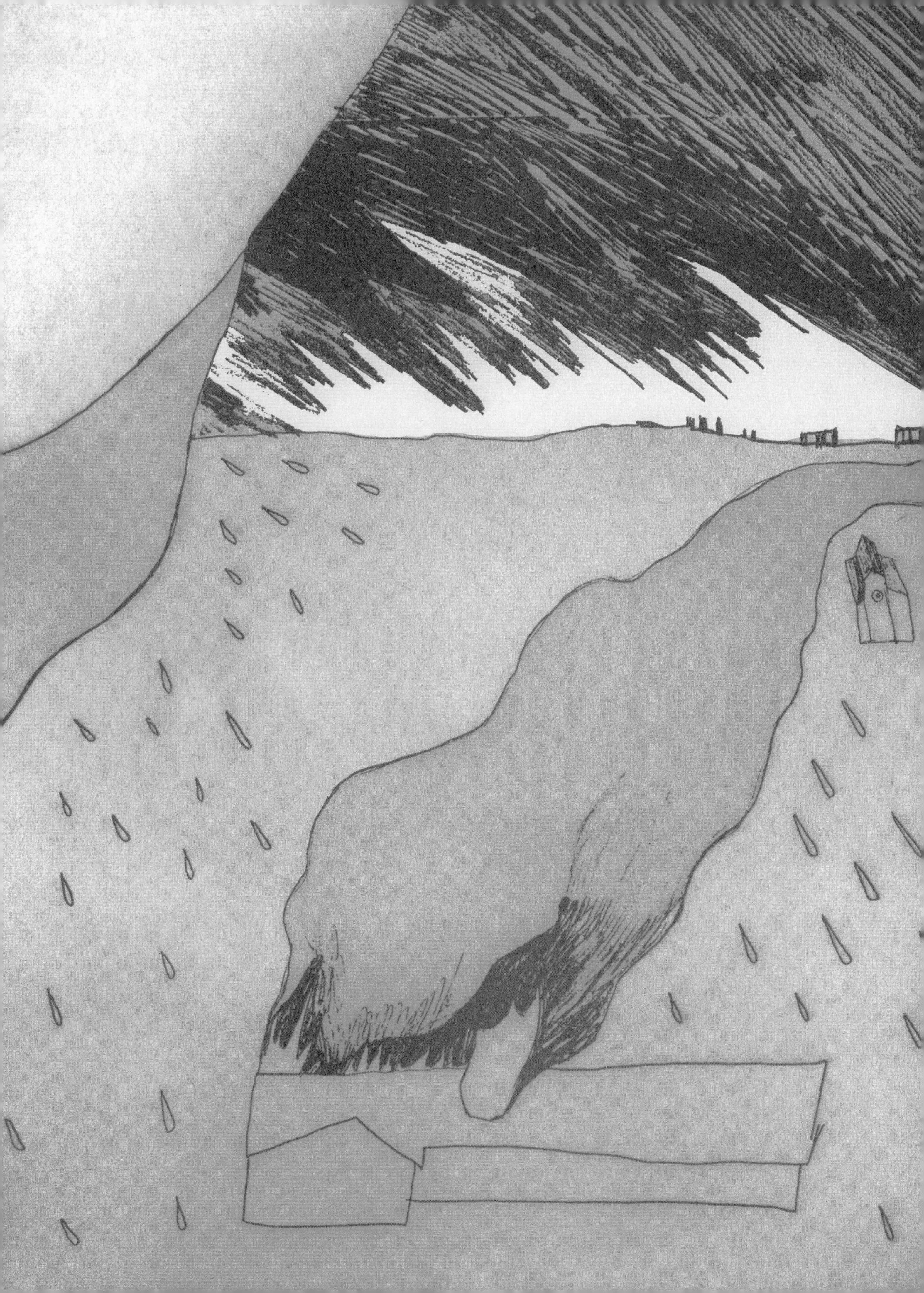

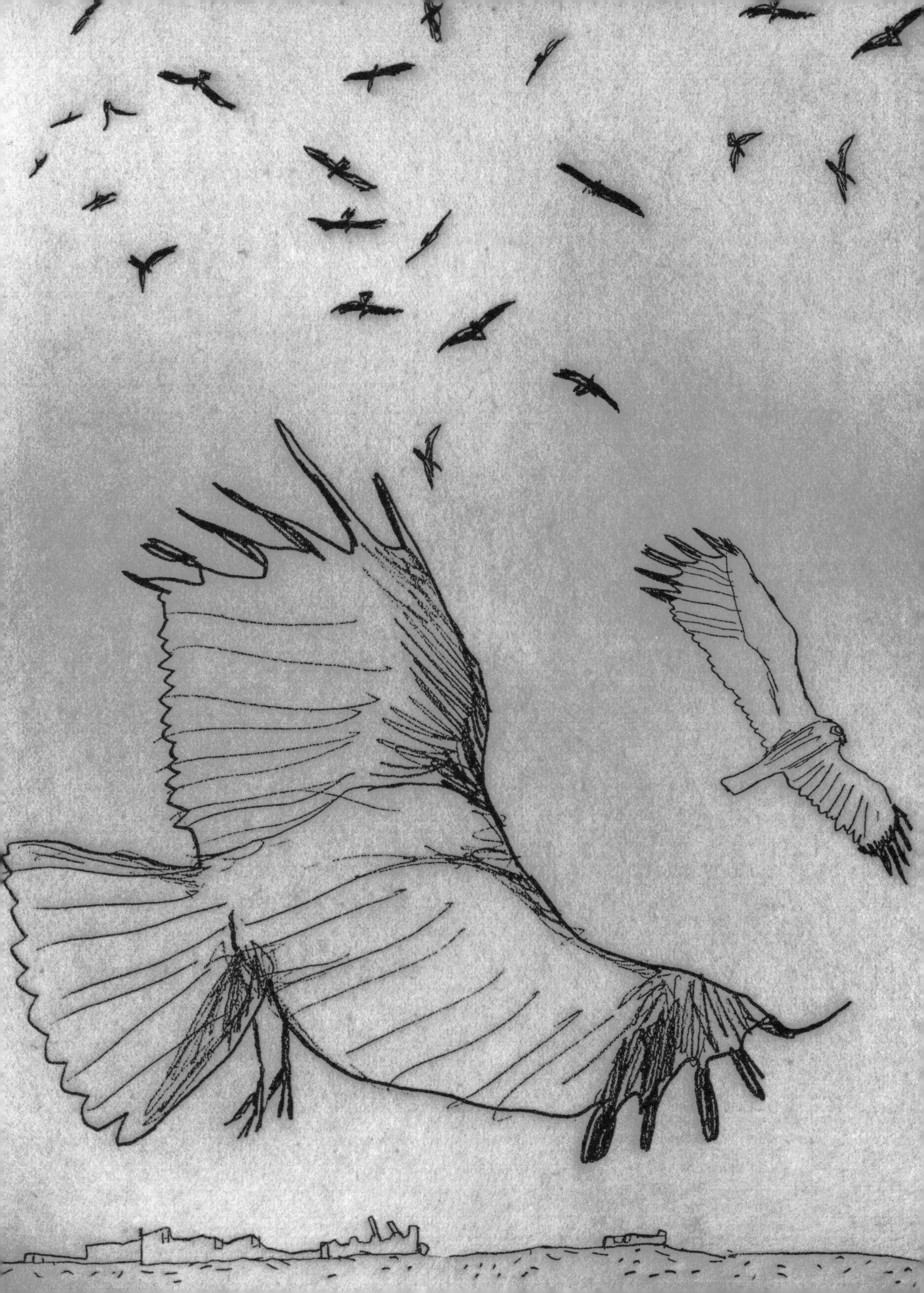

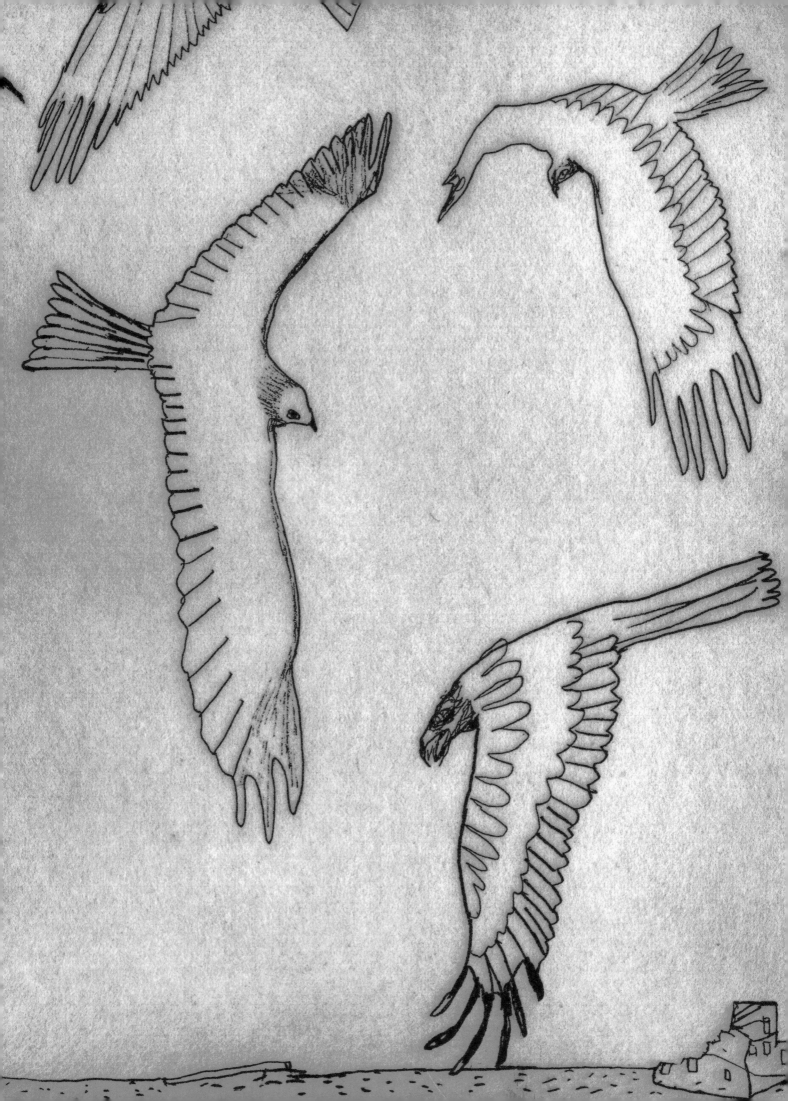

"We face the increased risk of fires almost everywhere," Chris Field, a Stanford forest ecology professor, told *The New York Times* in 2013. A recent Harvard study found that the probability of large fires in the American West will double or, depending on the region, triple by 2050. The fire season will be longer and the air smokier. In recent years, calamitous wildfires have burned in the United States, Europe, South America, South Africa. Around the world, more people are moving into "red zones" or the "wildland urban interface," where development meets uncultivated land and where wildfires can have their most deadly impact.

Even Siberia is on fire. In 2010, temperatures throughout Russia set record highs. There was drought. According to the Information Telegraph Agency of Russia (ITAR-TASS), nearly 2000 wildfires burned in Siberia in 2010. *The Siberian Times* reported that more than 50 people died and that a quarter of Russia's grain harvest was lost in the blazes. The Emergency Situations Ministry said that fire in some regions was traveling at 100 meters per minute. In August, Prime Minister Dmitry Medvedev traveled to Omsk in southwestern Siberia. "The wildfire situation," he said, "is abnormal."

CHAPTER 7:

SKY

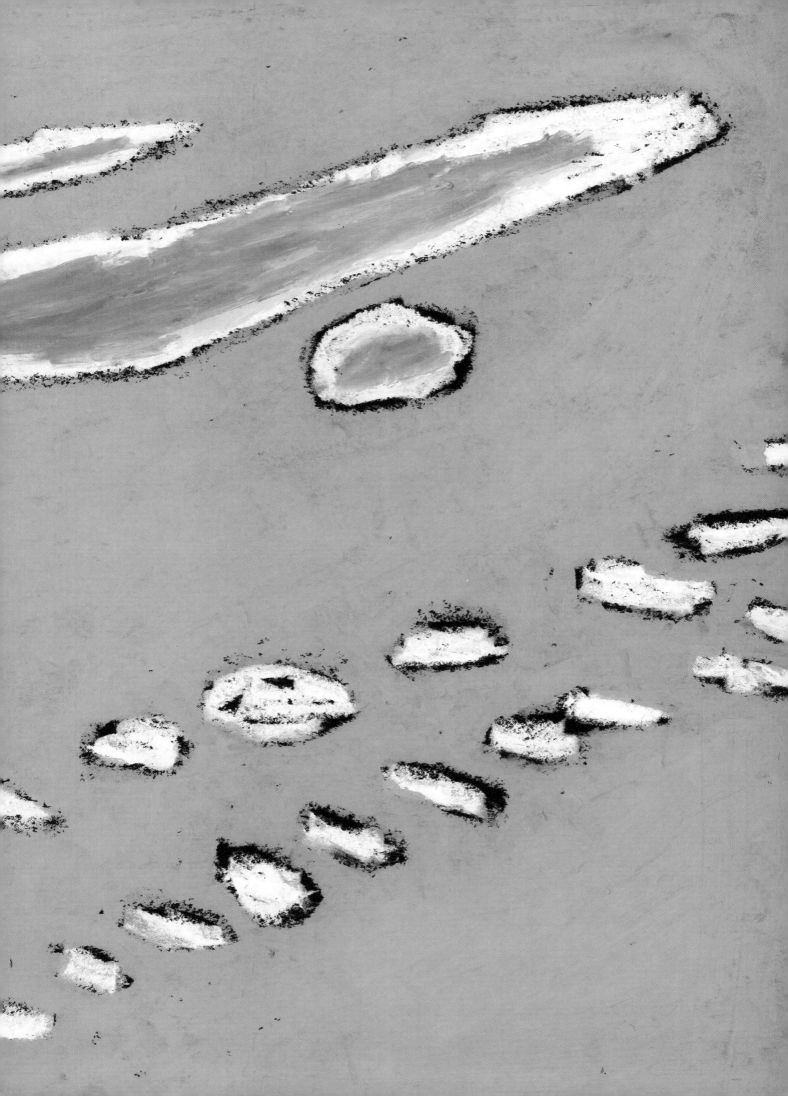

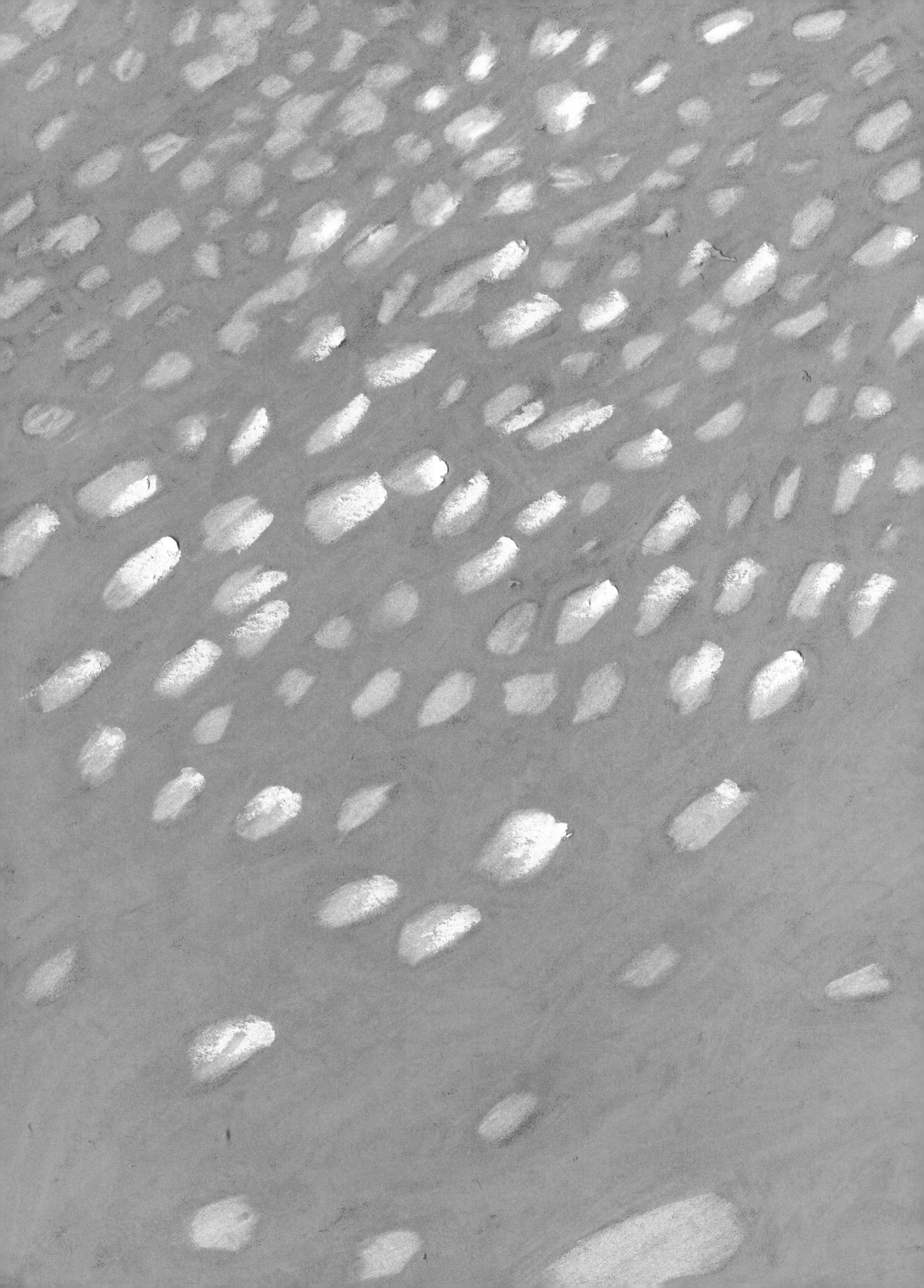

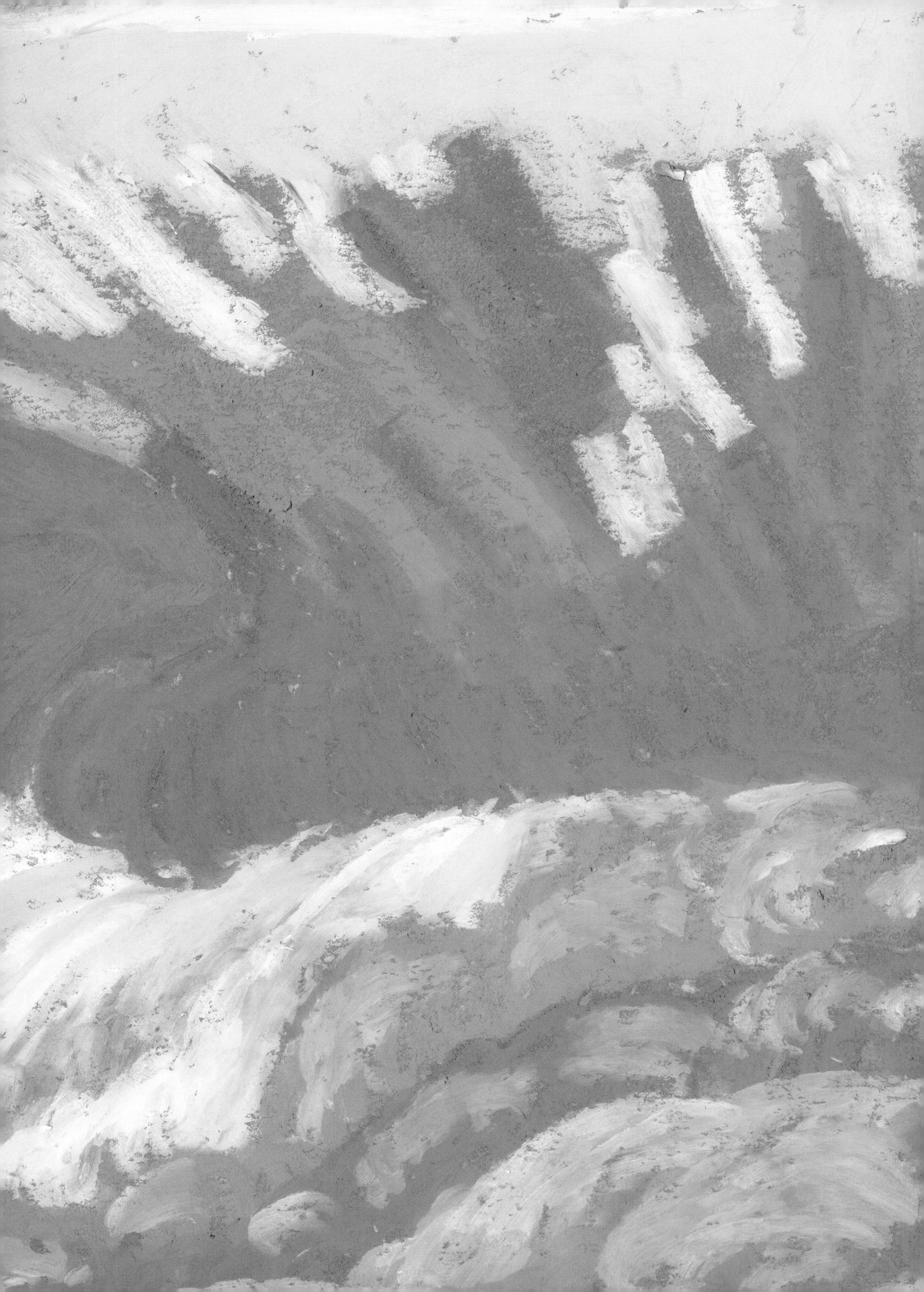

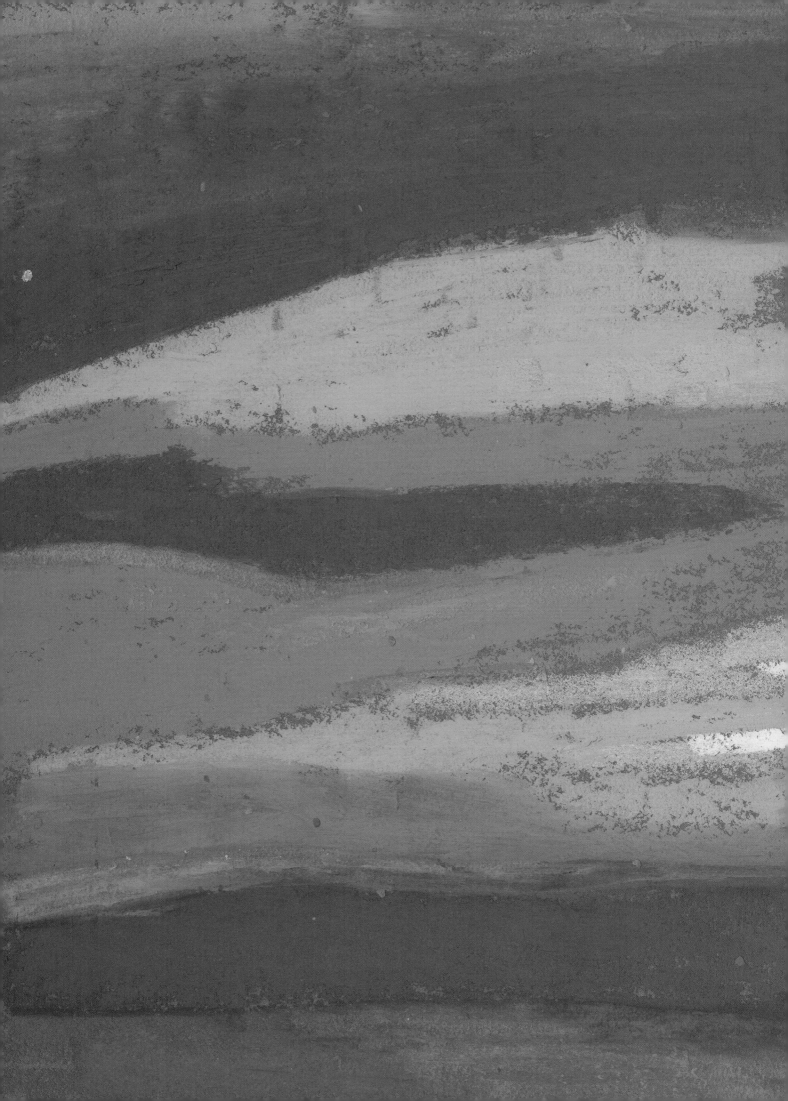

CHAPTER 8:

DOMINION

Two and a half weeks after the death of North Korean supreme leader Kim Jong Il on December 17, 2011, three ethereal photos were posted on the English-language home page of North Korea's most widely read newspaper, the state-controlled *Rodong Sinmun* ("Newspaper of the Workers"). Under the headline "Fascinating Frostwork," the paper ran images of a wintry scene in a village in Ryanggang, a northern province that borders China. The photographs were picturesque, like a Christmas card, showing spindly white birch branches and larch needles shimmering with ice, framed by cobalt blue skies. The "wonderful natural beauty" of the wintry weather, unusual in the region, was attributed to Kim Jong Il. Residents, the article stated, spoke "in one voice," saying, "it seems as if leader Kim Jong Il had unfolded such an amazing scene...telling them always to hold [son and successor] Kim Jong Un in high esteem and do farm work well to bring a bumper harvest this year."

North Korea's official news outlets linked other meteorological phenomena to the death of the Dear Leader. During the last days of Kim Jong Il's life, the Korean Central News Agency reported, winds were stronger, waves higher, and temperatures the coldest of the season — the coldest, in fact, in decades.

In national lore, Mount Paektu, a mountain that straddles the North Korea–China border, is the birthplace of the Korean people, and of Kim Jong Il himself. The majestic Lake Chon, or Heaven Lake, lies in a volcanic crater atop the mountain. On the morning of December 17, ice over Lake Chon fractured with a "loud roar." A group of state researchers testified to the unprecedented volume of the cracking. The ground rumbled and color flooded the sky. "At the view of an unusual glow tinging the sky with deep and clear color, people said in excitement that even nature, unable to forget the heaven-born man, unfolded in the sky a red flag associated with the life of Kim Jong Il." Elsewhere, snow was said to fall from cloudless space, prompting local residents to pronounce, "Kim Jong Il was the heaven-born man and so the sky shed tears at the news of his demise."

For millennia, people have found meaning, and divinity, in weather.

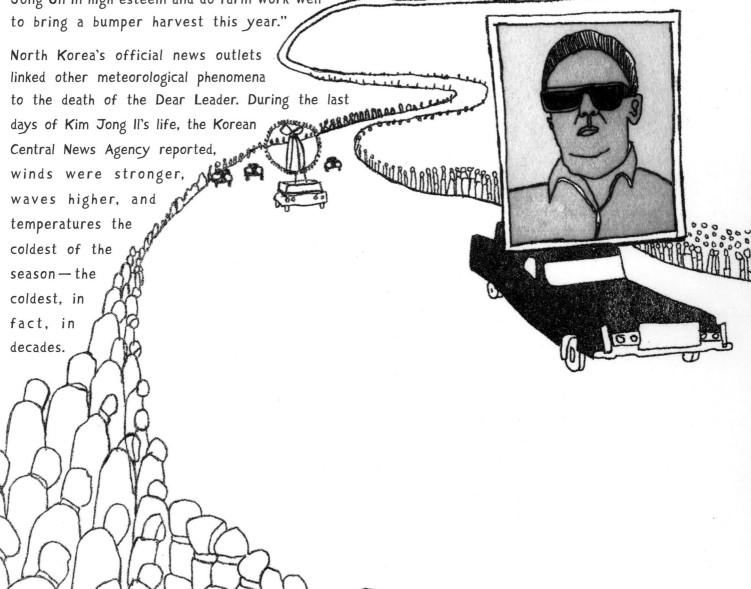

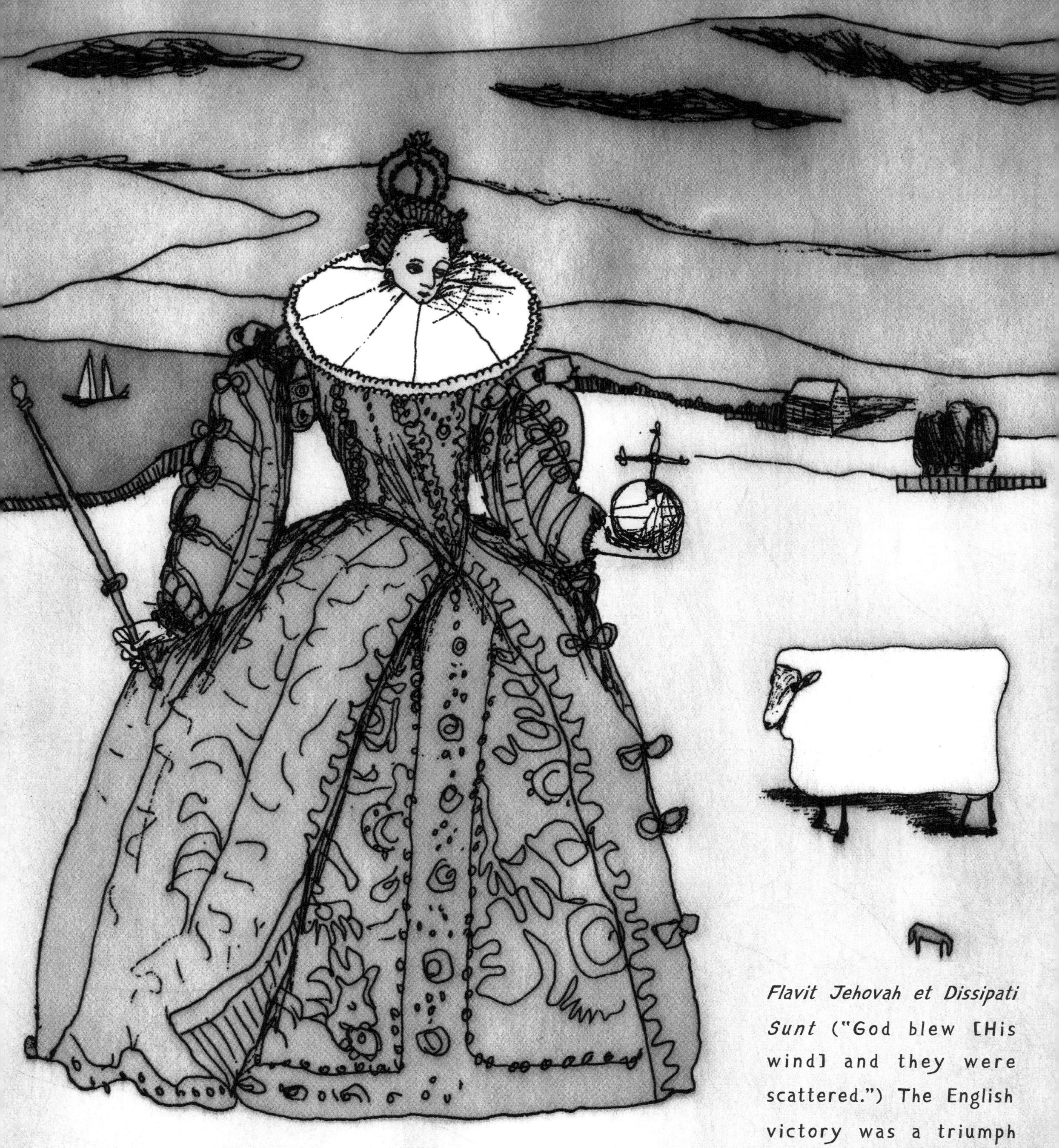

In 1588, the outnumbered and outgunned English fleet defeated the invading Spanish Armada, in part due to favorable tides and winds. Even in retreat the Spanish ships were ravaged by storms. A vanquished King Philip II of Spain lamented, "I sent the Armada against men, not God's winds and waves." In England, celebratory medals were minted with the phrase *Flavit Jehovah et Dissipati Sunt* ("God blew [His wind] and they were scattered.") The English victory was a triumph over the Catholic Church; the weather was evidence of God's own allegiance, and the air currents were dubbed a "Protestant Wind."

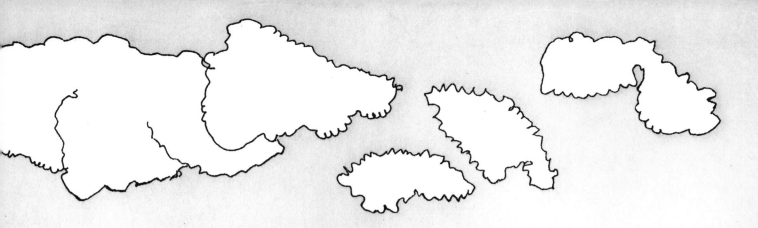

The typhoons that twice saved Japan from

Mongol invasion in the 13th century came to be known

as "kamikaze," or "divine wind." Some said Raijin, the Shinto

god of thunder and lightning and storms, had risen to

protect the nation. Later, the term gave an

ideological spin to World War II suicide missions.

Lyrics from what came to be known as the kamikaze

farewell song compared the pilots to cherry

blossoms, romanticizing their sacrifice:

"Blossoms know they must blow in

the wind someday/Blossoms in the

wind, fallen for their country."

Native American peoples have long practiced weather-conjuring rituals. The Hopi, the Navajo, the Mojave, and other tribes of the arid Southwest perform rain dances with rhythmic footwork and singing. An 1894 issue of *National Geographic* catalogued various other traditions: the Muskingum of Pennsylvania employed old men and women as jugglers to bring rain; the Mandan tribe had both rain callers and rain interrupters, who threatened the skies with bow and arrow. In times of drought, the Choctaw bathed with a fish to summon rain; they roasted sand in a pan to clear the skies. When the Moqui wanted rain, they wrapped wild honey in a husk of maize, chewed it, and spat it out on the parched earth. The Omaha tribes of Kansas would "flap their blankets" to call forth the wind. To stop a blizzard, an Omaha boy was painted red and rolled in the snow. The Omaha had a strategy for fog dispersal, too: tribesmen "drew the figure of a turtle on the ground with its face to the south. On the head, tail, middle of the back, and on each leg were placed small pieces of a (red) breech cloth, with some tobacco."

In the Bible, God speaks to man through meteorological events. In Genesis, grieving over the "wickedness of mankind," God floods the earth. "And I, lo, I am bringing in the deluge of waters on the earth to destroy all flesh" (Genesis 6:17). After wiping out nearly all living creatures — sparing only those refugees on Noah's ark — God ends the rain and bends a rainbow through the sky, a promise to never again destroy life on earth.

Through weather, God shows anger: "Then the Lord rained down burning sulfur on Sodom and Gomorrah — from the Lord out of the heavens" (Genesis 19:24); "The Lord will cause men to hear his majestic voice and will make them see his arm coming down with raging anger and consuming fire, with cloudburst, thunderstorm and hail" (Isaiah 30:30); "And I will rain on him, and on his hordes, and on the many peoples who are with him, an overflowing shower, and great hailstones, fire, and sulfur" (Ezekiel 38:22).

He threatens: "The Lord will send wasting disease, and burning pain, and flaming heat against you, keeping back the rain till your land is waste and dead; so will it be till your destruction is complete" (Deuteronomy 28:22).

He promises to nurture and protect: "The Lord will open to you his good treasure in the sky, to give the rain of your land in its season, and to bless all the work of your hand" (Deuteronomy 28:12).

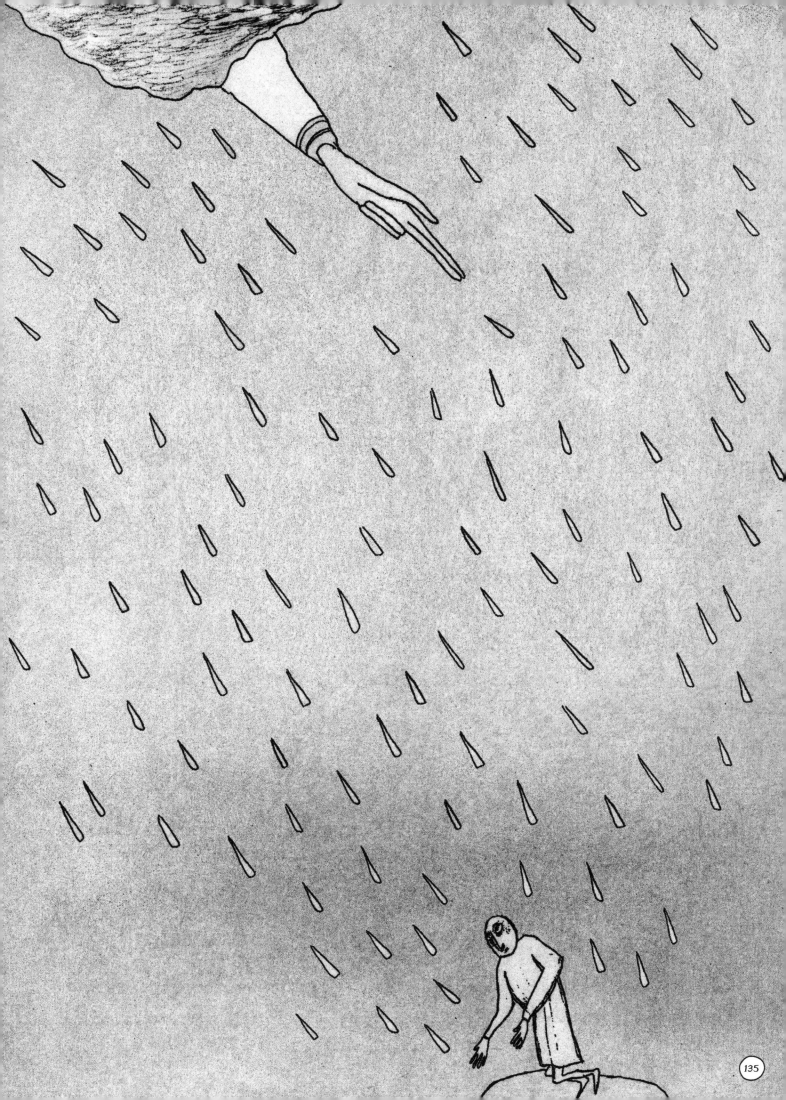

For a stretch of several centuries beginning in about 1300, global temperatures sank and weather became erratic. Europe faced bitter, wet winters, increased snowfall and hailstorms, droughts, floods, and irregular spikes in temperature. Unpredictable weather contributed to crop failures, ailing livestock, food shortages, famine, and disease. Drought, floods, and famine also ravaged Asia. This period is often referred to as the Little Ice Age.

Today, climatologists suggest possible causes for the weather vagaries of the Little Ice Age. A series of volcanic eruptions, dusting the atmosphere with ash and deflecting some of the sun's warming rays from reaching earth, may have played a role. A dip in solar activity may also have played a role. A swing in the North Atlantic Oscillation, in which a usually persistent pattern of low atmospheric pressure over Iceland and high pressure over the Azores reversed, ushering in cold air from the north, was, according to historian Brian Fagan, a "major player."

But no one was offering these theories at the time. People grew hungry and desperate. Some scholars link the unusual weather with a rise in witchcraft trials. Between the 13th and 19th centuries, a million accused witches, mostly poor women and widows, were put to death.

Brian Fagan: "A frenzy of prosecutions coincided with the coldest and most difficult years of the Little Ice Age, when people demanded the eradication of the witches they held responsible for their misfortunes."

In 1484, Innocent VIII issued a Papal Bull: "Many persons of both sexes...give themselves over to devils...and by their incantations, charms, and conjurings, and by other abominable superstitions and sortileges, offences, crimes, and misdeeds, ruin and cause to perish the offspring of women, the foal of animals, the products of the earth, the grapes of vines, and the fruits of trees."

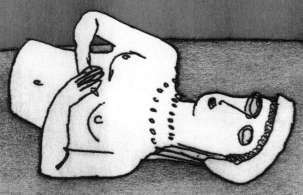

A 15 century witchcraft trial is documented in the *Malleus Maleficarum*, a tract on witchcraft by German church inquisitor Heinrich Kramer. Chapter Fifteen of *Malleus*, "How they Raise and Stir up Hailstorms and Tempests, and Cause Lightning to Blast both Men and Beasts," details the trial of two women near Salzburg. After "a violent hailstorm destroyed all the fruit, crops, and vineyards in a belt one mile wide, so the vines hardly bore fruit for three years," citizens demanded an inquiry, "many...being of the opinion that [the weather] was caused by witchcraft."

After a two week investigation, "a bath-woman," identified only by her first name, Agnes, and a second defendant, Anna von Mindelheim, were charged. "These two were taken and shut up separately in different prisons." Agnes was brought first before a panel of judges. Initially, she claimed innocence, demonstrating "that evil gift of silence." But eventually, she caved. Agnes had "performed coitus with an Incubus devil (...she had been most secret)," she admitted. The author relays Agnes's testimony: she met the devil in a field, under a tree. He commanded her to dig a hole in the ground, to fill it with water, and to stir it with her finger. She had just enough time to rush home before the skies opened up.

In court the following day, Anna von Mindelheim confessed to similar crimes.

On the third day, both women were burned.

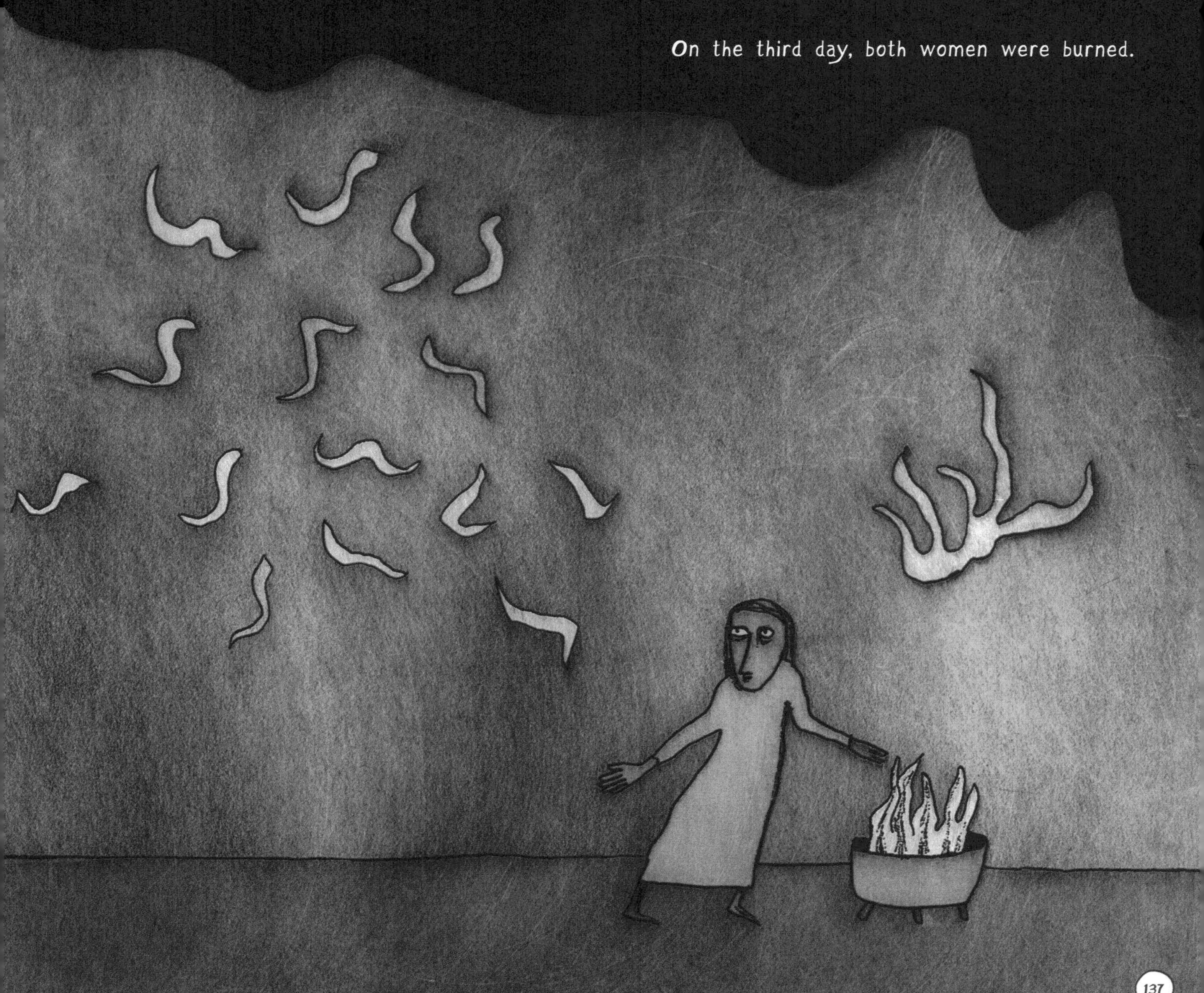

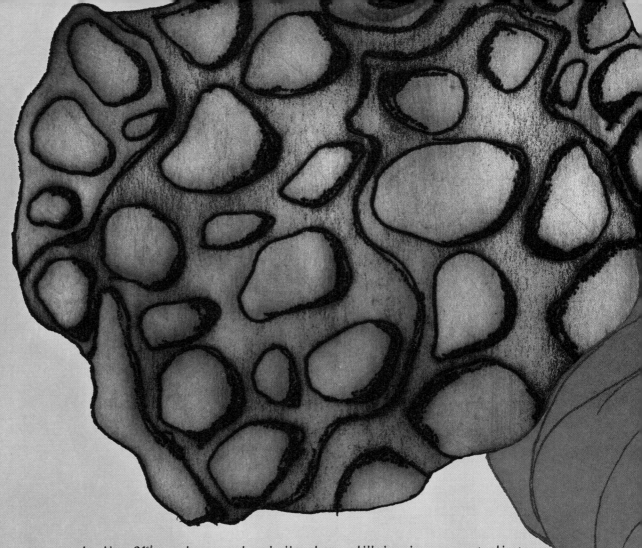

In the 21st century, natural disasters still inspire scapegoating.
Gays and lesbians have been a target. In 1998, Pat Robertson
warned the city of Orlando not to fly rainbow flags "in God's face."
"It'll bring about terrorist bombs, it'll bring earthquakes, tornadoes, and possibly
a meteor." John McTernan, founder of Defend and Proclaim the Faith ministries,
blamed Hurricane Sandy in 2012 on homosexuality. Rabbi Noson Leiter of Torah
Jews for Decency said Sandy was "divine justice" for New York State's
legalization of gay marriage, and that the flooding Lower Manhattan experienced
was because it is "one of the national centers of homosexuality."

Accusations of witchcraft continue, too.

Edward Miguel, professor of economics at the University of California, Berkeley,
sees a pattern in modern Tanzania. According to Miguel, during years in which
flooding or drought cause meager harvests, the resulting famine
conditions are accompanied by a doubling of "witch" murders.

"It would be a mistake to think that belief in witchcraft is confined to the rural areas or to people who are not well educated," says Estelle Trengrove, a lecturer at the University of the Witwatersrand in Johannesburg, who studies mythologies that surround lightning. By telephone from Johannesburg, Trengrove described a conversation with three fourth-year engineering students from Zulu families:

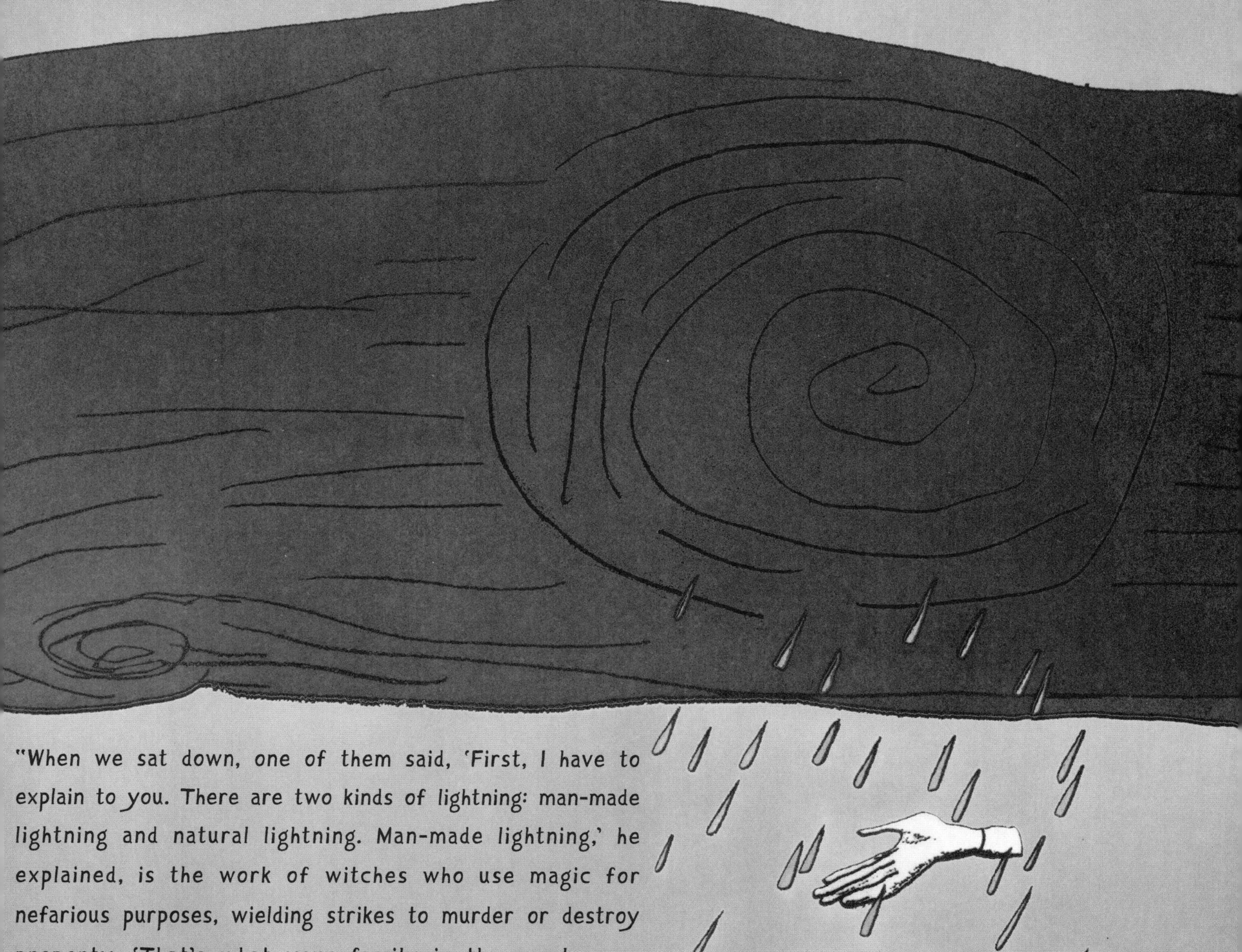

"When we sat down, one of them said, 'First, I have to explain to you. There are two kinds of lightning: man-made lightning and natural lightning. Man-made lightning,' he explained, is the work of witches who use magic for nefarious purposes, wielding strikes to murder or destroy property. 'That's what your family in the rural areas believe,' I replied. And he said, 'No. I'm explaining it to you. I'm explaining it to you because I can see you don't understand.' All the science he learned applied to 'natural' lightning, but 'man-made' lightning in his opinion fell into a completely different category, not governed by physics."

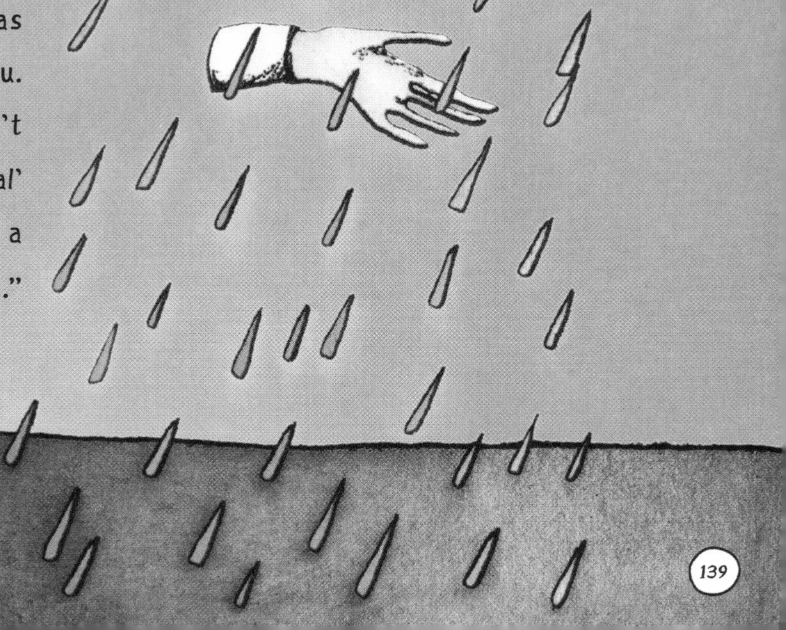

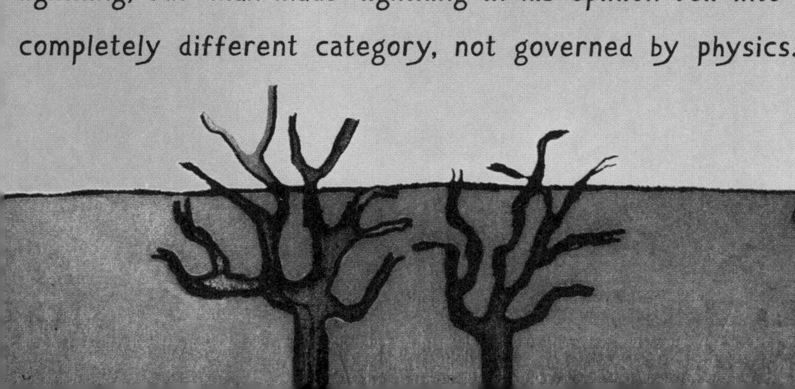

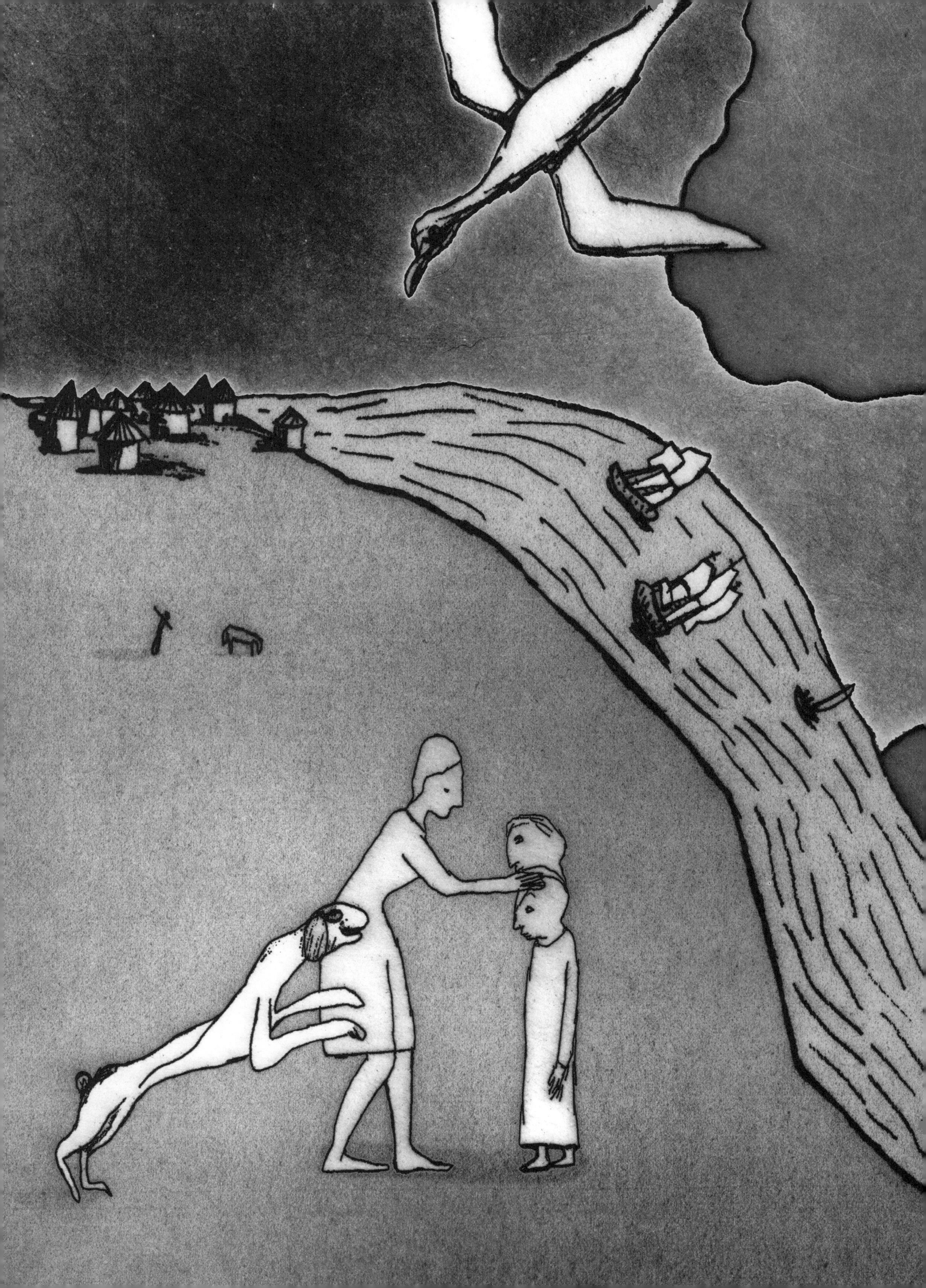

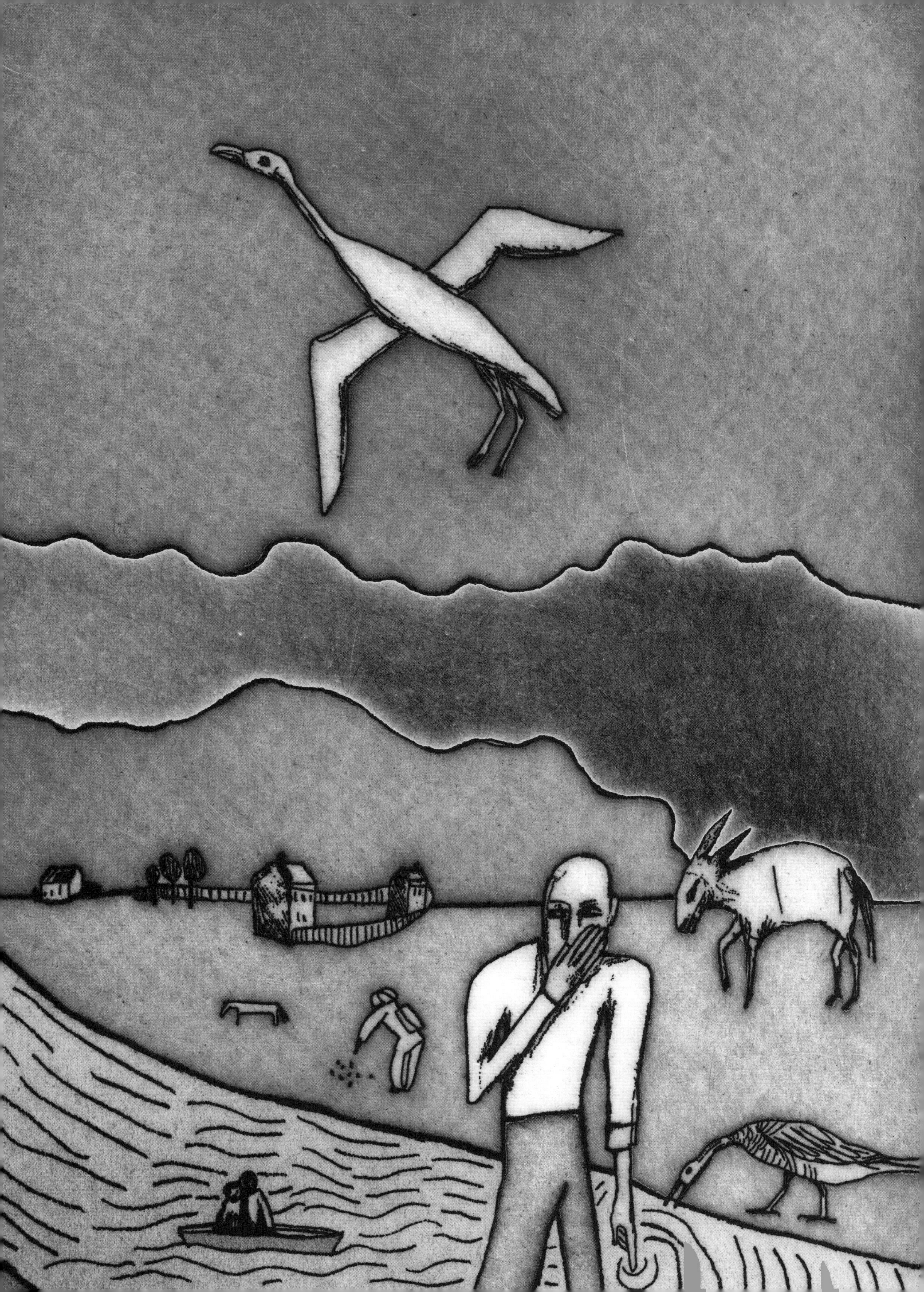

"Weather" is the state of the atmosphere at a given moment: temperature, precipitation, humidity, wind speed and direction, cloud cover, barometric pressure. "Climate" is a summary picture of the prevailing weather patterns in a certain region over an extended period of time — that is, climate describes the *usual* weather in a particular place. "We dress for the weather and build houses in accordance with the climate," writes scientist Edmond Mathez. Any one weather event can deviate from normal averages without necessarily signaling a larger shift in climate, but changes to climate, by definition, mean changes in the weather.

Scientists agree that we are living in an age of global climate change.

The Intergovernmental Panel on Climate Change has declared, "Warming of the climate system is unequivocal." This warming, the IPCC says, is due principally to greenhouse gas emissions.

Human activities are transforming the planet. The consequences, scientists contend, include warmer temperatures, extreme events, wildfires, floods and droughts, rising sea levels, and species extinction.

Harold Brooks, meteorologist at the National Severe Storms Lab at the National Oceanic and Atmospheric Administration: "The planet is warming, and the planet will continue to warm. That's almost an uninteresting statement, it's so obvious."

Many researchers, governments, and military strategists see climate change as a potential "threat multiplier." The Pentagon's 2010 Quadrennial Defense Review stated: "Climate change could have significant geopolitical impacts around the world, contributing to poverty, environmental degradation, and the further weakening of fragile governments. Climate change will contribute to food and water scarcity, will increase the spread of disease, and may spur or exacerbate mass migration." The 2014 QDR reiterated the concerns, adding that climate change will aggravate "conditions that can enable terrorist activity and other forms of violence."

In 2014, the Military Advisory Board of CNA Corporation, a government-funded military research non-profit, published "National Security and the Accelerating Risks of Climate Change." The report stated that climate change is already a catalyst "for instability and conflict" across the globe. Within the United States, the report forecast, climate change will "place key elements of our National Power at risk and threaten our homeland security." The report continued: "When it comes to thinking about how the world will respond to projected changes in the climate, we believe it is important to guard against a failure of imagination."

Some scientists and researchers are envisioning radical solutions: deliberate intervention in earth's systems, projects commonly referred to as geoengineering. Geoengineers seek to dim the sun, churn the oceans, cool things down.

If human activity is already inadvertently affecting climate, couldn't we, shouldn't we, take conscious steps to counterbalance the negative effects?

Can mankind and technology replace God and magic to claim dominion over the weather?

For over a decade Nathan Myhrvold was chief technology officer at Microsoft, where he established Microsoft's research division. In 1999, Myhrvold left Microsoft to found Intellectual Ventures, an "idea factory." Myhrvold holds Ph.D.s in mathematical economics and theoretical physics from Princeton University. He worked with Stephen Hawking researching "quantum field theory in curved space time." He has hunted for dinosaur fossils in Montana and Mongolia. He co-authored *Modernist Cuisine*, a six-volume opus on molecular gastronomy. In 1991, he won first prize at the World Championship of Barbecue. Intellectual Ventures has a proposal to counter the warming of the planet. It is a device Myhrvold calls the Stratoshield.

Nathan Myhrvold: "At our current course and speed, we're going to wind up cooking the planet. Lots of reasonable argument can occur as to how likely that is, or on what time frame that is, but a lot of it is just a question of when. And because we've done essentially nothing about global warming so far — nothing measurable — I just don't see how we escape it. If we do escape it, great. In the meantime, let's prepare for it."

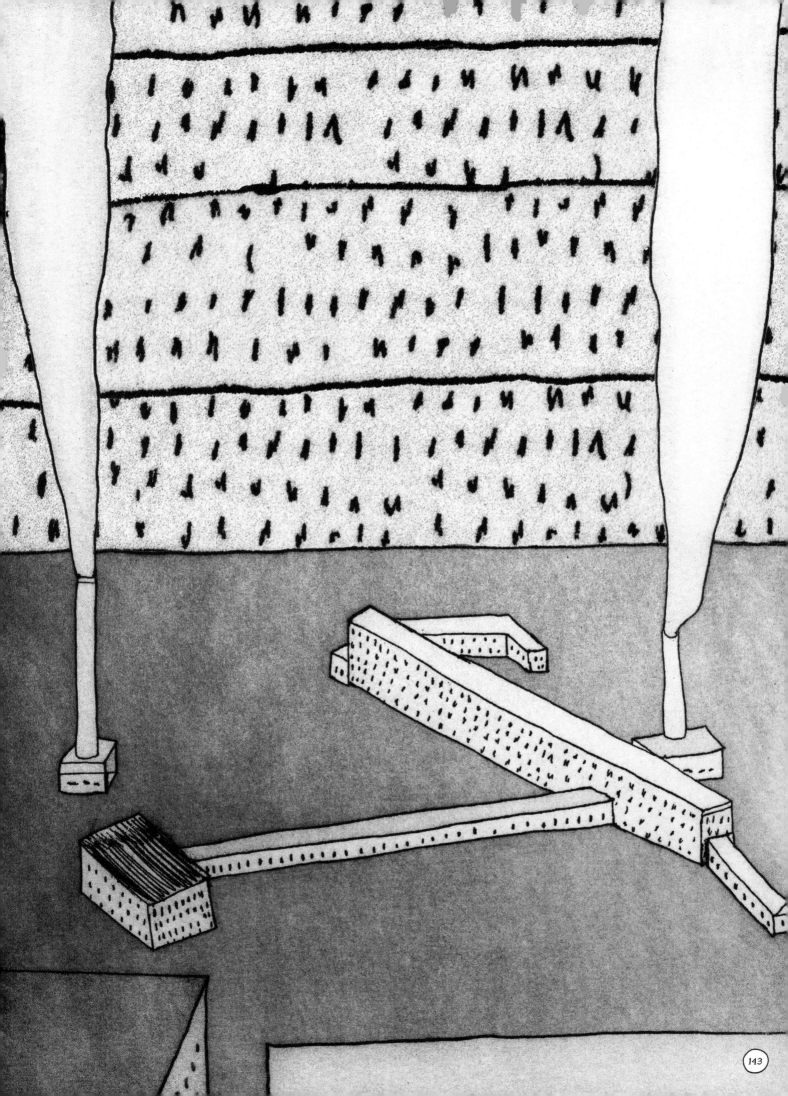

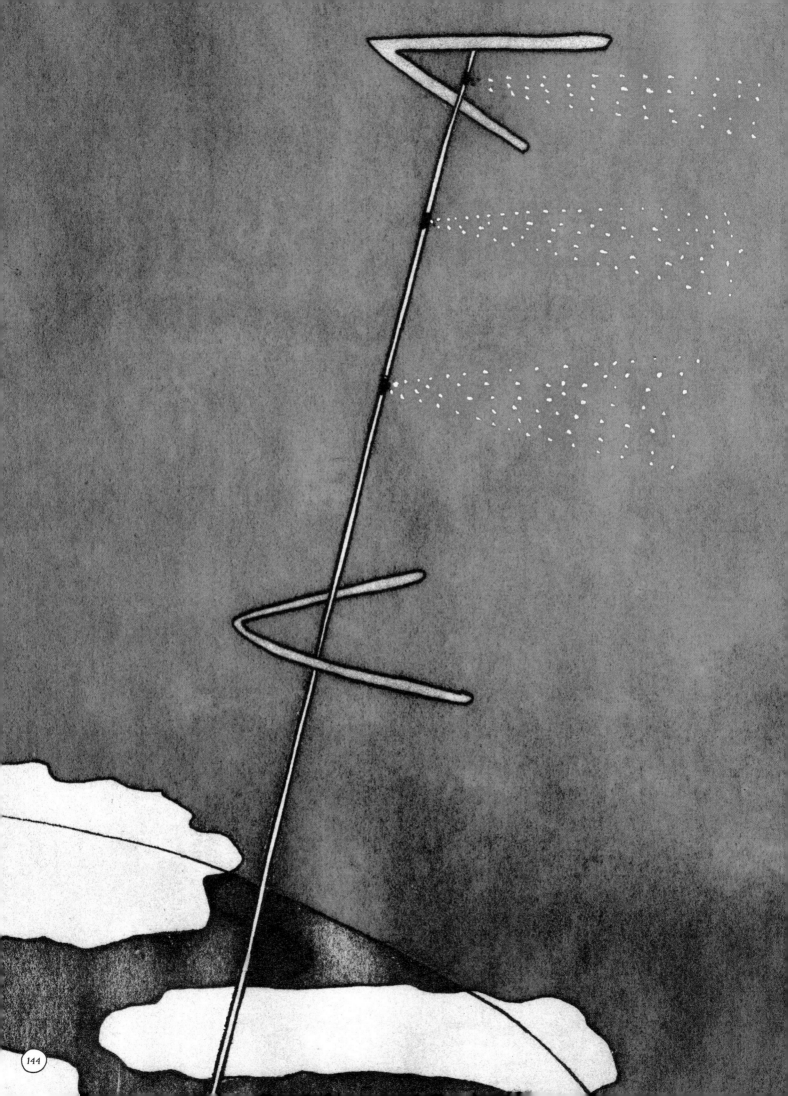

Geoengineering strategies are generally divided into two categories. One, carbon dioxide removal, seeks to mitigate the heat-trapping effect of carbon dioxide by pulling CO_2 out of the atmosphere. A second method, solar radiation management, attempts to prevent a certain amount of sunlight from penetrating the atmosphere, or to increase the amount of sunlight reflected back into space, in order to lower global temperatures.

Nathan Myhrvold: "One approach is called solar radiation management. SRM. There are acronyms for all this shit because people love acronyms, I guess. What it means is: let's bounce some of the sunlight back out into space. Our Stratoshield method is, in our view, the cheapest, most practical way anyone's come up with so far of doing that. We're trying to make the sun one percent dimmer.

"The first ideas to do this occurred in the 1960s to a guy named Mikhail Budyko, who is a Soviet scientist. Budyko realized that sulfate materials, which come in natural form from volcanic eruptions, dampen the sunlight. We know that when Mount Pinatubo erupted in 1991, it caused global temperatures to drop by about one degree for about 18 months. So all we need is one Pinatubo a year and we'd be fine. So the question is: how do you do one Pinatubo a year? And people came with a variety of ideas. People thought, Let's take artillery guns and shoot the artillery guns straight up in the air with artillery shells that would explode. Let's send it up in rockets. Let's pack the stuff into 747s. Those schemes would probably work, but they're very expensive.

"We came up with the simplest approach, which is: you run a hose up to the sky. It sounds really dumb, but it's very simple and it's very cheap. So we designed this in a bunch of detail. You take a series of balloons that hold up a pipe. The pipe has a bunch of little electric pumps on it. The inside diameter of the pipe is somewhere between one inch and two inches in diameter. It's like a big garden hose.

"It's made out of an ordinary pipe material. You don't need anything super strong. And you'd have a whole string of balloons that would hold this up into place, and a whole string of little electric pumps. The cool ones that we've done are V-shaped. But you could use a round balloon. The V-shaped ones are better in the wind. Every hundred meters of tubing has a little balloon that holds it up. We call that the string-of-pearls design.

"Then you need the material to put up there. The simplest material to put up there is sulfate, which is a sulfur dioxide. It's completely natural.

"Anyway, it's a couple of garden hoses. If you do a calculation of how many of these units you need, you need one per hemisphere. They're long garden hoses, and they're going up into the sky. One of them could negate global warming for an entire hemisphere.

"The best scheme we've come up with is to put it in the high Arctic, or the low Antarctic. We would put it somewhere near the Arctic Circle. There are a couple of ideal places in Canada.

"So you're spraying that stuff up there. At the end of the hose there are some spray nozzles that spray it out as a fine mist. You've got a bunch of ways of monitoring it so that you can adjust it very accurately, and you could decide to stabilize the temperature, or stabilize the climate, at any temperature you wanted. So you could say, 'Let's stabilize it at today's temperature.' That's probably the best thing to do. But you could also say, 'Let's take it back to the preindustrial climate. Let's just negate all global warming.'"

Many people find geoengineering initiatives frightening. Some, for instance, dread the predicted end of blue skies — the color dulled by sulfate particles clouding the atmosphere. (Myhrvold says that any lightening of the sky would be imperceptible to the naked eye.) Until recently, geoengineering research was, Jeff Goodell writes in *How to Cool the Planet*, "the scientific equivalent of a porn habit, something you thought about and explored in the privacy of your own lab but did not discuss in polite company."

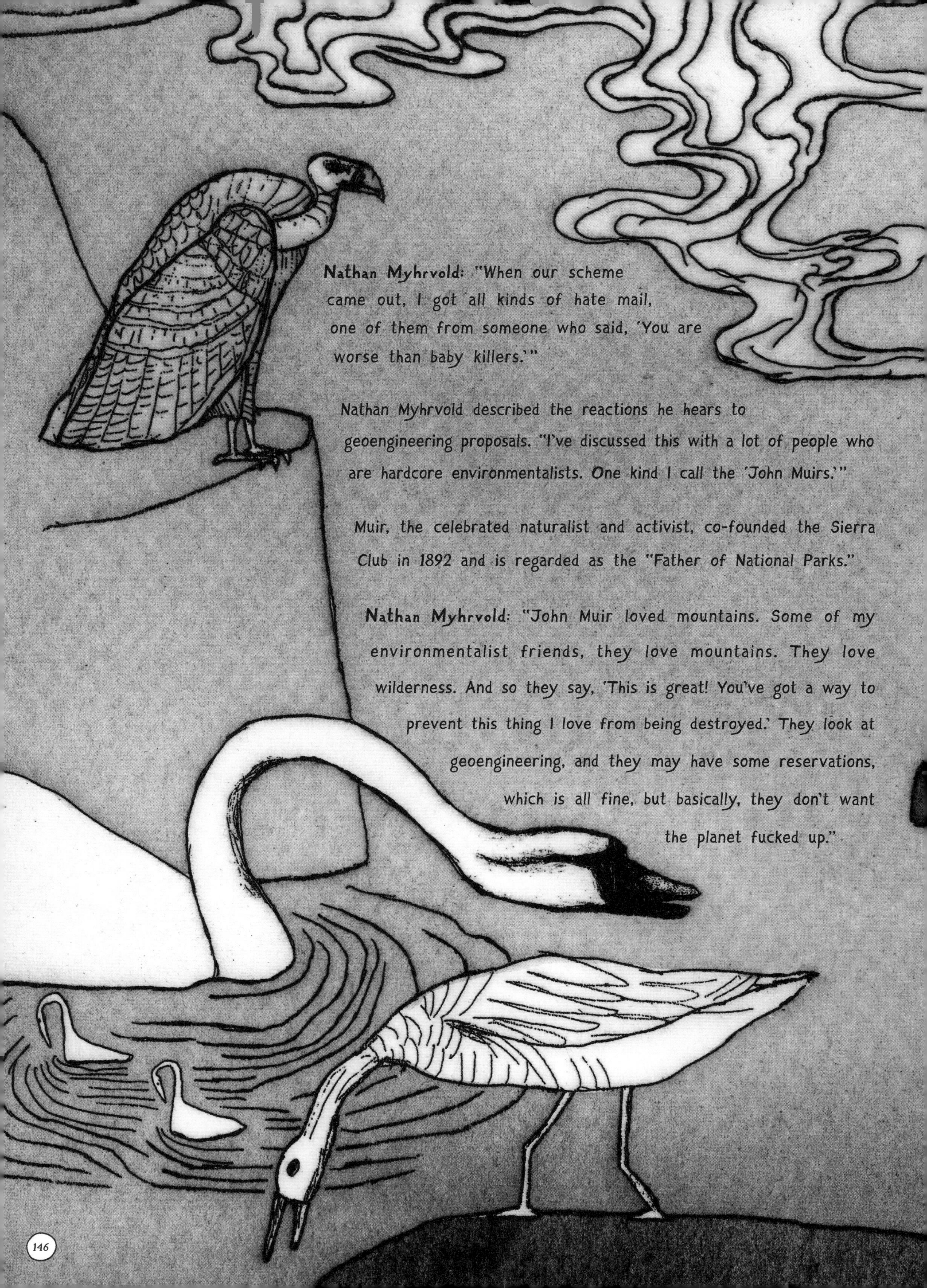

Nathan Myhrvold: "When our scheme came out, I got all kinds of hate mail, one of them from someone who said, 'You are worse than baby killers.'"

Nathan Myhrvold described the reactions he hears to geoengineering proposals. "I've discussed this with a lot of people who are hardcore environmentalists. One kind I call the 'John Muirs.'"

Muir, the celebrated naturalist and activist, co-founded the Sierra Club in 1892 and is regarded as the "Father of National Parks."

Nathan Myhrvold: "John Muir loved mountains. Some of my environmentalist friends, they love mountains. They love wilderness. And so they say, 'This is great! You've got a way to prevent this thing I love from being destroyed.' They look at geoengineering, and they may have some reservations, which is all fine, but basically, they don't want the planet fucked up."

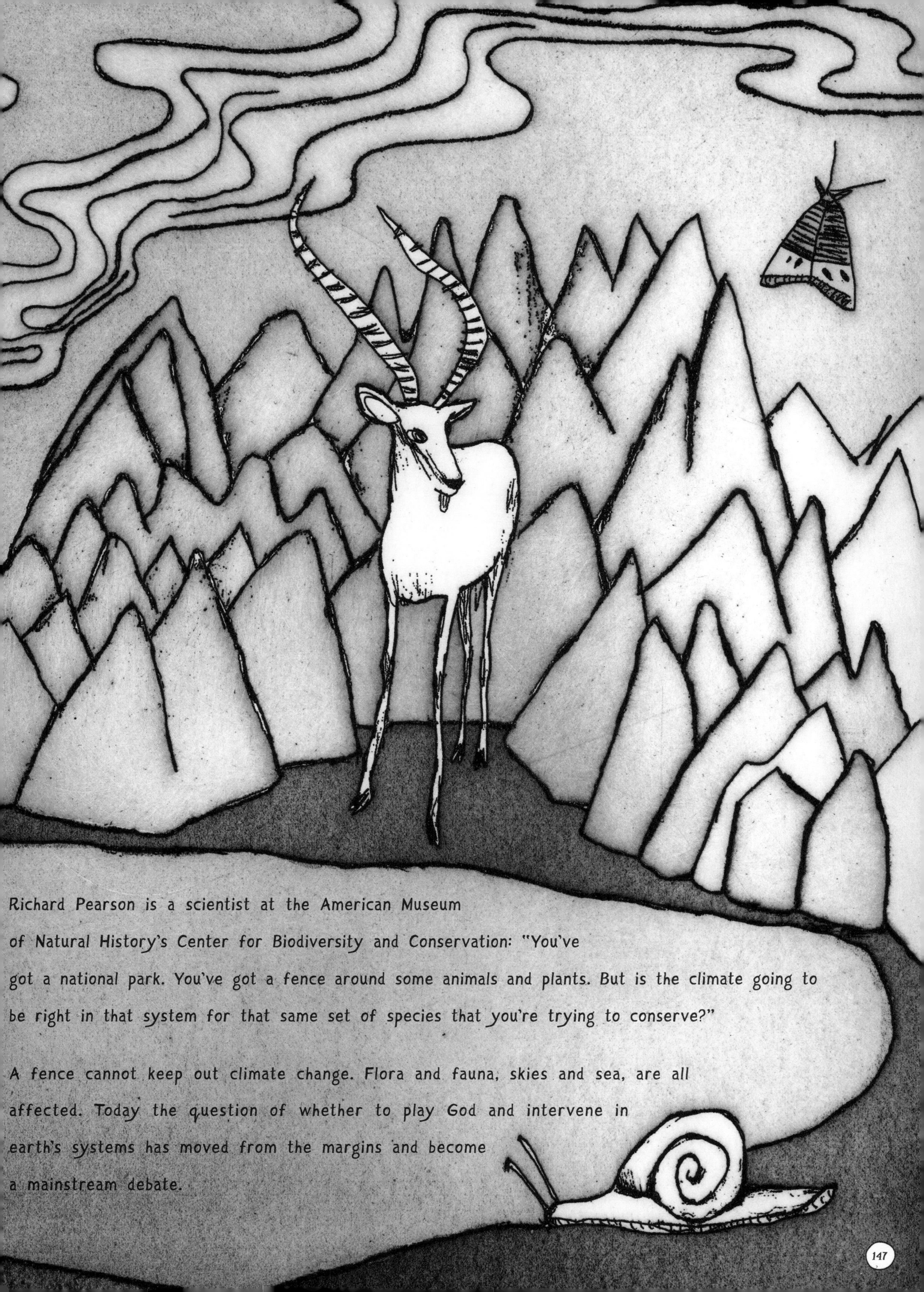

Richard Pearson is a scientist at the American Museum of Natural History's Center for Biodiversity and Conservation: "You've got a national park. You've got a fence around some animals and plants. But is the climate going to be right in that system for that same set of species that you're trying to conserve?"

A fence cannot keep out climate change. Flora and fauna, skies and sea, are all affected. Today the question of whether to play God and intervene in earth's systems has moved from the margins and become a mainstream debate.

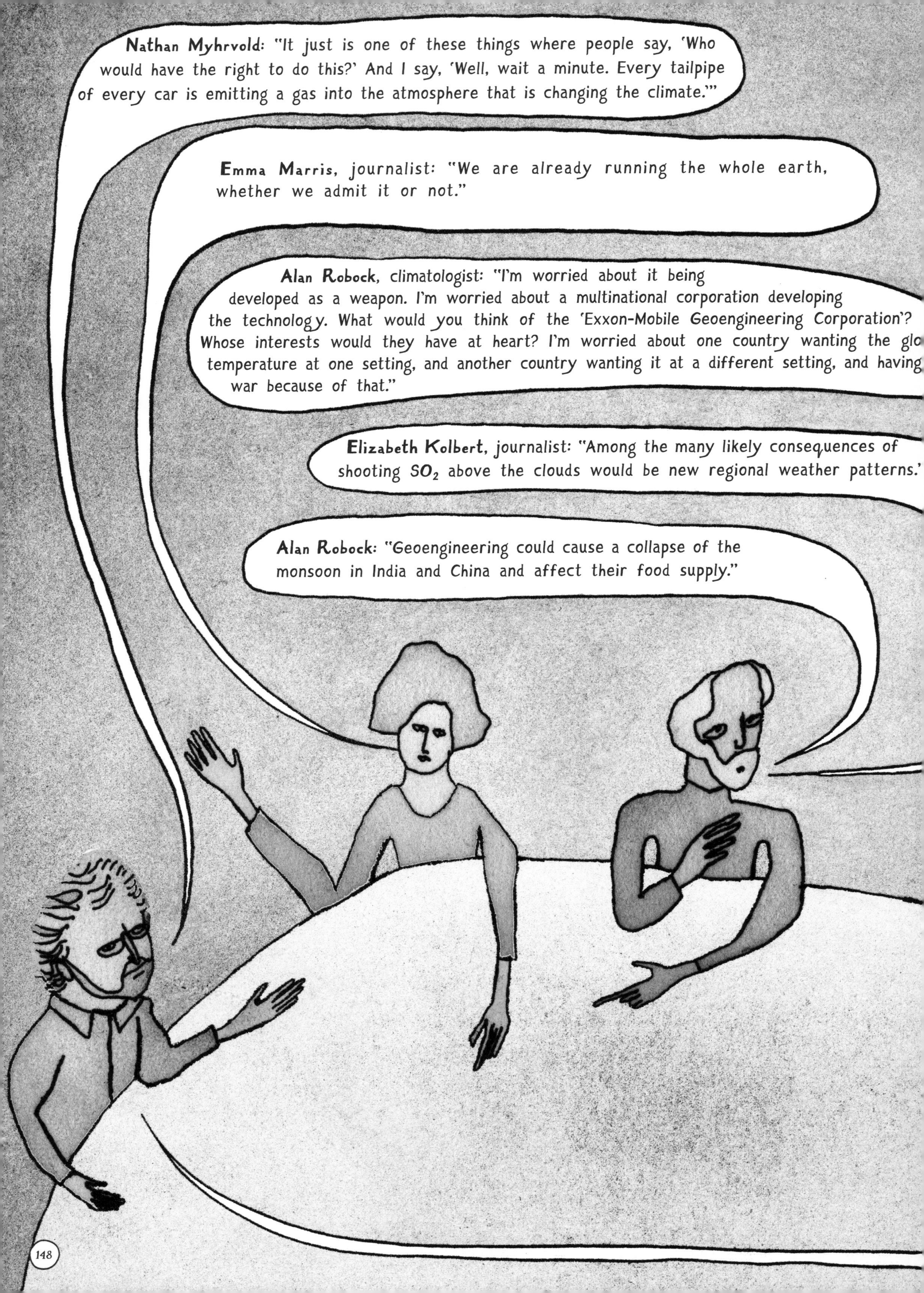
Nathan Myhrvold: "It just is one of these things where people say, 'Who would have the right to do this?' And I say, 'Well, wait a minute. Every tailpipe of every car is emitting a gas into the atmosphere that is changing the climate.'"
Emma Marris, journalist: "We are already running the whole earth, whether we admit it or not."
Alan Robock, climatologist: "I'm worried about it being developed as a weapon. I'm worried about a multinational corporation developing the technology. What would you think of the 'Exxon-Mobile Geoengineering Corporation'? Whose interests would they have at heart? I'm worried about one country wanting the glo temperature at one setting, and another country wanting it at a different setting, and having war because of that."
Elizabeth Kolbert, journalist: "Among the many likely consequences of shooting SO_2 above the clouds would be new regional weather patterns.'
Alan Robock: "Geoengineering could cause a collapse of the monsoon in India and China and affect their food supply."

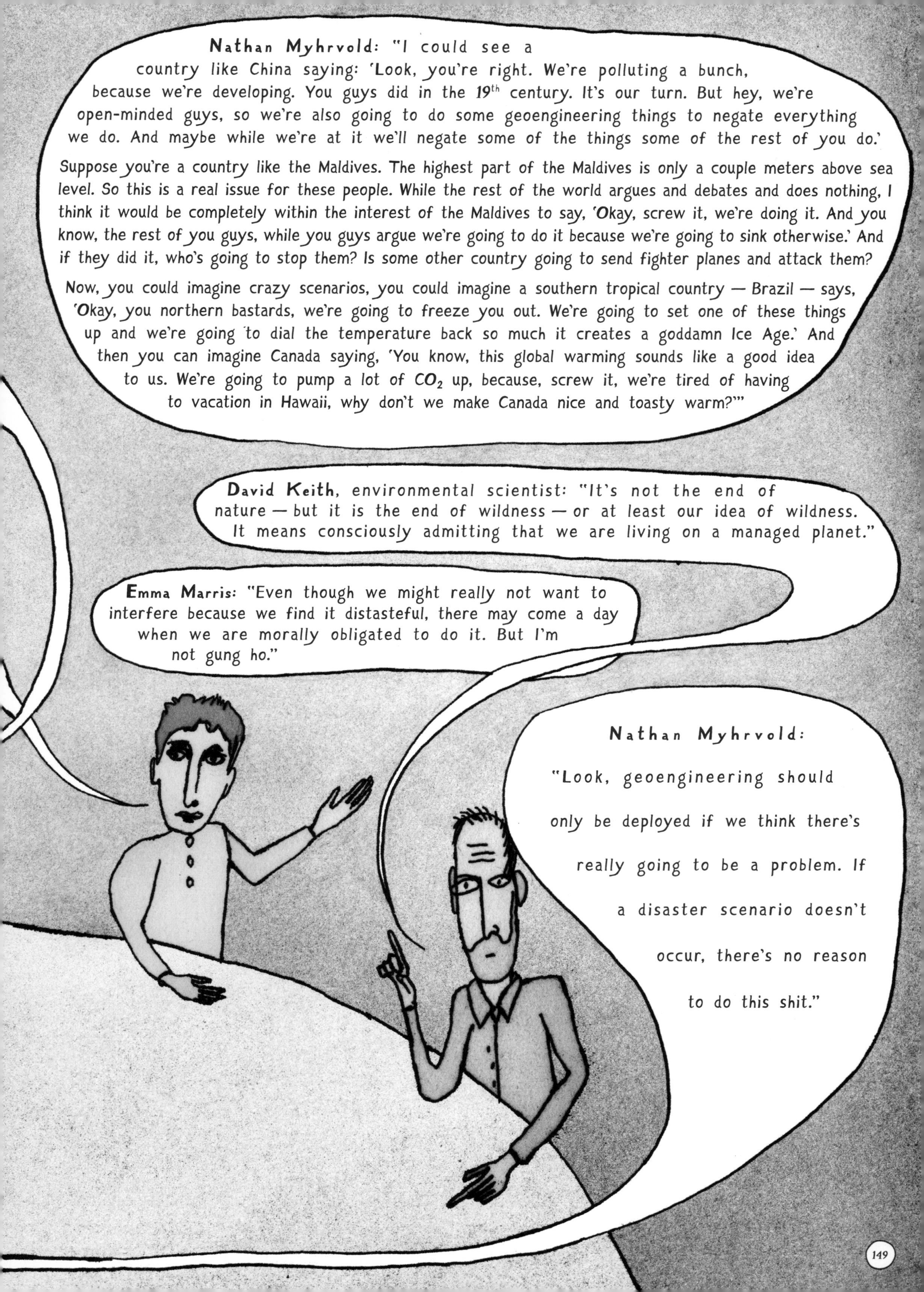
Nathan Myhrvold: "I could see a country like China saying: 'Look, you're right. We're polluting a bunch, because we're developing. You guys did in the 19th century. It's our turn. But hey, we're open-minded guys, so we're also going to do some geoengineering things to negate everything we do. And maybe while we're at it we'll negate some of the things some of the rest of you do.'
Suppose you're a country like the Maldives. The highest part of the Maldives is only a couple meters above sea level. So this is a real issue for these people. While the rest of the world argues and debates and does nothing, I think it would be completely within the interest of the Maldives to say, 'Okay, screw it, we're doing it. And you know, the rest of you guys, while you guys argue we're going to do it because we're going to sink otherwise.' And if they did it, who's going to stop them? Is some other country going to send fighter planes and attack them?
Now, you could imagine crazy scenarios, you could imagine a southern tropical country — Brazil — says, 'Okay, you northern bastards, we're going to freeze you out. We're going to set one of these things up and we're going to dial the temperature back so much it creates a goddamn Ice Age.' And then you can imagine Canada saying, 'You know, this global warming sounds like a good idea to us. We're going to pump a lot of CO_2 up, because, screw it, we're tired of having to vacation in Hawaii, why don't we make Canada nice and toasty warm?'"
David Keith, environmental scientist: "It's not the end of nature — but it is the end of wildness — or at least our idea of wildness. It means consciously admitting that we are living on a managed planet."
Emma Marris: "Even though we might really not want to interfere because we find it distasteful, there may come a day when we are morally obligated to do it. But I'm not gung ho."
Nathan Myhrvold: "Look, geoengineering should only be deployed if we think there's really going to be a problem. If a disaster scenario doesn't occur, there's no reason to do this shit."

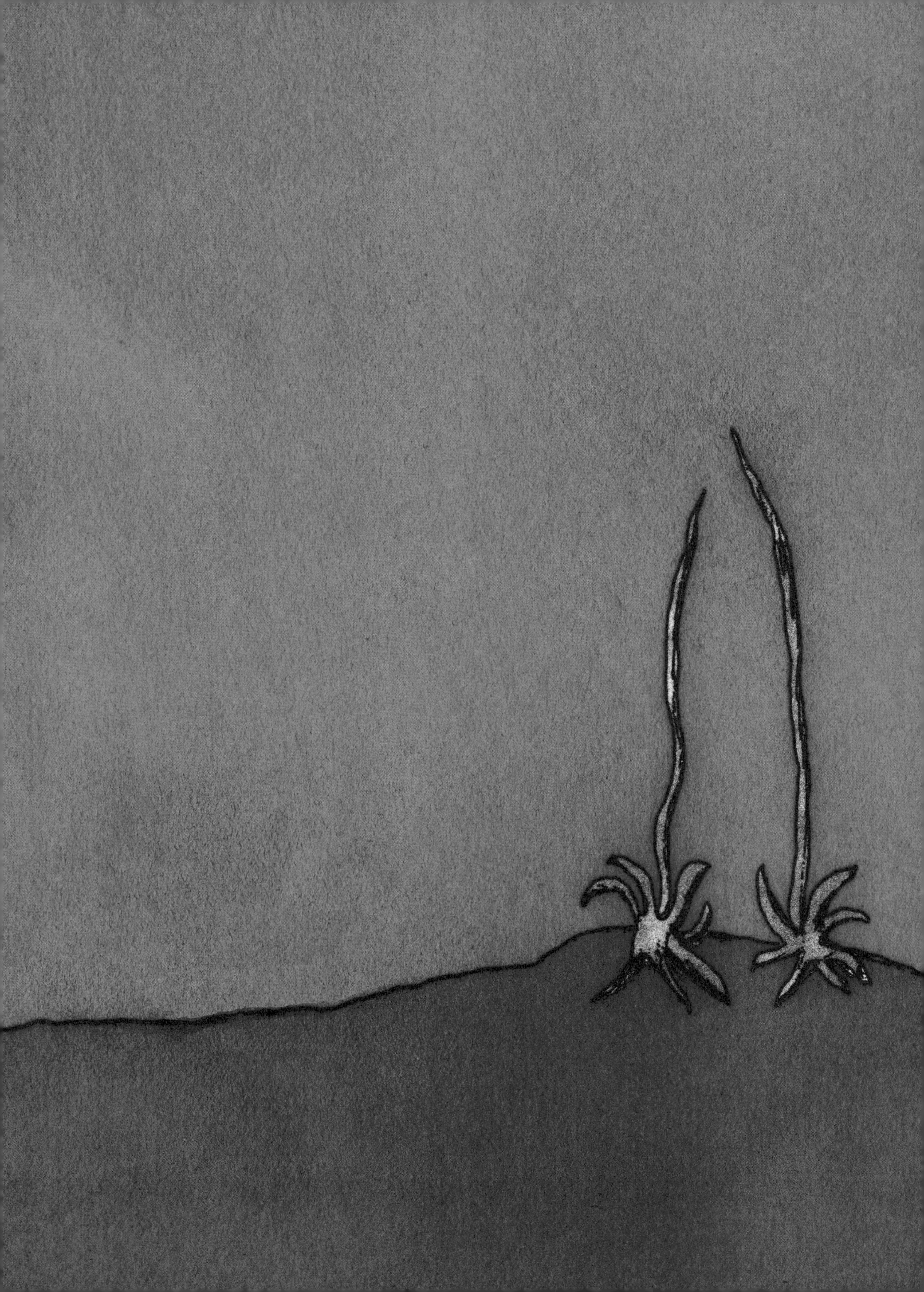

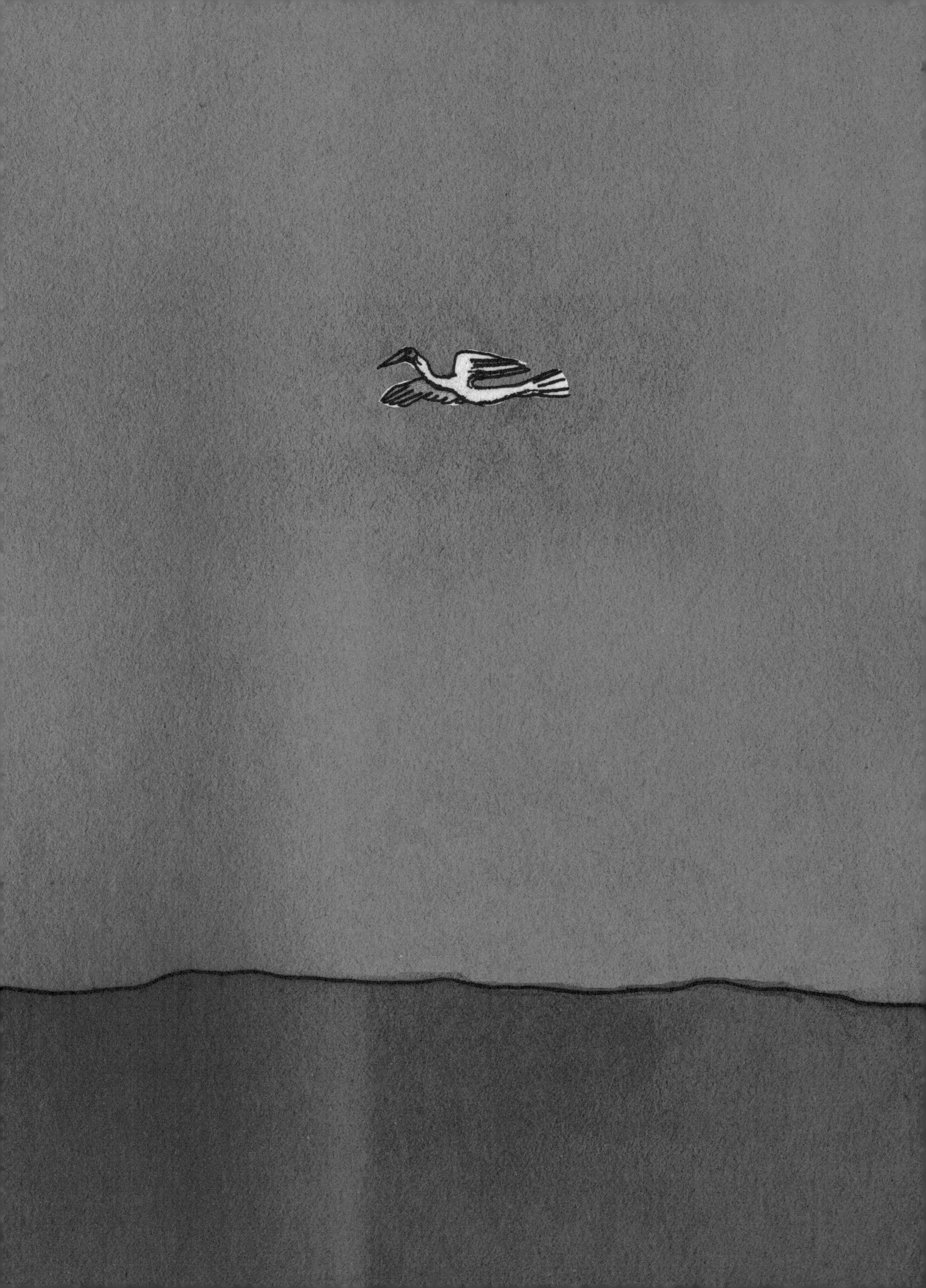

CHAPTER 9:

WAR

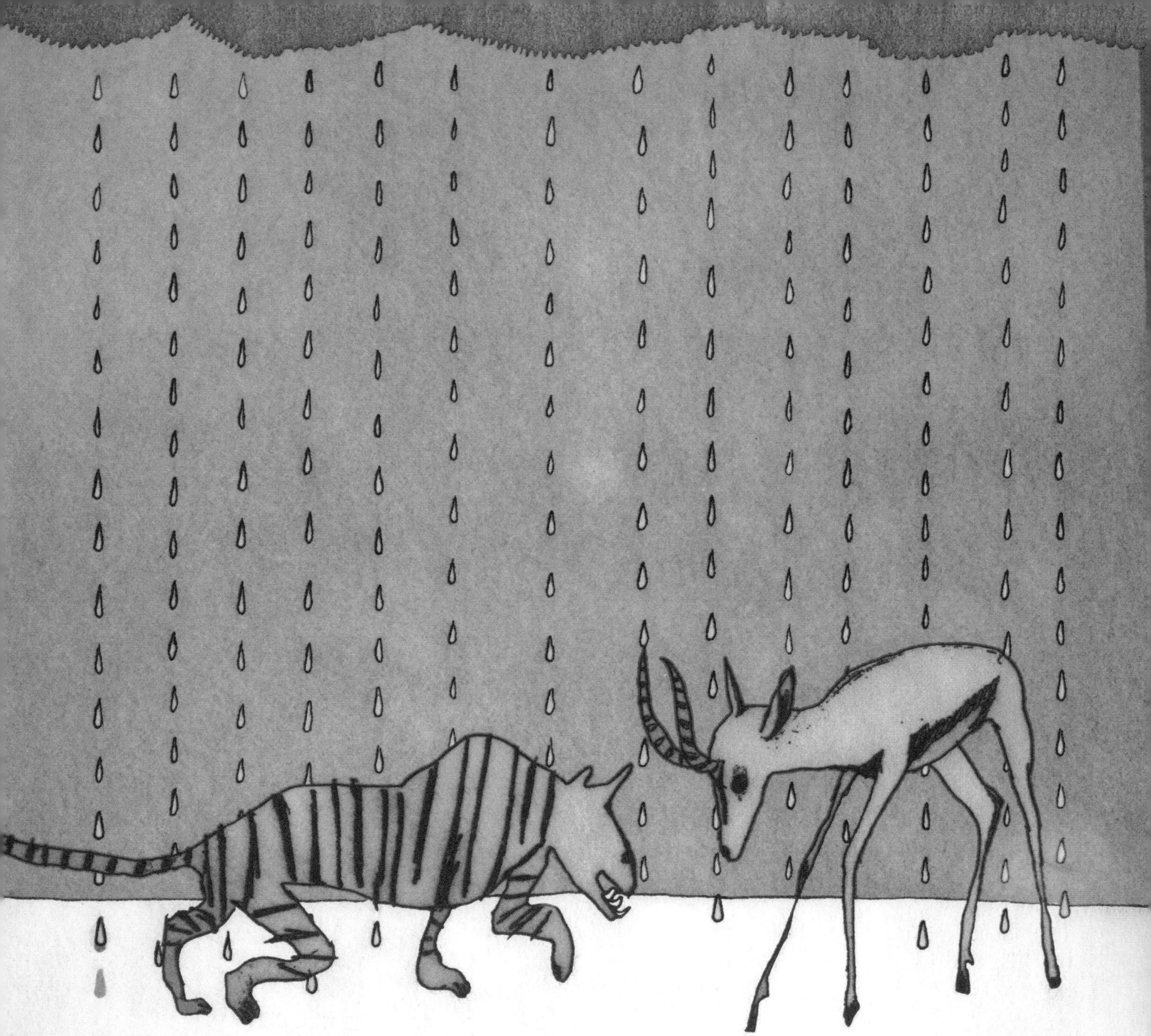

In 1963, South Vietnam's Catholic president Ngo Dinh Diem faced Buddhist monks protesting his government's repressive religious policies. In May, Diem's security forces killed nine monks demonstrating against the ban on raising a Buddhist flag. In June, the army poured chemicals on religious protesters. In July, Associated Press photographer Malcolm Browne caught monk Thich Quang Duc's self-immolation on film, focusing worldwide attention on the situation. The strife in South Vietnam caused tension with the United States, which supported Diem's anti-Communist government in its struggle with Communist North Vietnam. A CIA agent who witnessed protests in August of '63 described the monks' response to government aggression: "They would just stand around during demonstrations when the police threw tear gas at them." But, the agent noted, when it rained, the monks scattered.

This gave the CIA an idea: make it rain. A rain shower could forestall the protests, and there would be no violent government response; the U.S. would no longer find itself in the position of supporting a regime that attacked its own citizens. According to the CIA agent, "The agency got an Air America Beechcraft and had it rigged up with silver iodide," the key ingredient for "seeding" clouds to induce precipitation.

Seventeen years earlier, three men—a high school dropout, a Nobel laureate, and Kurt Vonnegut's older brother—had developed cloud-seeding techniques at the General Electric Company in Schenectady, New York.

Vincent Schaefer left school in 1921, when he was 15. That year he got a job at GE, working first as a drill press operator, and later in research. Industrial chemist Irving Langmuir, who had advanced the technology of the light bulb, furthered understanding of atomic structure, and, in 1932, won the Nobel Prize in chemistry, directed GE's research division and became Schaefer's mentor. During the Second World War, the two collaborated to aid the war effort, making improvements to the gas mask and naval sonar for submarine detection. They also explored weather phenomena, studying the problem of icing on airplane wings and inventing a cloud generator to mask military maneuvers.

During the summer of 1946, Schaefer lined a home freezer with black velvet and proved he could make artificial snow. A GE promotional film from the period shows Schaefer, with tousled hair, a dark tie, and white shirtsleeves rolled up above his elbows, narrating the process.

He leans in and exhales into the freezer's void: "The moist air from the breath condenses and forms a cloud." The camera lingers on the swirling mist, tinted neon blue by artificial light. It looks like cigarette smoke in a nightclub. The cloud, Schaefer tells us, is "supercooled." That is, though the droplets of moisture are at a temperature below freezing, they remain liquid. Schaefer holds up a chunk of dry ice and flicks a few shards into the cloud: "A few long streaks develop, like the vapor trails of an airplane. These contain millions and millions of tiny snow crystals which grow

very fast." On screen we see a toy snowstorm begin to whirl. "They grow about a billionfold in volume in a few seconds." The camera zooms in. The flakes have become a blizzard, catching and reflecting light in pink, violet, and yellow flashes. "The ice crystals are scintillating. The colors are produced by the fact that they are tiny prisms and separate light into its various colors."

When Schaefer created the first replicable, artificially generated weather event, GE put out a press release: "Man-made snow, every bit as real as that which makes for a 'white Christmas' has been produced for the first time." The New York Times wrote, "A step that might lead to some human control over snow clouds was announced today by the General Electric Company....It might be possible to keep snow from falling on cities or cause it to fall on farms." At higher temperatures, this man-made precipitation would become rain. Soon after the initial experiments, another GE researcher, physicist Bernard Vonnegut—whose 23-year-old brother Kurt began working in the company's publicity department the next year—made a further discovery: silver iodide was an even more effective seeding agent than dry ice.

Irving Langmuir saw military applications. "'Rainmaking' or weather control can be as powerful a war weapon as the atom bomb....In the amount of energy liberated, the effect of thirty milligrams of silver iodide under optimum conditions equals that of one atomic bomb."

In Vietnam, the U.S. tested the technology. When in South Vietnam in 1963 Buddhist monks continued religious demonstrations, the CIA acted. "We seeded the area," a CIA agent told The New York Times. "It rained." It was, the Times wrote, "the first confirmed use of meteorological warfare."

Soon after, the Air Force began its own weather modification efforts in Southeast Asia. The project's focus shifted from dispersing religious protesters to hindering the flow of goods and weapons from North to South Vietnam. "We were trying to arrange the weather pattern to suit our convenience," a government official said later. The program was kept secret.

Ben Livingston was a cloud physicist for the U.S. Navy during the Vietnam War. He flew dozens of cloud-seeding missions in Southeast Asia in 1966 and 1967. Photos of Livingston from the time show a tall man with a long neck, thick features, and a lopsided smile. In one photo he is shirtless, holding open the door to a small airplane. In others he smokes a cigarette. In later pictures, Livingston wears black-rimmed glasses. Today he lives with his wife, Bettie, and grown son, Jim, in a ranch house in Midland, Texas. George W. Bush's childhood home, now a museum, is five blocks away.

Ben Livingston: "I was a young kid in West Texas, cutting weeds out of cotton field. You'd see clouds coming along and hope that would get you in the shade for a little bit, then hope in the afternoon it might even turn into a shower. I guess I never quit wondering why you couldn't take a cloud—since sometimes it rains anyway — and get that rain out of it when you wanted to. That was in my head before I was 10 years old."

Waylon Alton Livingston, called Ben, was born on August 17, 1928, to Addie Floyd and Ernest Livingston in Fisher County, Texas. After high school, Ben volunteered for the U.S. Navy. He studied meteorology and Japanese. He became a pilot and by 1958 was a flight meteorologist in Guam, training pilots to make "low level typhoon eye penetrations." In the 1960s he became involved with Project Stormfury, a government effort to manage hurricanes with cloud seeding. In August of 1966, he was deployed to Da Nang, South Vietnam.

Ben Livingston: "The object of the cloud seeding in Vietnam was to make the monsoon season start sooner and last longer. So that's what I did. I went over there to seed the clouds and make them rain."

According to a Department of Defense chart submitted to the Senate Foreign Relations Committee in 1974, the project aimed to "increase rainfall sufficiently in carefully selected areas to deny the enemy use of roads by: 1) softening road surfaces, 2) causing landslides along roadways, 3) washing out river crossings, 4) maintain[ing] saturated soil conditions beyond the normal time span." The silver iodide flares—aluminum

cartridges holding the seeding material — were mounted in rows on the wing of an airplane. The pilot would trigger a time-delayed firing mechanism and release the flares, dropping them into a cloud.

Ben Livingston: "You got a cloud of a certain dimension and you want that thing to grow, you might just fly, say, 200 or 300 yards to the outside of it, put your cloud-seeding material in that little part of it—in which case you fly in the cloud for a few seconds and you're out in the clear blue sky again. Or, if you're working on a really big cloud, you get inside and you stay there until you're through, which may be 30 or 40 minutes. You take your cues from your radar. You're inside a cloud.

"It was pretty simple. I usually went out every day to a certain place over North Vietnam to take a look. If it turned out that they weren't developing the way they needed to, I would turn around and come home, you know? I mean, back to Thailand or wherever."

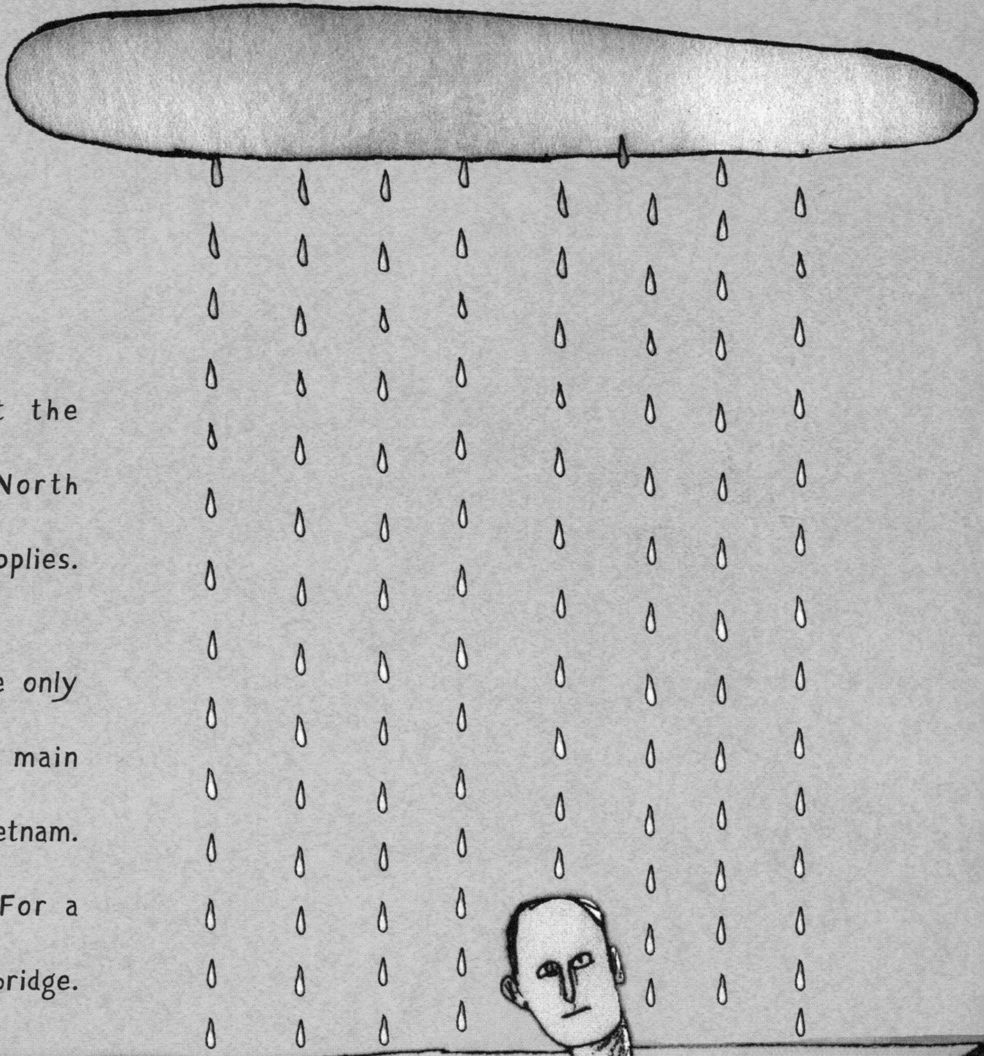

Livingston remembers washing out the Mu Gia Pass Bridge, used by the North Vietnamese for the transportation of supplies.

"That was a hell of a bridge. It was the only bridge on Highway 1—that was the main thoroughfare between North and South Vietnam. Everything went across that bridge. For a couple hundred miles there wasn't another bridge.

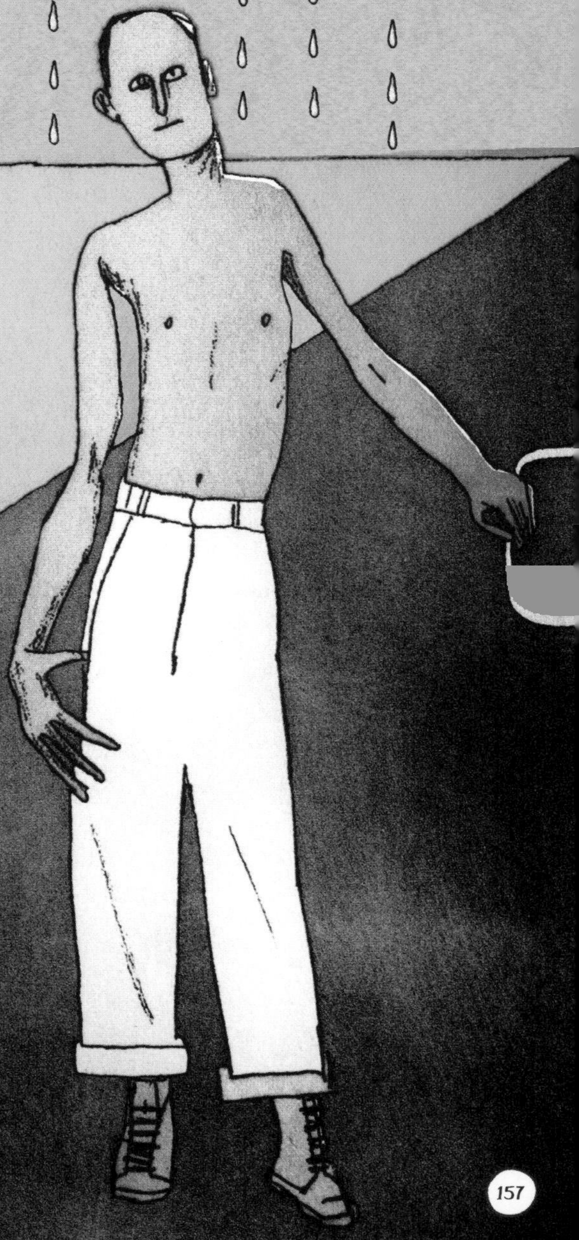

"It's my understanding that the bombers had tried to bomb that bridge out for day after day after day and hadn't had any luck. We put so much water through the valley that it just took the bridge with it. We made a hell of a lot of water that day.

"I remember of course we reported several times that we had killed a lot of people. We flooded them and they drowned. That was just part of what went on every day in making it rain."

In October 1966, Livingston met with Lyndon Johnson in the Oval Office.

Ben Livingston: "I was called to Washington to report on how we were doing and what I was doing in Vietnam. Those meetings involved, of course, briefing the president of the United States. He was certainly interested in knowing what in the heck we were doing to change the weather and so forth. Boy, he was pleased to find that we were able to do something without putting troops on the ground."

On March 18, 1971, journalist Jack Anderson published a nine-paragraph item in *The Washington Post* exposing "the hush-hush project," though with minimal detail.

"Air Force rainmakers, operating secretly in the skies over the Ho Chi Minh trail network," Anderson wrote, "have succeeded in turning the weather against the North Vietnamese." The Pentagon Papers, leaked that March to *The New York Times* by RAND Corporation military analyst Daniel Ellsberg, confirmed the existence of a weather modification program, referring to it as "Operation Pop Eye." In July 1972, *Times* reporter Seymour Hersh wrote a more extensive article, sparking public debate. One letter to the editor pointed out that the ambiguous "legal status of clouds" would need to be defined in order to determine nations' jurisdictional claims of "ownership and control." Another *Times* reader called rainmaking a "weapon of mass destruction."

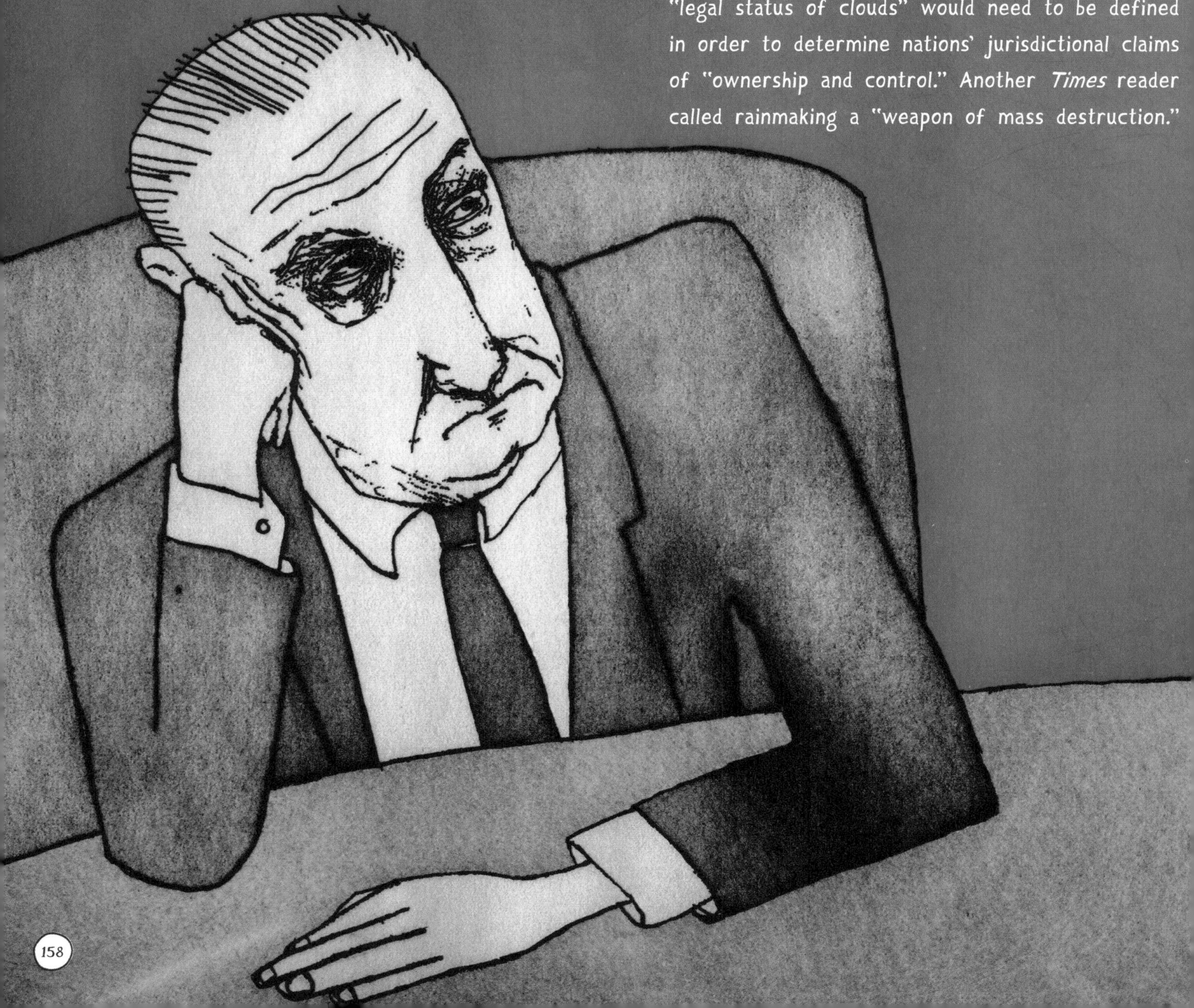

On March 20, 1974, the Senate Foreign Relations Subcommittee on Oceans and International Environment convened a secret hearing on weather-control efforts in Vietnam. Senator Claiborne Pell, Democrat from Rhode Island, presided. Dennis J. Doolin, deputy assistant secretary of defense (East Asia and Pacific Affairs), and Lieutenant Colonel Ed Soyster of the Joint Chiefs of Staff testified. The transcripts were made public two months later. Soyster and Doolin were asked about the effectiveness of the project:

Lieutenant Colonel Ed Soyster: "It was one of the most difficult parts of the project to try and quantify how well we were doing."

Dennis Doolin: "In my own mind on the basis of the material that I have seen, I am not convinced that it had anything more than marginal effect, but that is something that even the experts disagree on."

Nevertheless, Doolin endorsed the program.

Dennis Doolin: "If an advisory wanted to stop me from getting from point A to point B so I could do something at point B, I would rather he stopped me with a rainstorm than with bombs. Frankly, I view this in that context as really quite humane, if it works."

Not everyone agreed. Within the Johnson administration, a contingent of State Department officials objected, worrying that weather modification might cause "unusual suffering" and unforeseen ecological damage.

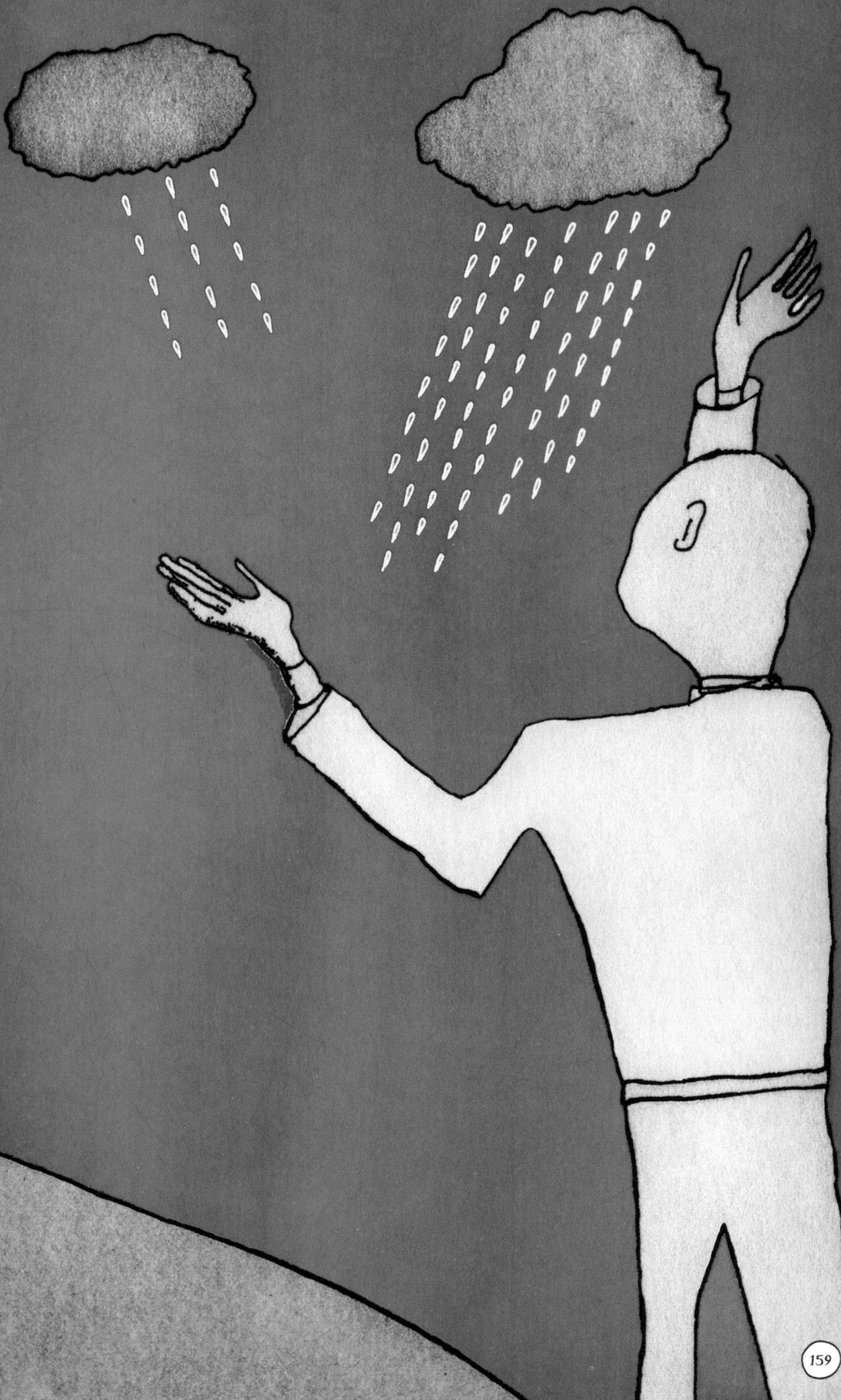

After the exposure of the American program in Vietnam, weaponized weather became subject to an international treaty, the Convention on the Prohibition of Military or Any Other Hostile Use of Environmental Modification Techniques, or ENMOD. The treaty, which went into effect in 1978, bans aggressive environmental modification activity "on the scale of several hundred square miles" that lasts "for a period of months," and that involves "serious or significant disruption or harm to human life, natural and economic resources, or other assets." Some have criticized ENMOD for its implied acceptance of smaller-scale, shorter-term environmental manipulation. It has not prevented military strategists from considering a future in which weather is again brandished as a weapon.

One document that circulates on the Internet is a 1996 study entitled "Weather as a Force Multiplier: Owning the Weather in 2025." The paper was produced, it states, in compliance with "a directive of the chief of staff of the Air Force." The authors imagine a future in which "weather modification can provide battlespace dominance to a degree never before imagined."

The paper begins with a scenario from the future:

"Imagine that in 2025 the U.S. is fighting a rich, but now consolidated, politically powerful drug cartel in South America. The cartel has purchased hundreds of Russian- and Chinese-built fighters that have successfully thwarted our attempts to attack their production facilities....Meteorological analysis reveals that equatorial South America typically has afternoon thunderstorms on a daily basis throughout the year. Our intelligence has confirmed that cartel pilots are reluctant to fly in or near thunderstorms. Therefore, our weather force support element (WFSE), which

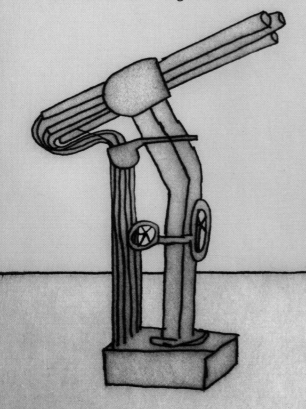

WILHELM REICH'S CLOUDBUSTER

AUSTRIAN PSYCHOANALYST AND FREUD PROTÉGÉ WILHELM REICH PRACTICED THERAPEUTIC TECHNIQUES THAT INCLUDED NAKED MASSAGE. HE ALSO INVENTED THE CLOUDBUSTER, A COSMIC ENERGY ACCUMULATOR MADE OF METAL PIPES AND TUBING THAT PROMISED TO CONTROL RAINFALL.

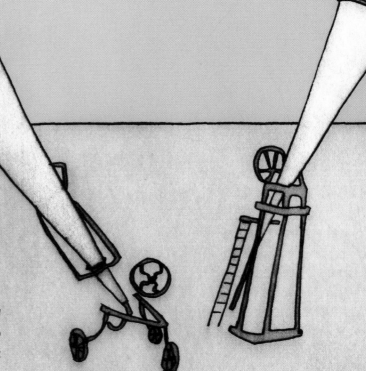

HAIL CANNONS

HAIL CANNONS PURPORT TO THWART HAIL BY GENERATING ATMOSPHERIC SHOCK WAVES. THEY HAVE LONG BEEN USED IN WINE-GROWING REGIONS, WHERE GRAPE HARVESTS ARE DAMAGED BY THE ICY HAILSTONES. THEIR EFFECTIVENESS IS DISPUTED.

is a part of the commander in chief's (CINC) air operations center (AOC), is tasked to forecast storm paths and trigger or intensify thunderstorm cells over critical target areas that the enemy must defend with their aircraft. Since our aircraft in 2025 have all-weather capability, the thunderstorm threat is minimal to our forces, and we can effectively and decisively control the sky over the target."

By 2025, the paper continues, the U.S. could wield a weaponized weather arsenal including lasers to dissipate fog, aircraft able to repel lightning strikes, and drones for seeding maneuvers. Virtual weather projections would confuse the enemy. "Precipitation enhancement" strikes could flood lines of communication and crush morale.

"While some segments of society will always be reluctant to examine controversial issues such as weather-modification, the tremendous military capabilities that could result from this field are ignored at our own peril. From enhancing friendly operations or disrupting those of the enemy via small-scale tailoring of natural weather patterns to complete dominance of global communications and counterspace control, weather-modification offers the war fighter a wide-range of possible options to defeat or coerce an adversary."

CHARLES HATFIELD'S RAINMAKING TOWERS

IN THE EARLY 20TH CENTURY, SEWING MACHINE SALESMAN CHARLES HATFIELD BUILT EVAPORATION TOWERS CONTAINING A SECRET MIX OF CHEMICALS AND MARKETED THEM AS RAINMAKING DEVICES. "I DO NOT FIGHT NATURE AS...OTHERS HAVE DONE BY MEANS OF DYNAMITE BOMBS AND OTHER EXPLOSIVES," HE SAID. "I WOO HER BY MEANS OF SUBTLE ATTRACTION." IN 1915, SAN DIEGO'S CITY COUNCIL HIRED HATFIELD TO FILL THEIR EMPTY RESERVOIR. SUBSEQUENT PRECIPITATION CAUSED FLOODING AND MILLIONS OF DOLLARS IN DAMAGES, PROVOKING THE FIRST WEATHER MODIFICATION LAWSUIT. HATFIELD'S STORY INSPIRED A 1956 FILM CALLED *THE RAINMAKER*. BURT LANCASTER PLAYED STARBUCK, A SMOOTH OPERATOR IN TIGHT PANTS WHO COMES TO A DROUGHT-STRICKEN TOWN IN THE SOUTHWEST, OFFERS TO MAKE RAIN FOR $100, AND SEDUCES A SPINSTER (KATHARINE HEPBURN).

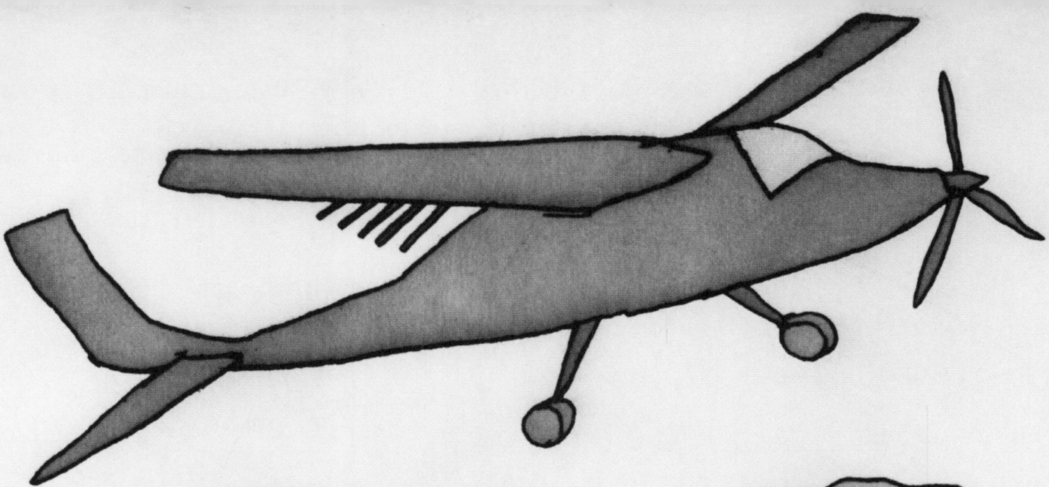

Returning from Vietnam, Ben Livingston was awarded the Navy Commendation Medal. The citation notes his involvement in "a weapon system under development." It praises Livingston as "relentless" and "unyielding" with an "unwavering devotion to duty," all of which contributed to the "outstanding success of the project and were instrumental in the development of a unique, major combat capability for the United States." The Air Force awarded him an Air Medal for "his outstanding airmanship and courage...in the successful accomplishment of important missions under extremely hazardous conditions including the continuous possibility of hostile ground fire."

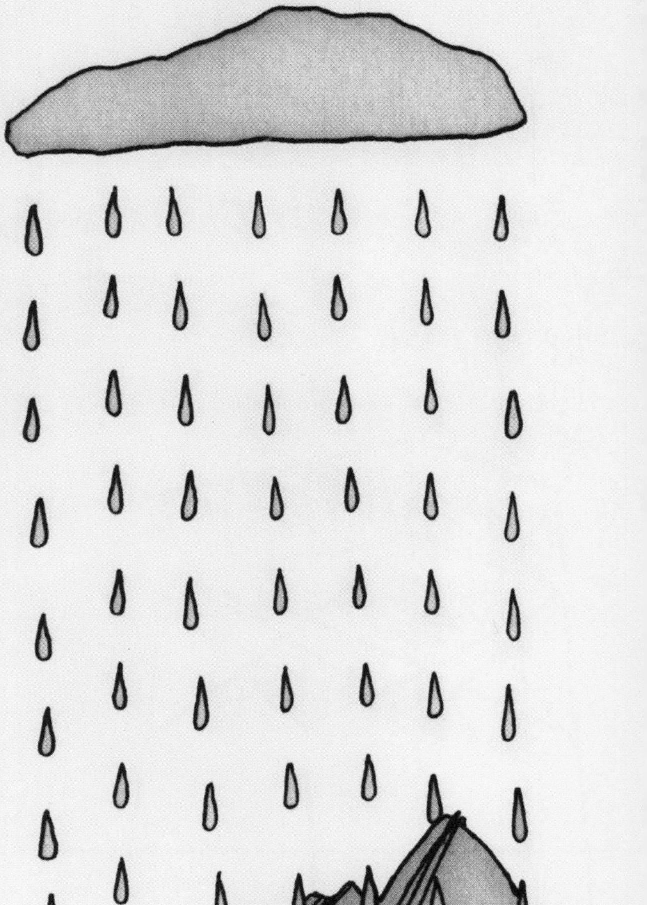

In 1969, Livingston retired from the Navy and began envisioning peacetime applications for weather modification. He moved to Alamosa, Colorado, where he designed a specially pressurized bovine treatment center to cure cows suffering from oxygen deprivation at high altitudes. He opened a commercial cloud seeding business, San Luis Valley Weather Engineering, Inc.

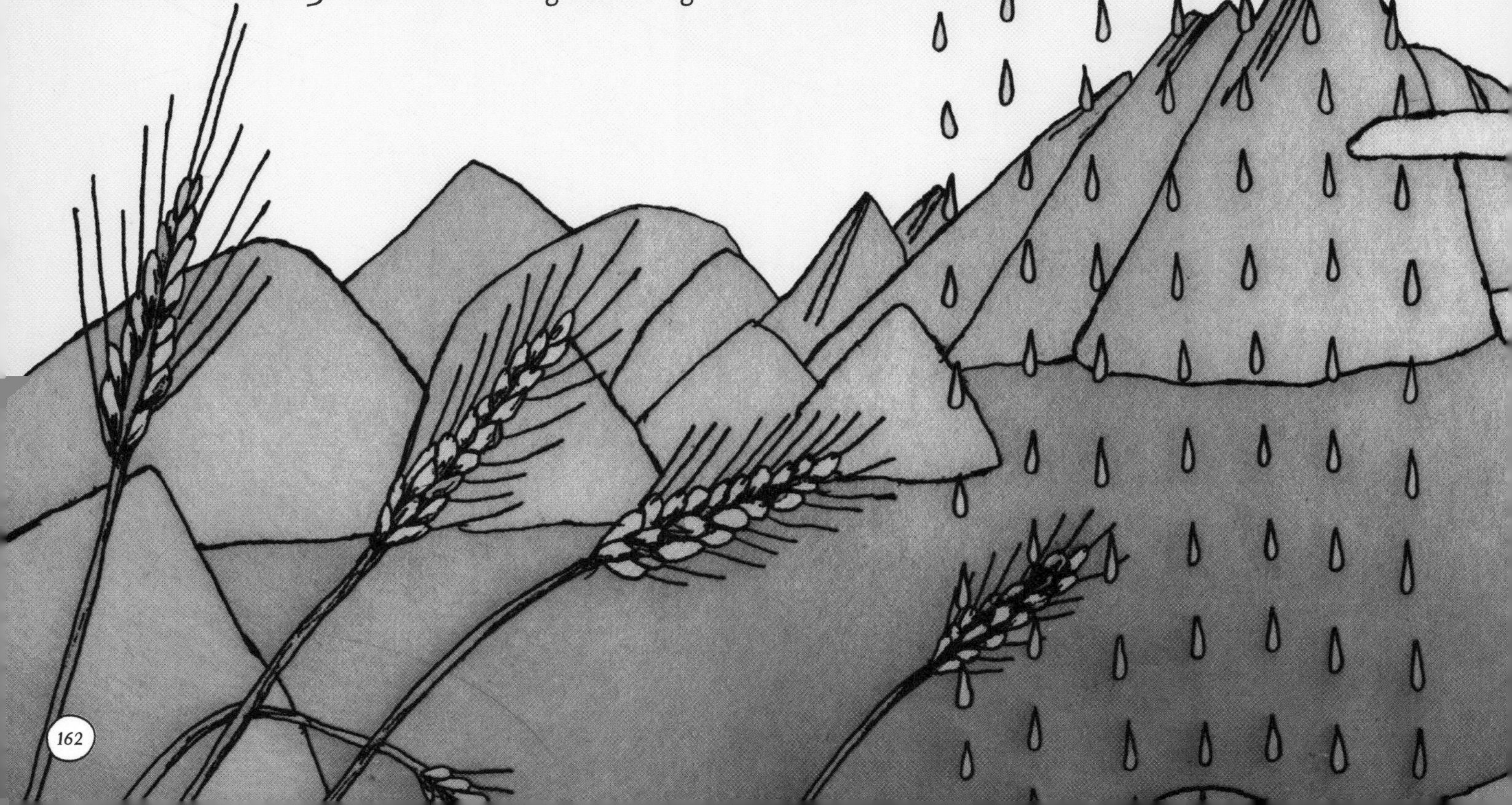

Ben Livingston: "I was flying for Coors Brewery. Our mission was to make it rain as much as possible during the growing season, and then, through the end of the growing season, to try and prevent it from raining so that the grain—the barley that they use for the Coors beer—could mature with a very bright amber sheen to it: make it rain until the Fourth of July, and then from the fifth day of July until the harvest, to kill all these clouds."

Some 40 countries around the world operate weather modification programs today. Thailand has a Bureau of Royal Rainmaking and Agricultural Aviation. The Greek National Hail Suppression Program has fought damage to agricultural crops in recent years. In 2008, the Beijing Meteorological Bureau announced China would use cloud seeding to ensure a rain-free opening ceremony for the Olympics. In 2013, *The Jakarta Globe* reported that Indonesia's Agency for the Assessment and Application of Technology would seed clouds to try and manage flooding in the capital. Scientists still debate the effectiveness—and ethics—of such programs.

Unlike geoengineering schemes designed to counteract changes to the climate across the globe, these weather modification efforts aim for short-term, localized results. In the U.S., individual states allocate funds towards weather modification and numerous private companies offer weather-making services, but the federal government no longer supports these efforts.

Ben Livingston believes we are missing an important opportunity to use weather modification to prevent catastrophic storms.

Ben Livingston: "I took the year 2004 off and went around to the people that I had known while in the military as being movers and shakers, and was trying to convince them that we ought to seed some clouds or control the weather down off the Gulf Coast or whatever to keep these hurricanes from damaging, you know, New Orleans, for instance. I made the rounds doing what I thought needed to be done. I went to the place where they manufacture seeding materials in a little town near Fargo, North Dakota—Ice Crystal Engineering. I had the aircraft all lined up, and people and pyrotechnics and everything ready to go. Then I went to Washington, D.C. I sent letters to every senator in Washington and told them what I was proposing to do. And boy, they wouldn't touch it. They can come up with all kinds of reasons why you shouldn't mess with nature, you know? The main reason they give is that you don't know for sure where your effects are going to be. And that's not true."

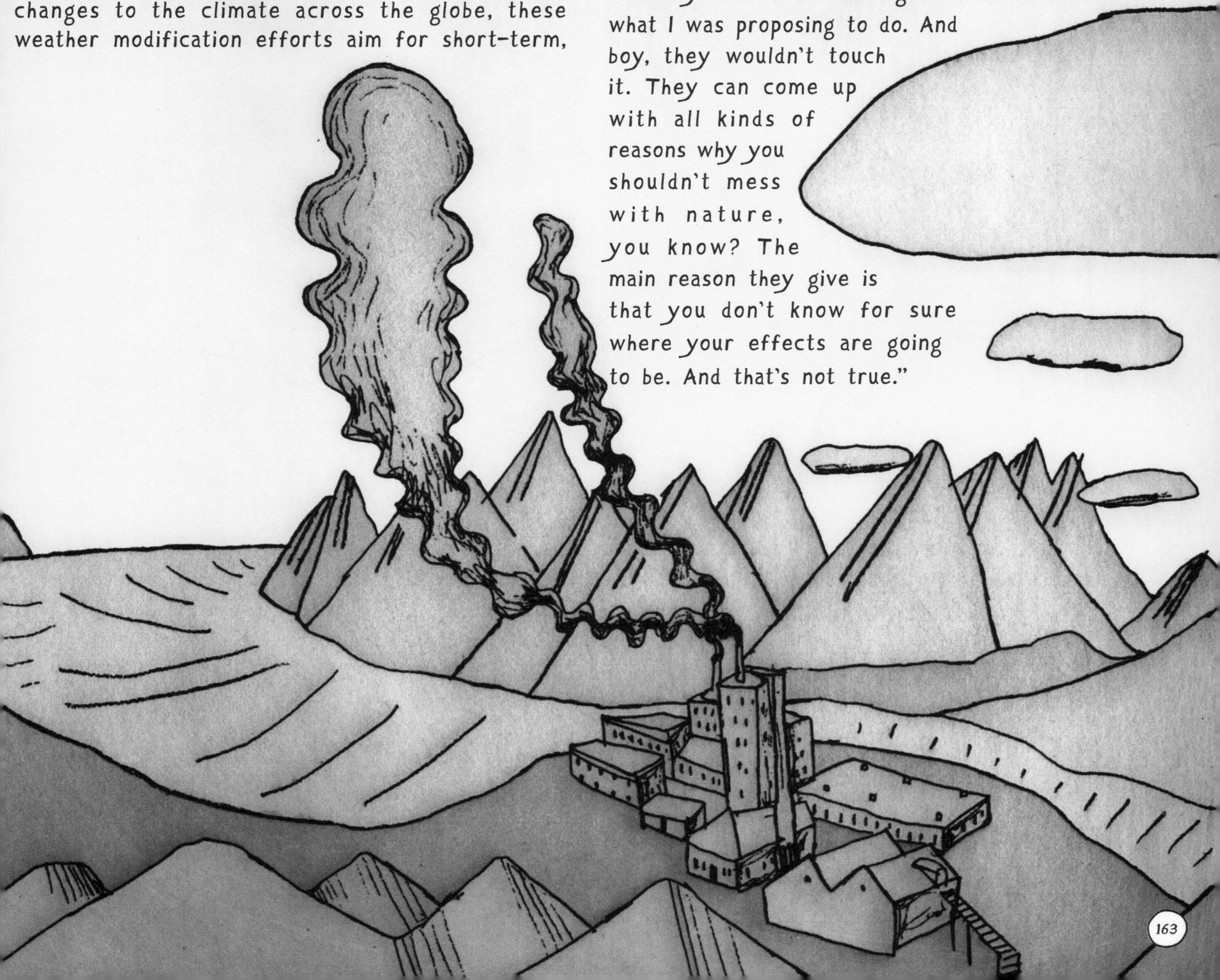

The same year he went to Washington, Ben Livingston self-published *Dr. Lively's Ultimatum*, a novel about weather control with a hero based on the author himself. ("I am Dr. Lively in that book," Livingston says.) It is a swaggering and pulpy tale. Dr. Ken Lively is a straight-talking former Navy cloud physicist who ran a U.S. rainmaking program during the Vietnam War. He has a sexy secretary and an extra tooth. The plot turns on Dr. Lively's top secret plan: manipulating the weather to save the world from a massive asteroid called **TOFU** and the toxic debris cloud it creates—an epic showdown in a war between man and nature.

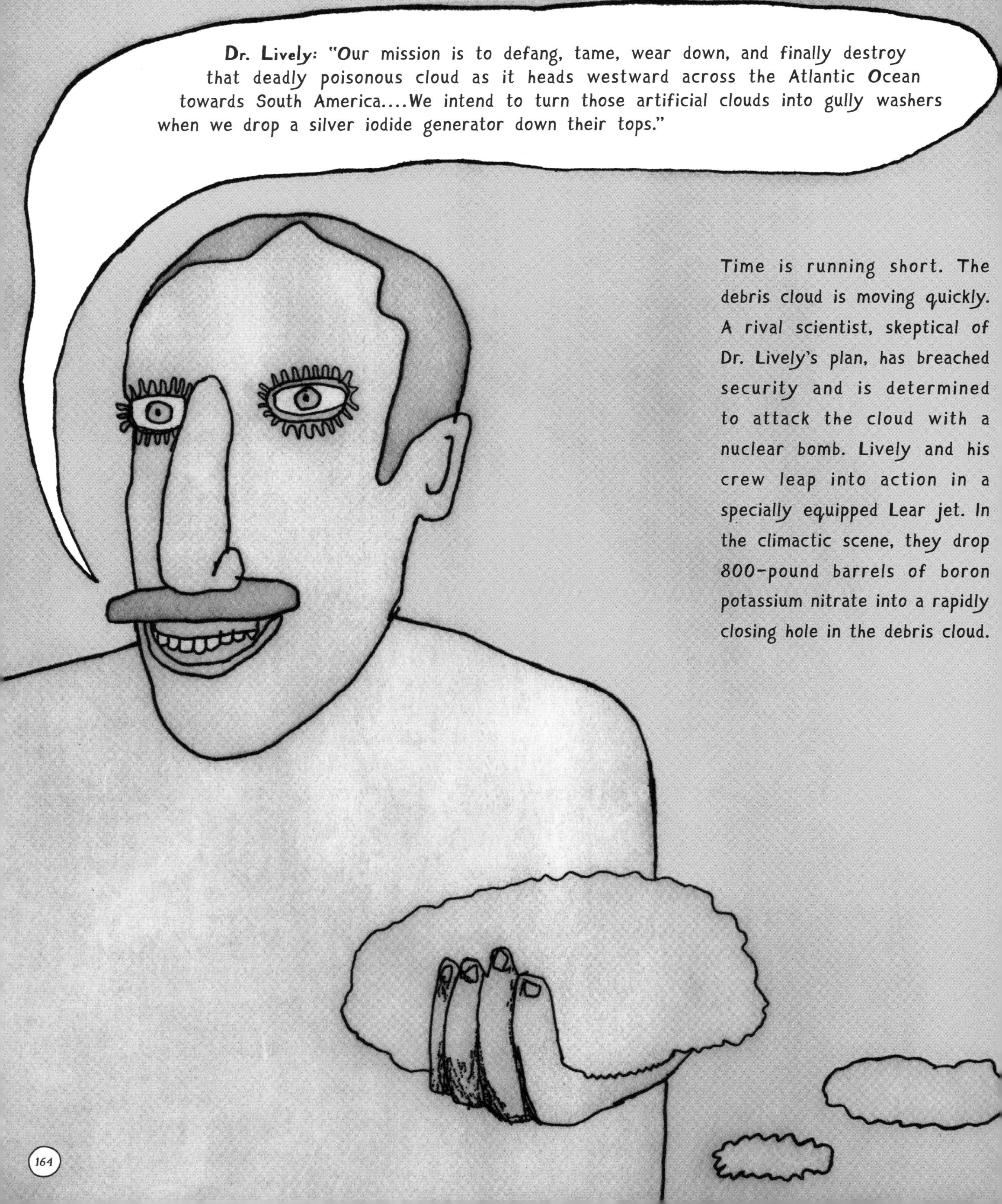
Dr. Lively: "Our mission is to defang, tame, wear down, and finally destroy that deadly poisonous cloud as it heads westward across the Atlantic Ocean towards South America....We intend to turn those artificial clouds into gully washers when we drop a silver iodide generator down their tops."

Time is running short. The debris cloud is moving quickly. A rival scientist, skeptical of Dr. Lively's plan, has breached security and is determined to attack the cloud with a nuclear bomb. Lively and his crew leap into action in a specially equipped Lear jet. In the climactic scene, they drop 800-pound barrels of boron potassium nitrate into a rapidly closing hole in the debris cloud.

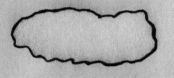

"In an instant of intense
light, every color in the
rainbow flashed and melted
into a blinding sun-like
fireball. The occupants
of the cockpit appeared
as filaments in a giant
camera bulb at the moment
of the flash. Through the
dark-green sun protective
visors in the helmets, the
three Lear crewmembers
saw the brown cloud turn
brighter than the sun and
then change to a bright
blue-green. Every inch of
the debris cloud's volume,
from top to bottom, must
have detonated, causing
the huge mass of gases
and the cremated residue
to be sucked towards the
center of the cloud....
The intense rush of
gases and debris flowing
towards the center of the
imploding cloud produced a
thunderous roar, followed
instantly by a sharp clap
as the renegade cloud
and the warm ocean air
collided and shot upwards."

Success.

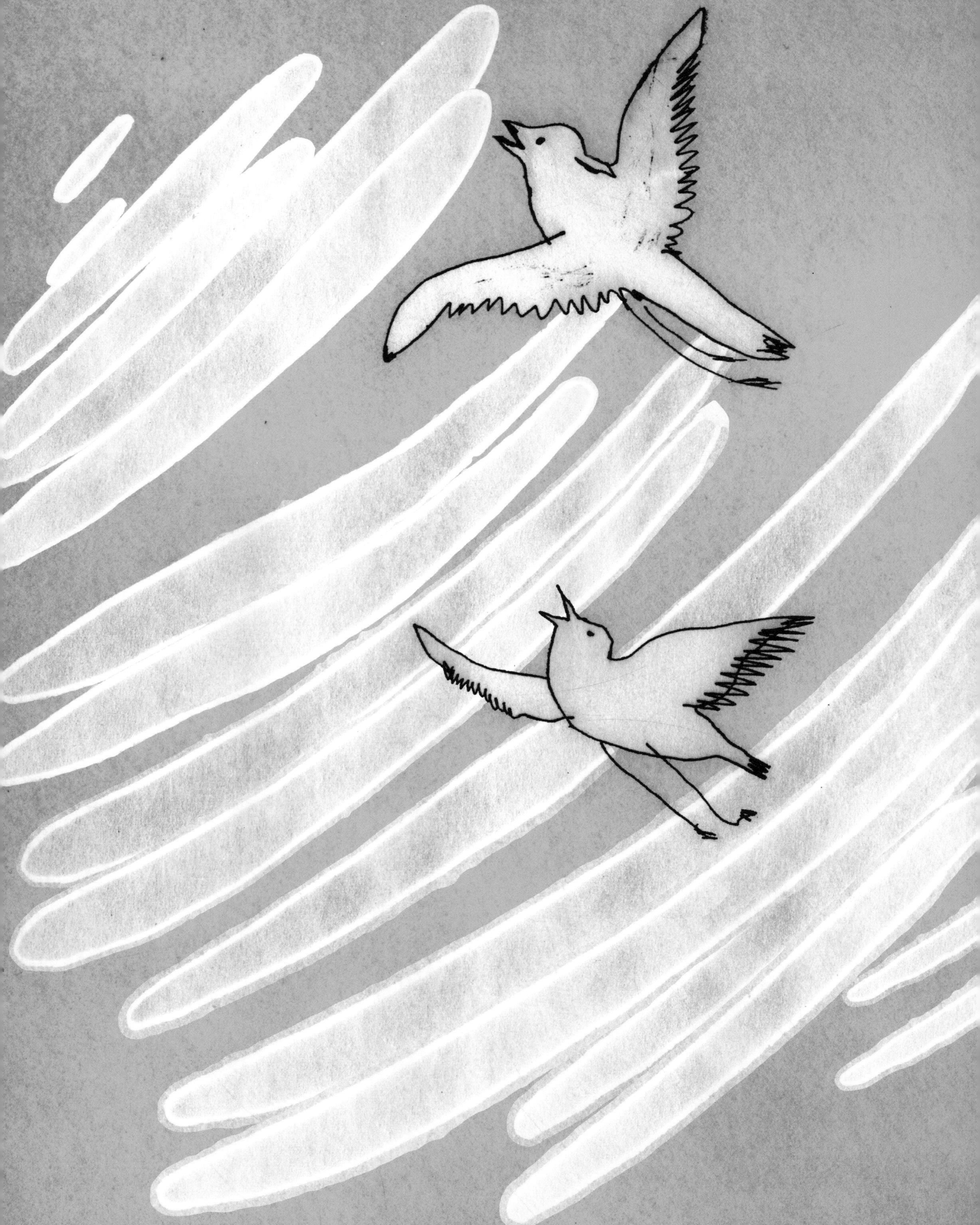

"When the sequence of intensely bright lights and strong winds ended,

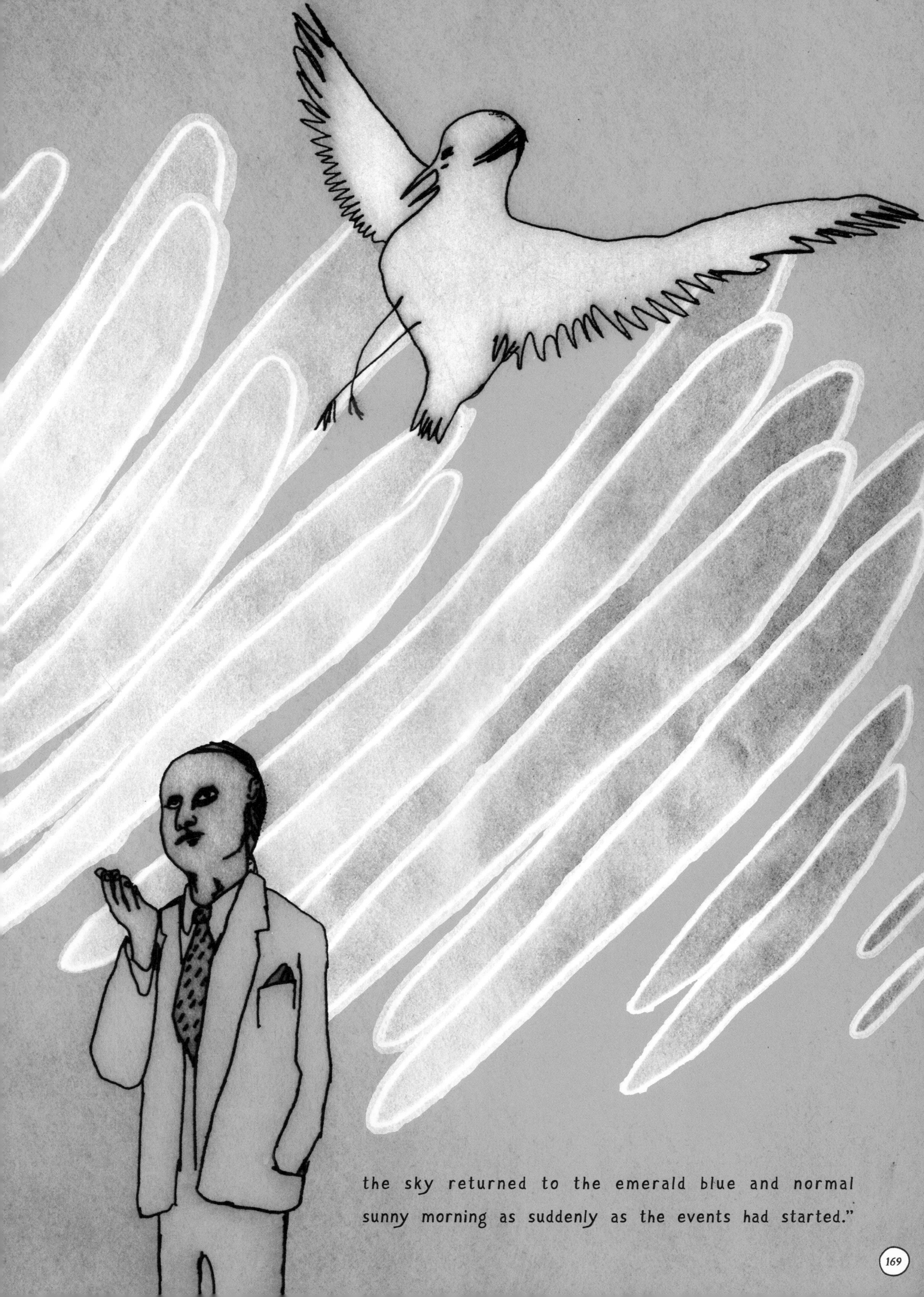

the sky returned to the emerald blue and normal sunny morning as suddenly as the events had started."

Ben Livingston: "On our farm growing up, my habit was to go with all the other guys and steal watermelons.

"We'd go in the night

and get watermelons out of somebody's patch.

"I told my dad,

'I think this is so dumb.

Why don't we just fix it so

that these people who want to

have watermelon can just get them?'

He thought that

was a good idea.

So we planted some

rows of watermelon

next to

the road so people could come

by and take them whenever they

wanted them.

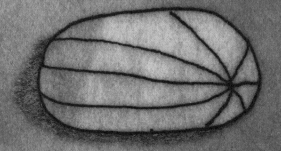

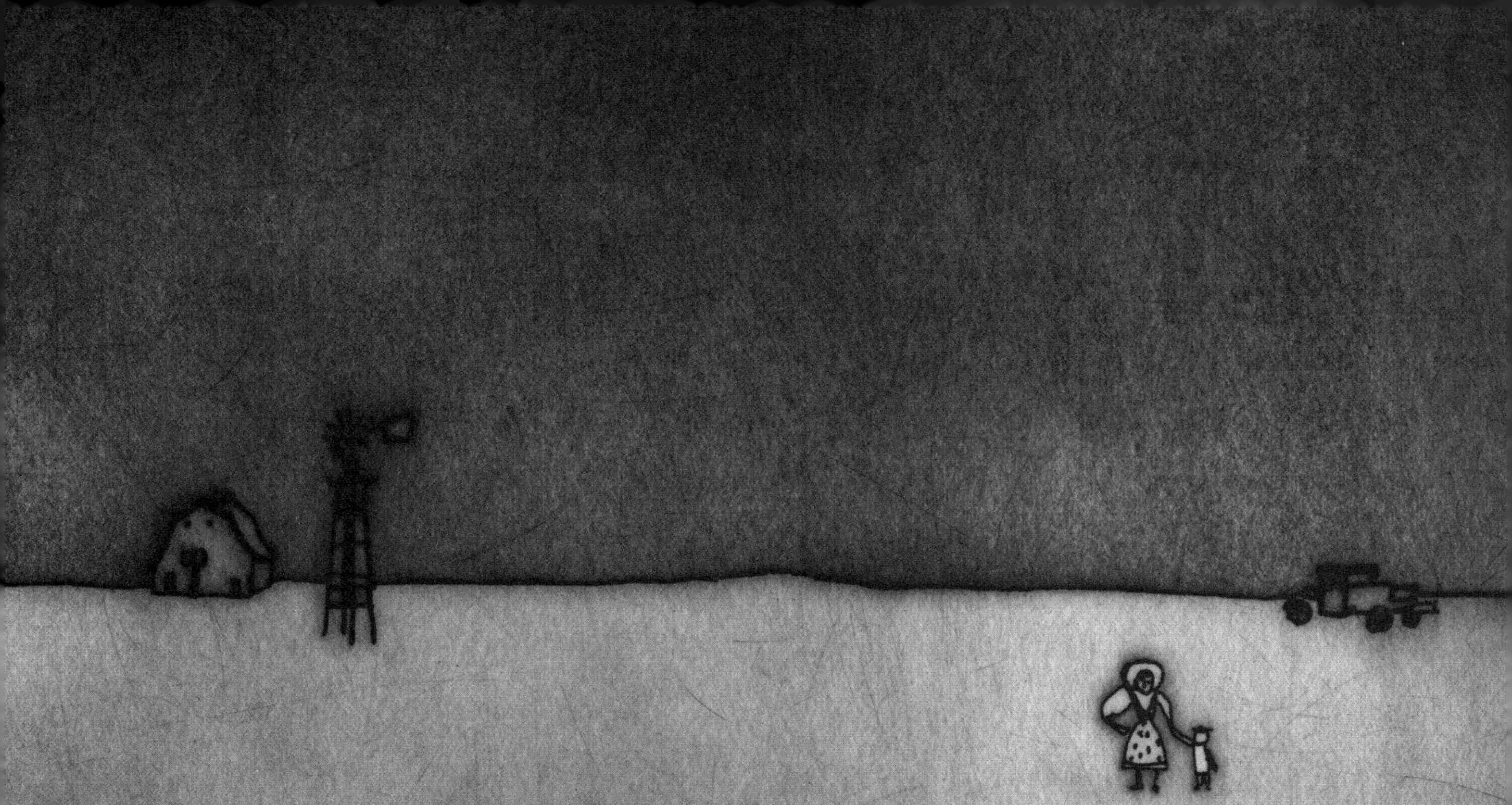

"In the growth of any plant, there's an optimum time when you get more good out of rain than other times. When you grew a lot of watermelon, if it didn't rain exactly when you needed it, those watermelons were always smaller, but they were much sweeter. So it wasn't always bad that it didn't rain when you wanted it to. I mean, as far as watermelon was concerned."

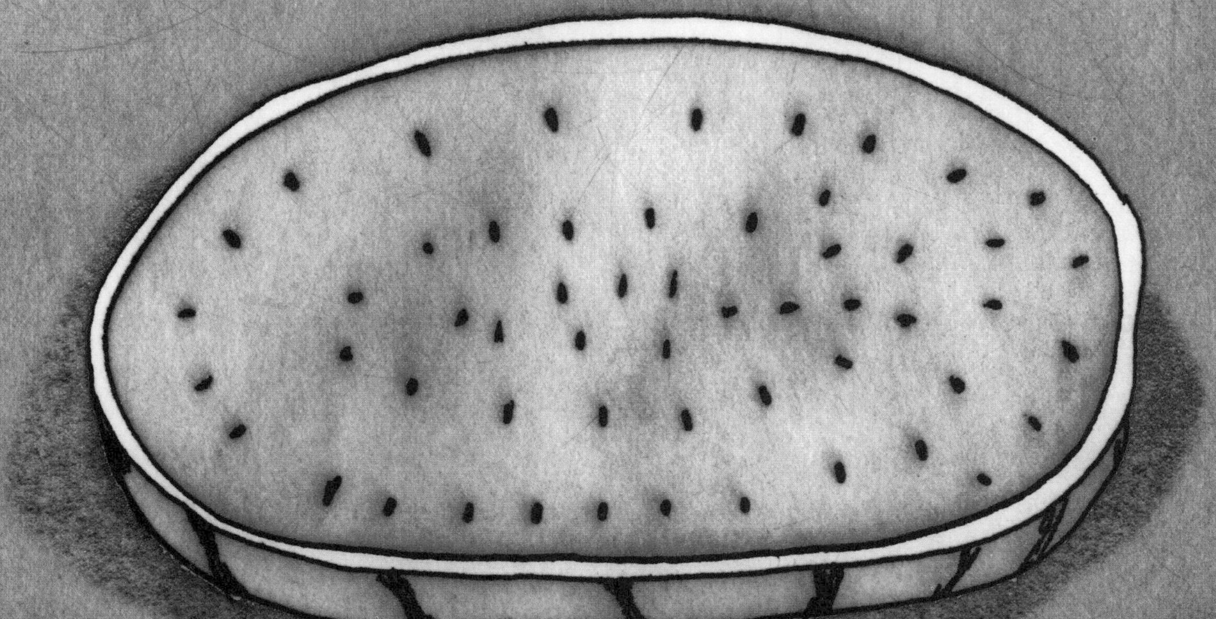

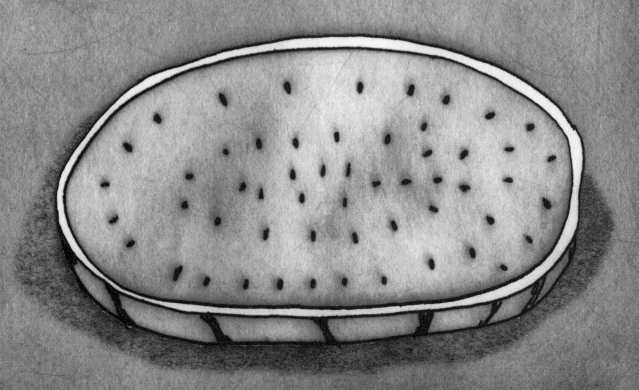

CHAPTER 10:

PROFIT

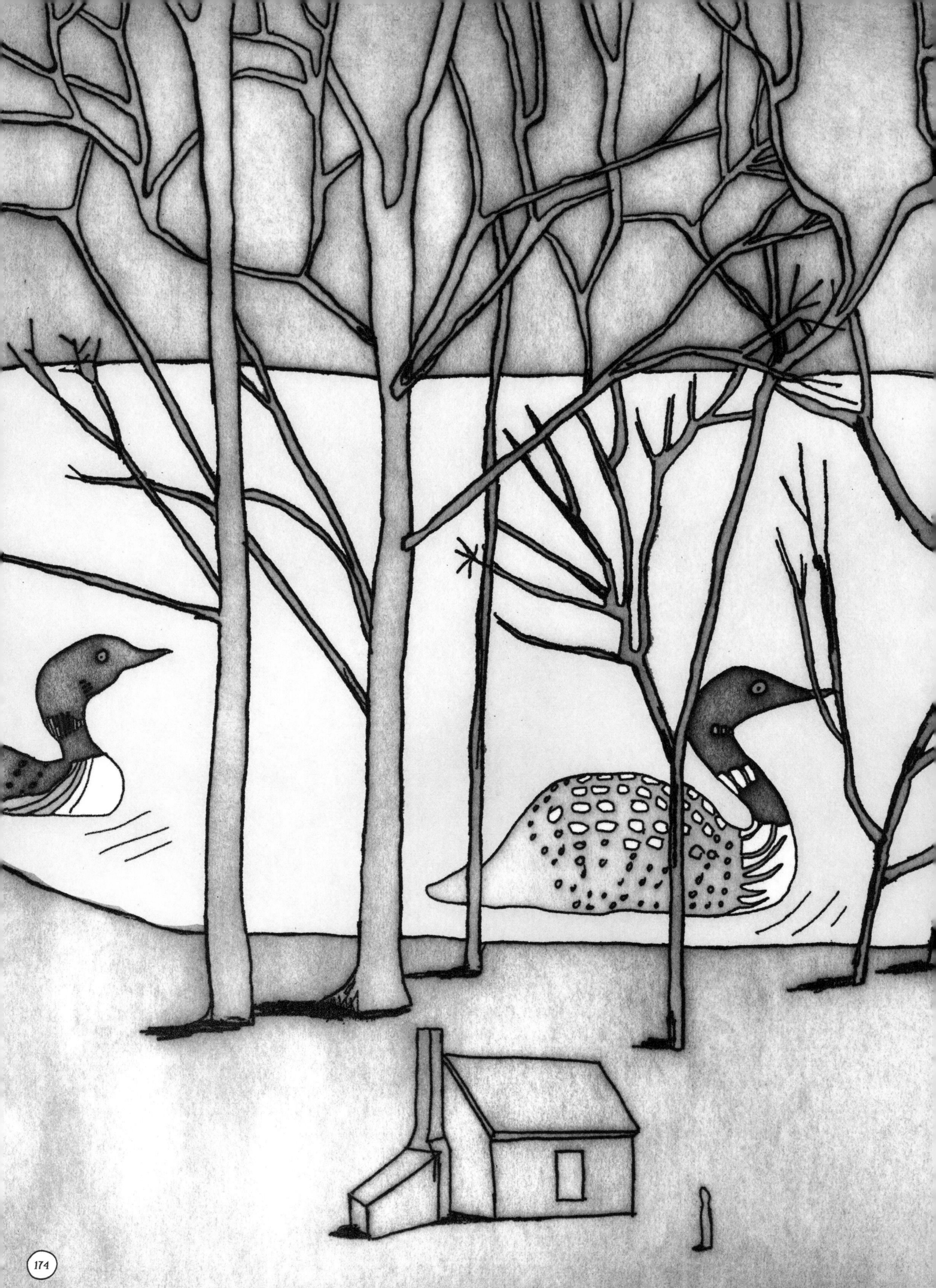

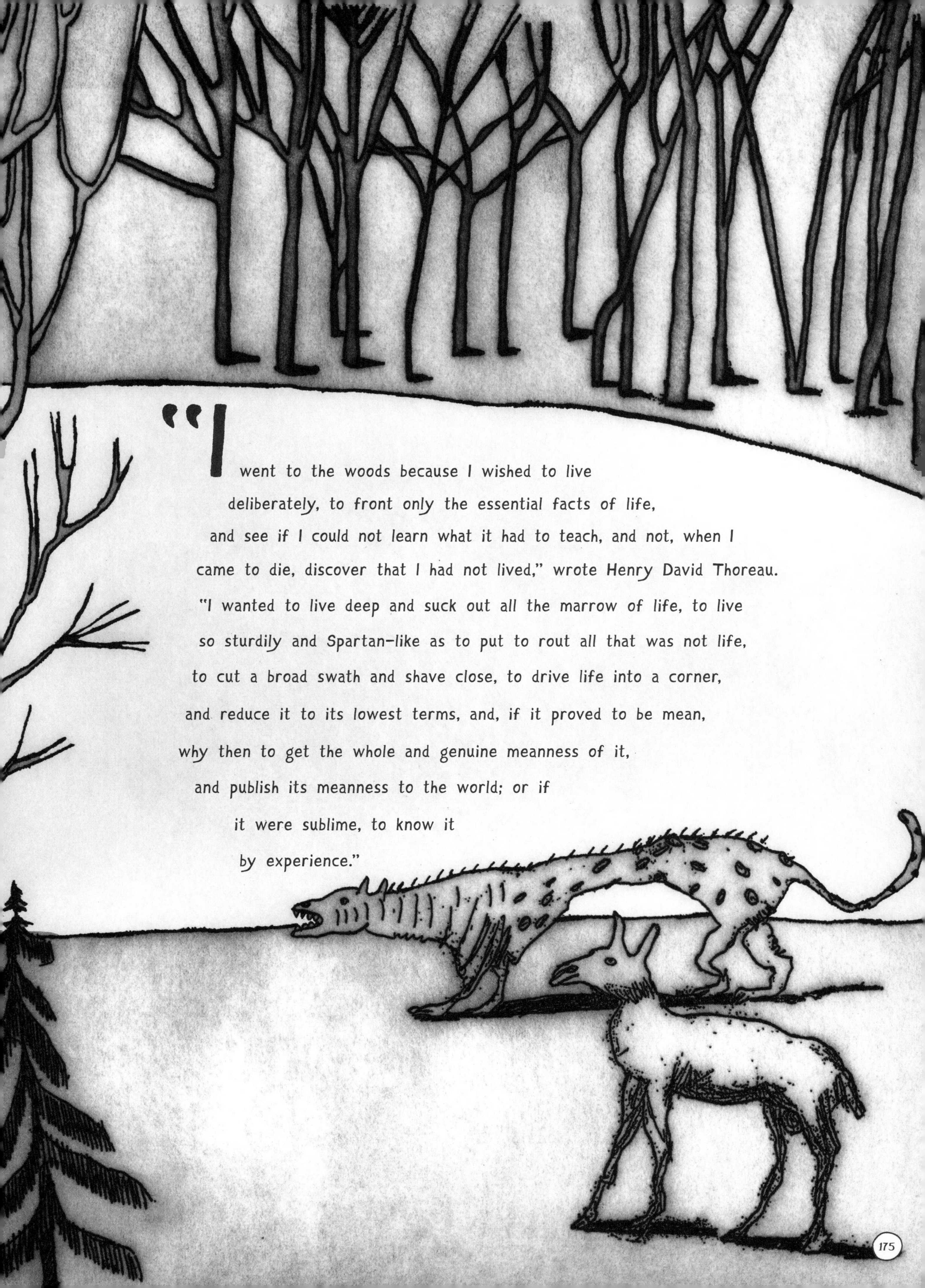

"I went to the woods because I wished to live deliberately, to front only the essential facts of life, and see if I could not learn what it had to teach, and not, when I came to die, discover that I had not lived," wrote Henry David Thoreau. "I wanted to live deep and suck out all the marrow of life, to live so sturdily and Spartan-like as to put to rout all that was not life, to cut a broad swath and shave close, to drive life into a corner, and reduce it to its lowest terms, and, if it proved to be mean, why then to get the whole and genuine meanness of it, and publish its meanness to the world; or if it were sublime, to know it by experience."

You may remember the lines from high school, or from the movie *Dead Poets Society*. To suck the marrow out of life, to live close to nature and cast off the burdens of society, in 1845 Thoreau built a cabin in the woods near Walden Pond in Concord, Massachusetts. For two years, he recorded his thoughts and observations. He listened to the trill of the whip-poor-wills, the grunts of the bullfrogs, the "wild laughter" of the loons. He noted the change of seasons: the "first signs of the infant year just peeping forth" in spring, the maple trees turning scarlet in autumn. He tracked dropping temperatures as winter approached.

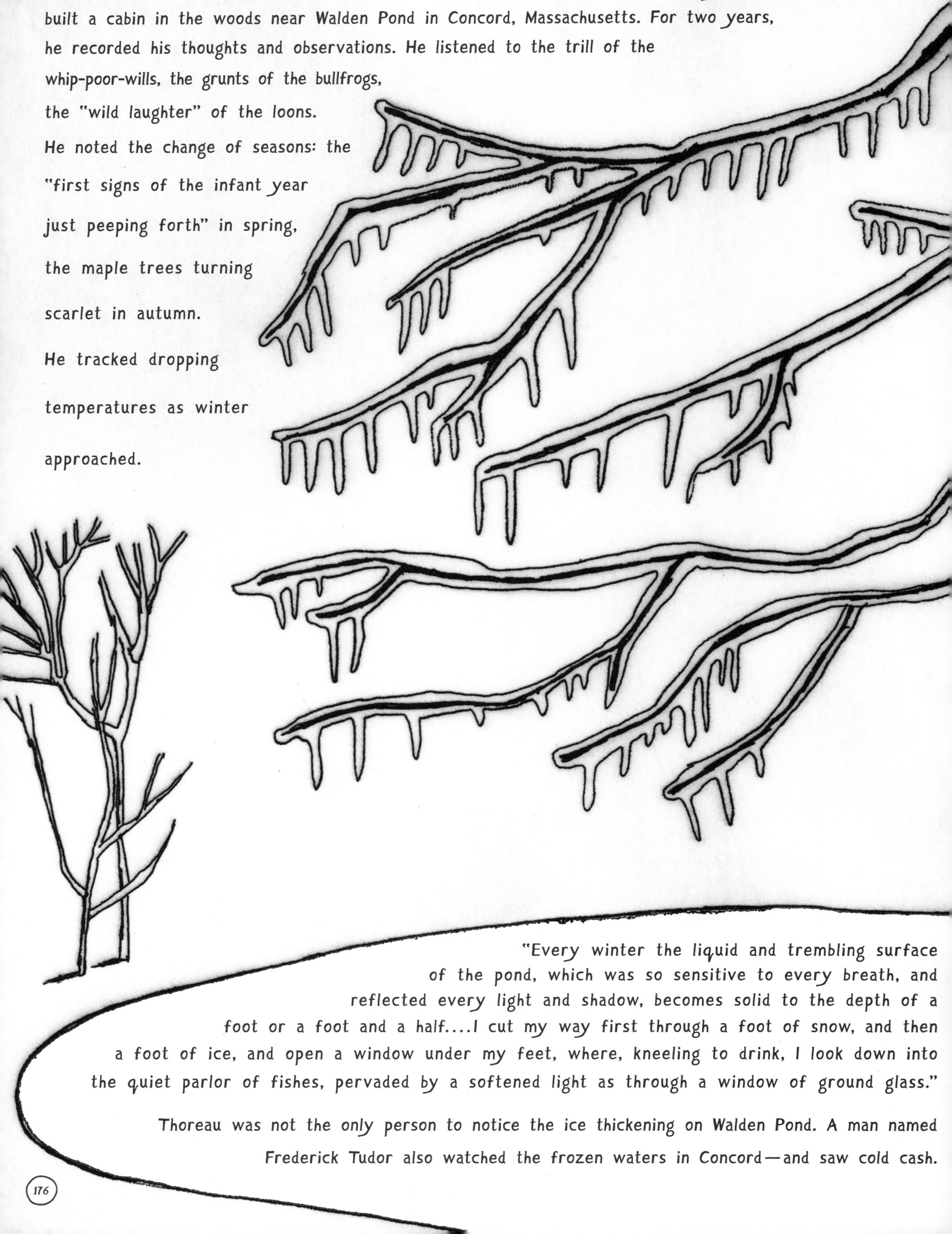

"Every winter the liquid and trembling surface of the pond, which was so sensitive to every breath, and reflected every light and shadow, becomes solid to the depth of a foot or a foot and a half.... I cut my way first through a foot of snow, and then a foot of ice, and open a window under my feet, where, kneeling to drink, I look down into the quiet parlor of fishes, pervaded by a softened light as through a window of ground glass."

Thoreau was not the only person to notice the ice thickening on Walden Pond. A man named Frederick Tudor also watched the frozen waters in Concord—and saw cold cash.

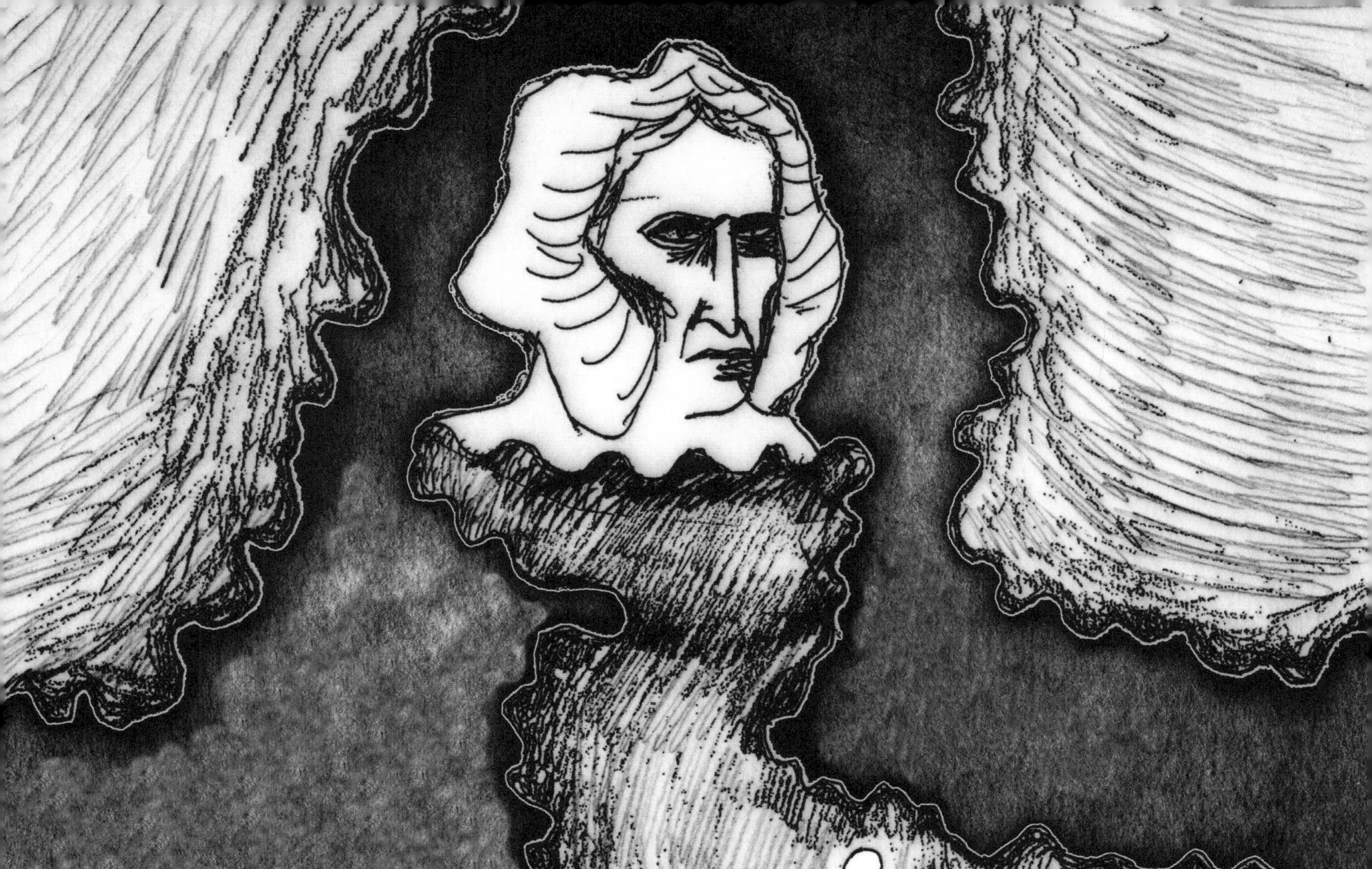

Tudor was the third son of a Boston Brahmin judge who had worked for John Adams and fought with George Washington. Instead of going to Harvard like his father and elder brothers, Frederick Tudor dropped out of school at 13 years old. By the time he was 21, in 1805, Tudor had a business plan. He would harvest the ice from frozen New England waters and ship it thousands of miles away to the tropics, where it would be sold in chunks as a delicacy and a medicine. Great profits were, he believed, "inevitable."

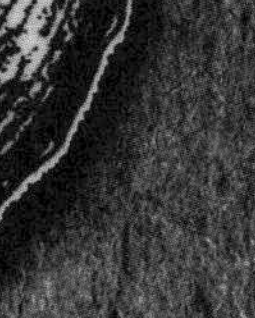

Tudor's scheme met with derision. Widespread use of mechanical cooling systems was more than a hundred years off. His father thought the idea "wild and ruinous." The *Boston Gazette* issued news of the first shipment with a disclaimer: "No joke. A vessel with a cargo of 80 tons of Ice has cleared out from this port to Martinique."

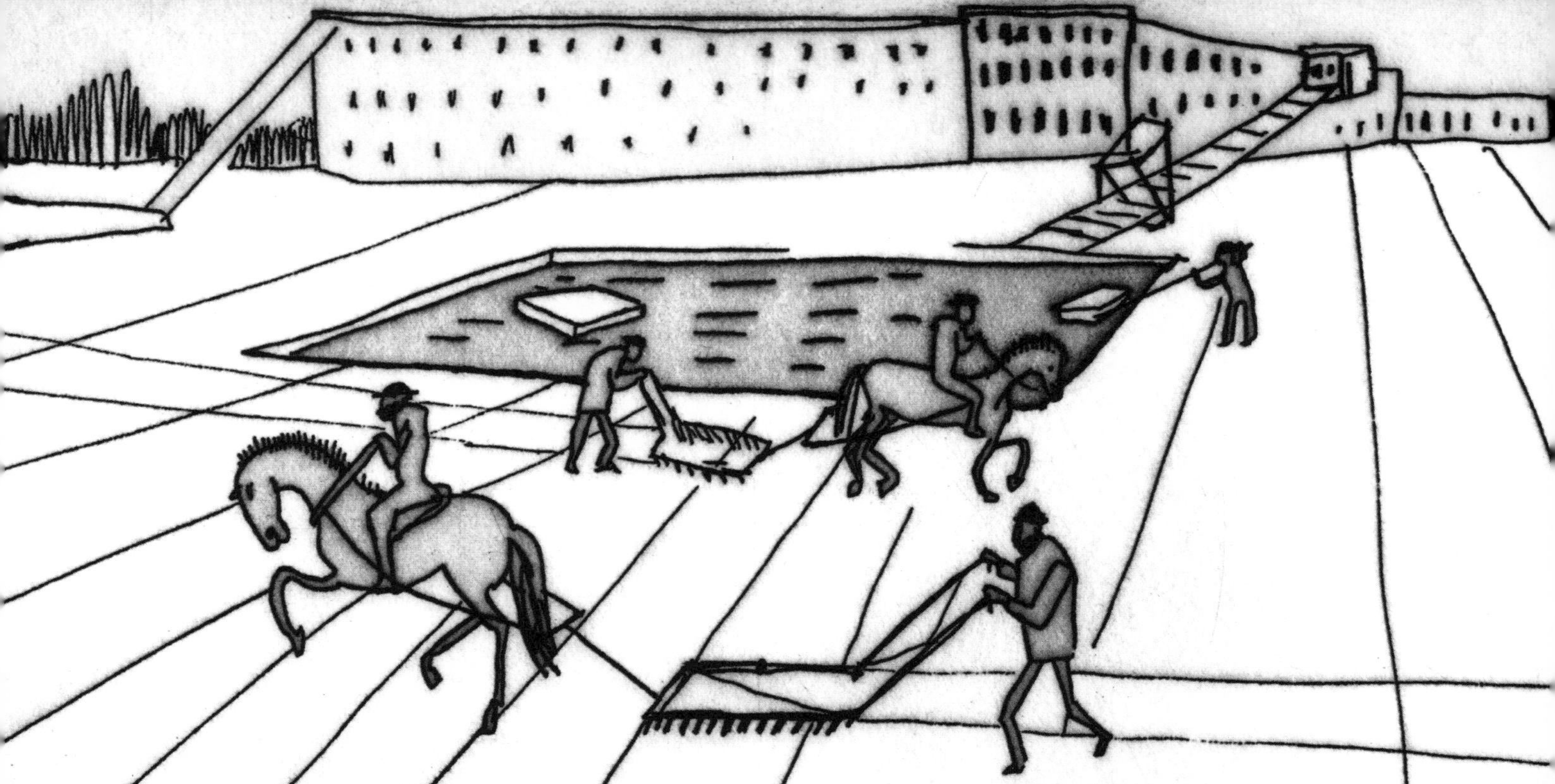

Tudor did suffer setbacks: stormy seas, corrupt local officials, distracted business partners, melting cargo. He lost thousands of dollars and was twice imprisoned for debt. But he persisted. Eventually, advances in cutting and harvesting techniques, improved methods of storage and insulation, and the cultivation of a reliable customer base made Tudor's business viable, and lucrative. Rivals entered the field. By 1840, you could buy New England ice in Calcutta, Bombay, Madras, Manila, Martinique, Singapore, Brazil, Cuba, China, Peru, New Orleans, Savannah, and Charleston.

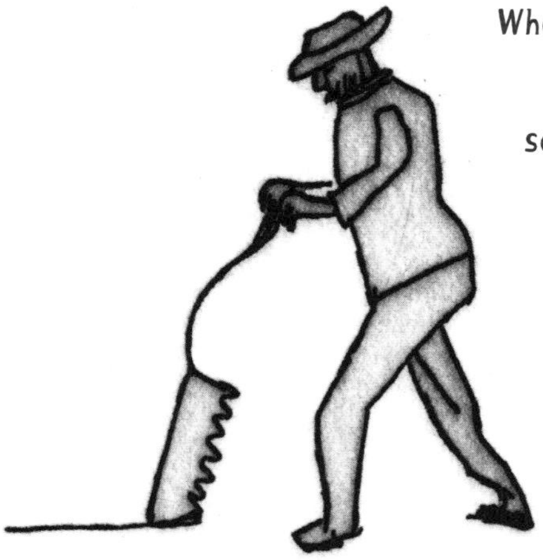

When Thoreau looked across Walden Pond, he saw a serene seasonal tableau—water distilled from "celestial dews" in which "the beholder measures the depth of his own nature." But he also saw burly workers, cutting ice.

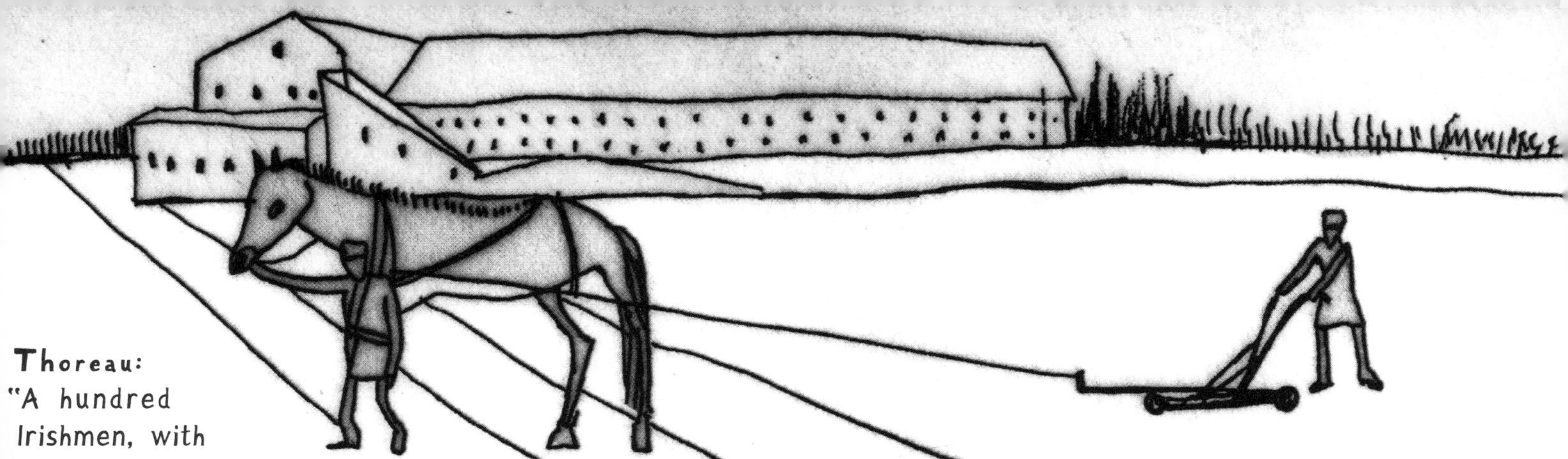

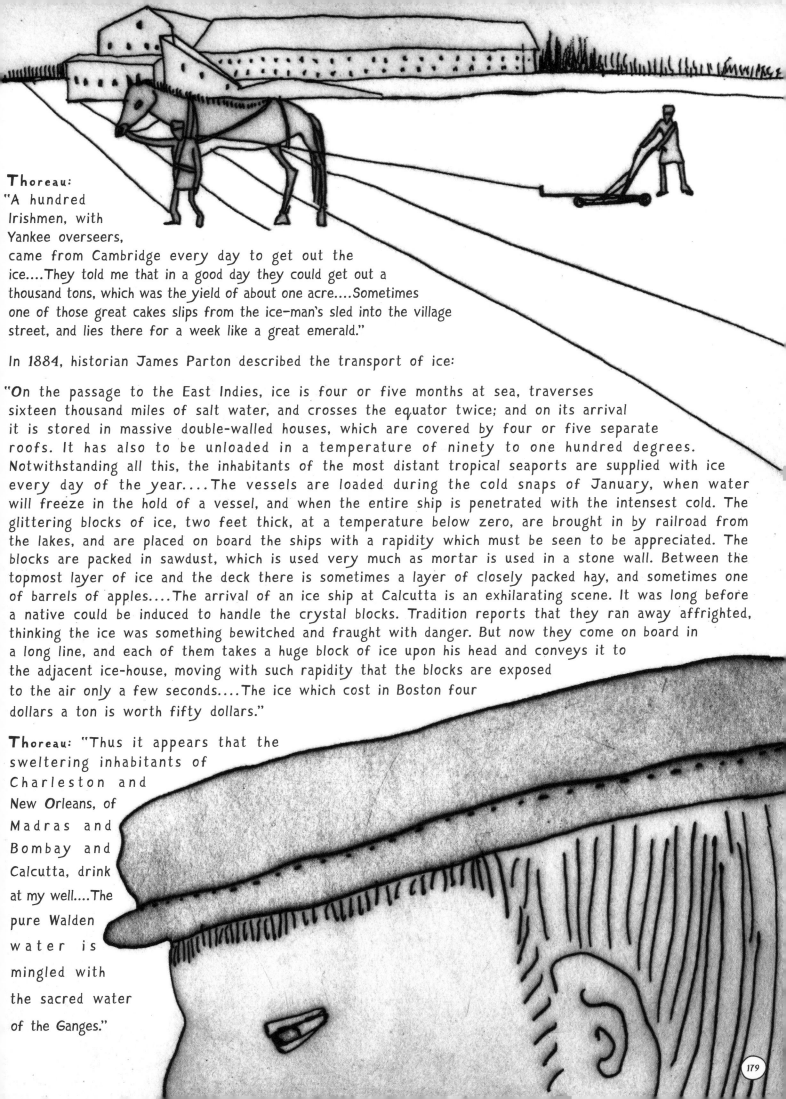

Thoreau: "A hundred Irishmen, with Yankee overseers, came from Cambridge every day to get out the ice....They told me that in a good day they could get out a thousand tons, which was the yield of about one acre....Sometimes one of those great cakes slips from the ice-man's sled into the village street, and lies there for a week like a great emerald."

In 1884, historian James Parton described the transport of ice:

"On the passage to the East Indies, ice is four or five months at sea, traverses sixteen thousand miles of salt water, and crosses the equator twice; and on its arrival it is stored in massive double-walled houses, which are covered by four or five separate roofs. It has also to be unloaded in a temperature of ninety to one hundred degrees. Notwithstanding all this, the inhabitants of the most distant tropical seaports are supplied with ice every day of the year....The vessels are loaded during the cold snaps of January, when water will freeze in the hold of a vessel, and when the entire ship is penetrated with the intensest cold. The glittering blocks of ice, two feet thick, at a temperature below zero, are brought in by railroad from the lakes, and are placed on board the ships with a rapidity which must be seen to be appreciated. The blocks are packed in sawdust, which is used very much as mortar is used in a stone wall. Between the topmost layer of ice and the deck there is sometimes a layer of closely packed hay, and sometimes one of barrels of apples....The arrival of an ice ship at Calcutta is an exhilarating scene. It was long before a native could be induced to handle the crystal blocks. Tradition reports that they ran away affrighted, thinking the ice was something bewitched and fraught with danger. But now they come on board in a long line, and each of them takes a huge block of ice upon his head and conveys it to the adjacent ice-house, moving with such rapidity that the blocks are exposed to the air only a few seconds....The ice which cost in Boston four dollars a ton is worth fifty dollars."

Thoreau: "Thus it appears that the sweltering inhabitants of Charleston and New Orleans, of Madras and Bombay and Calcutta, drink at my well....The pure Walden water is mingled with the sacred water of the Ganges."

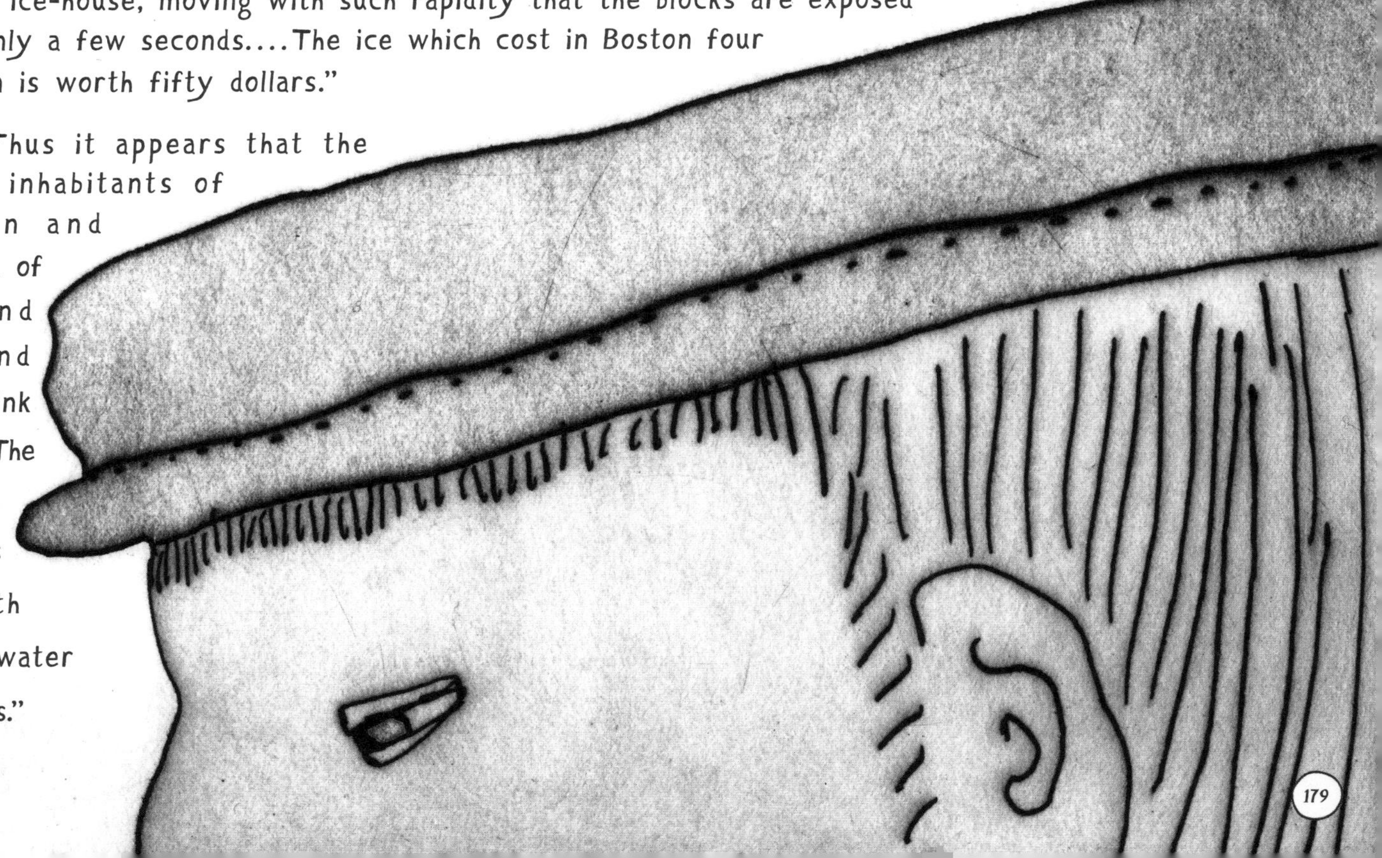

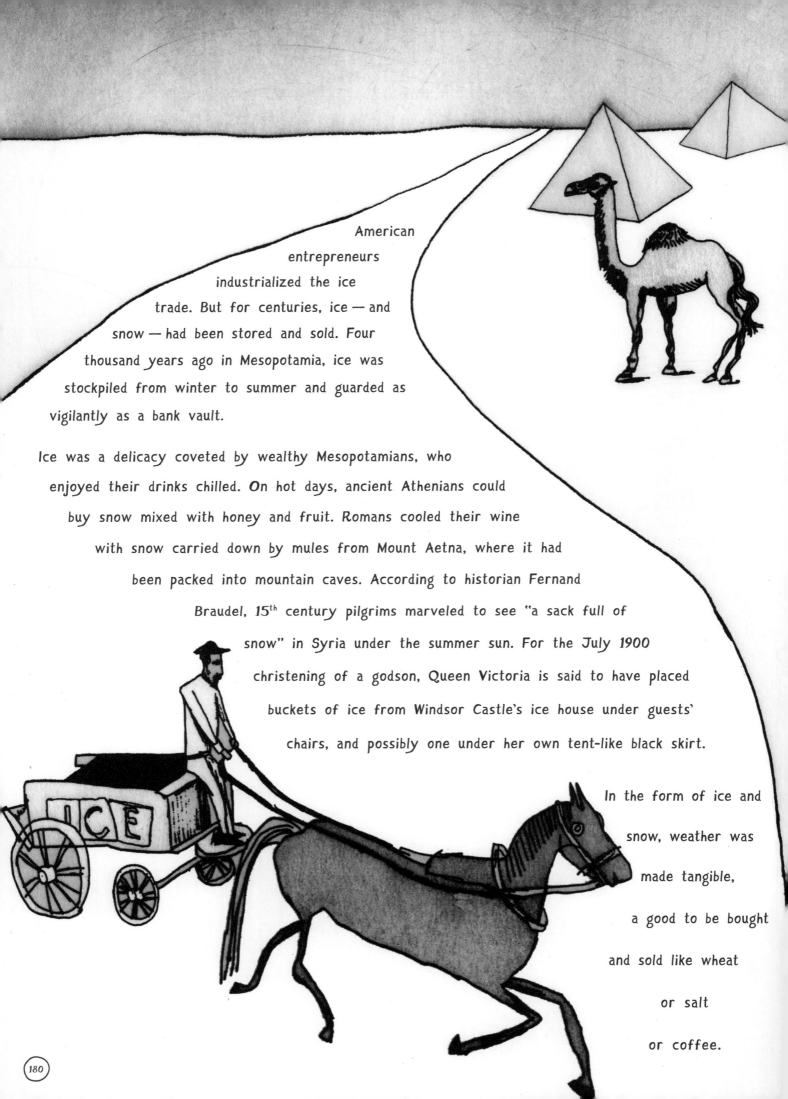

American entrepreneurs industrialized the ice trade. But for centuries, ice — and snow — had been stored and sold. Four thousand years ago in Mesopotamia, ice was stockpiled from winter to summer and guarded as vigilantly as a bank vault.

Ice was a delicacy coveted by wealthy Mesopotamians, who enjoyed their drinks chilled. On hot days, ancient Athenians could buy snow mixed with honey and fruit. Romans cooled their wine with snow carried down by mules from Mount Aetna, where it had been packed into mountain caves. According to historian Fernand Braudel, 15th century pilgrims marveled to see "a sack full of snow" in Syria under the summer sun. For the July 1900 christening of a godson, Queen Victoria is said to have placed buckets of ice from Windsor Castle's ice house under guests' chairs, and possibly one under her own tent-like black skirt.

In the form of ice and snow, weather was made tangible, a good to be bought and sold like wheat or salt or coffee.

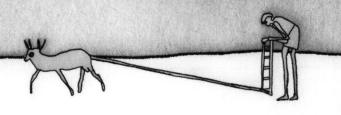

In 1997, the Enron Corporation made the first weather derivative deal, a hedge against temperatures unfavorable to a utility company. Hypothetical weather had become a commodity. Today, weather derivatives are a $12 billion business. In a speculation-fueled environment, the uncertainty of future weather creates value, offering risk and reward.

A 2011 paper published by the American Meteorological Society tallied weather's impact on the U.S. economy at $485 billion in 2008, or 3.4 percent of that year's GDP. Other estimates have placed the estimate of weather's economic importance far higher, reaching one third of annual GDP.

Brad Davis is president of MSI GuaranteedWeather LLC, a weather risk management company based in Overland Park, Kansas. The company sells weather insurance and weather derivatives.

Brad Davis: "Financially, some people love bad weather. People that plow snow, they want snowstorms. People that sell umbrellas, they want rain.

"A company that sells electricity in the summer wants it to be hot. Say you wanted average temperatures of 100 degrees in the month of July. You might purchase an option that would allow you to collect a payment if the average temperature in the month of July was only 85 degrees.

"A construction company in the summertime, they probably don't care about the heat because people work when it's hot, but they probably have a concern about rain. They might experience delays in their construction business. Folks in that industry have purchased contracts that would pay them if there was an excessive rainfall during that time period.

"The sky is the limit of what's impacted by weather and what can be covered by a weather derivative. You're buying or selling the right to receive—or the obligation to make—a payment if a weather event does or does not occur. Technically, we can't guarantee the weather. But we can structure a contract for you that will make you feel a lot better financially if the weather goes in a direction that's not to your liking."

Weather-related insurance has traditionally been used to protect against disaster—floods, tornadoes, hurricanes. Weather derivatives promise to mitigate the impact of less dramatic circumstances, to insulate a business from mundane temperature deviations that can impact its bottom line. They can also be used as instruments of speculation: while insurance offers compensation for damages, purchasers of weather derivatives may seek returns greater than mere restitution: a profit off of weather, good or bad.

Social critics have identified other ways people take advantage of weather events. Writer-activist Naomi Klein coined the term "disaster capitalism" to describe, for example, the profiteering of entrepreneurs who swooped in to exploit development opportunities in New Orleans in the aftermath of Hurricane Katrina.

The corporate campus of Planalytics, Inc., is in Berwyn, Pennsylvania. It sits just down the road from Valley Forge National Historical Park, where, in 1777–78, General George Washington's Continental Army endured a bitter cold winter.

The grounds around the parking lot are shaded by tulip, birch, and weeping willow trees.

Three fountains burble in a green pond.

Planalytics offers its clients "Business Weather Intelligence," "the actionable information companies need to understand and optimize the impact of weather on their business."

Planalytics cross references sales data and weather records, searching for relationships that may not be intuitive or obvious. Once recognized, a correlation between temperature or precipitation—or any other meteorological phenomenon—and consumer behavior may be translated into business strategy. Planalytics's client list includes Coca-Cola, Pepsico, Dow Chemical, Bayer International, Bloomberg, Caterpillar, ConAgra Foods, Dunkin' Donuts, Equitable Gas Company, Hanes, Heinz, Johnson & Johnson, John Deere, Levi Strauss Corporation, Payless Shoes, Pet Smart, Rite Aid, Starbucks, and the United Farmers Cooperative. Frederick Fox is the company's co-founder and CEO.

Frederick Fox: "We have a supermarket client with a lot of stores down in Florida.

When a hurricane is forecast,

what is the number one product that they sell?

The *number one product*.

It's not water, or candles, or

matches, or batteries. It's not

canned goods."

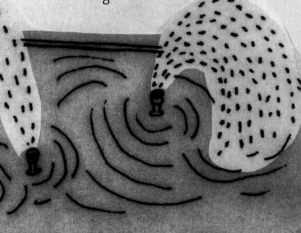

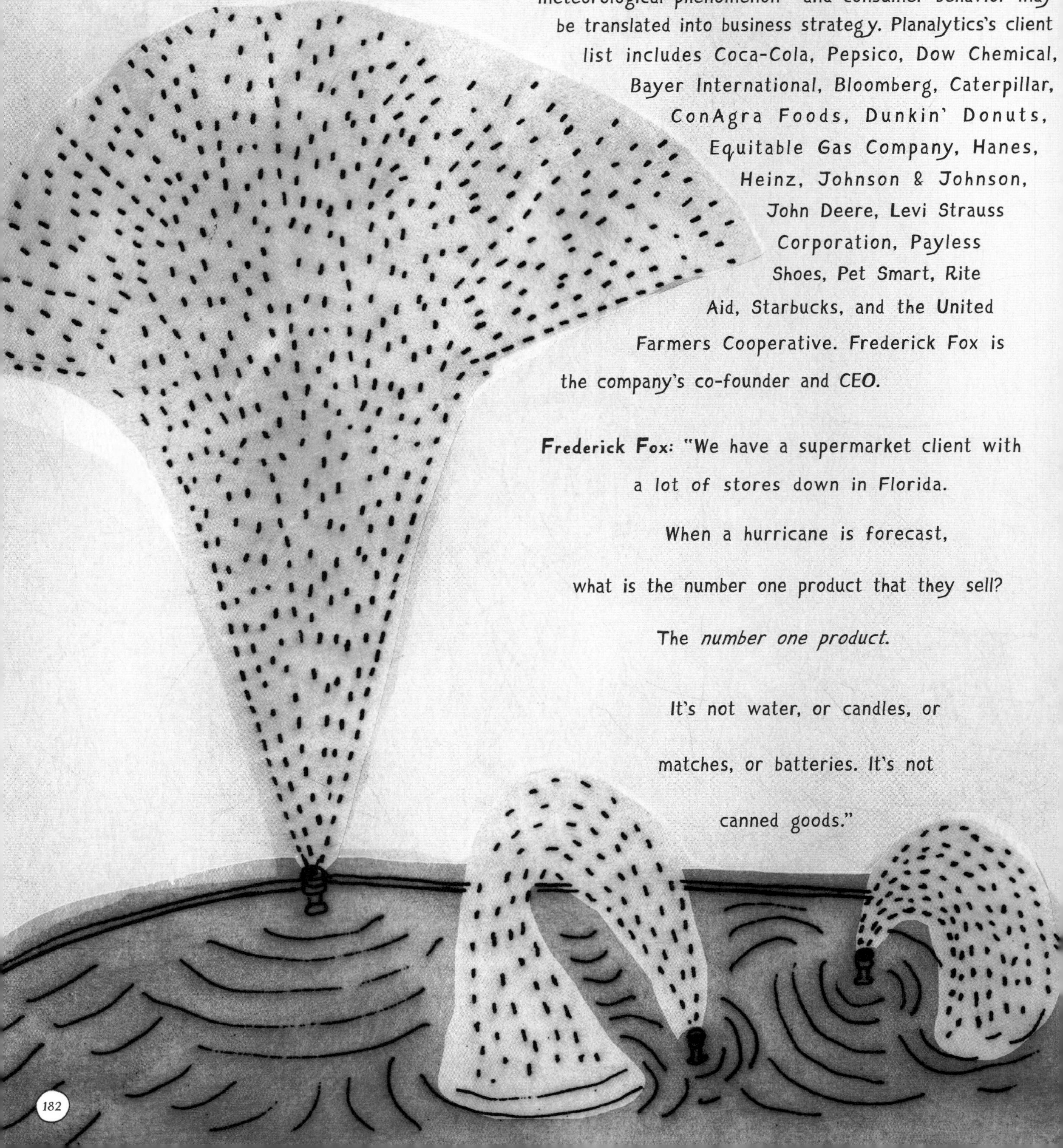

Frederick Fox: "It's fried chicken. It's their number one. Water obviously sells out, but fried chicken? Really? But that's what their data show. So they'd like a few days' heads-up before a storm. They want to get a jump on the chicken houses in Georgia and the Carolinas, to get those orders in so they can make enough fried chicken.

"Supermarkets sell out on the *forecast* of a storm. No one wants to lose those actual traffic days due to snow piling up. So *that's* the perfect storm. Same with a hurricane threatening to hit the East Coast: sales go up, and if the storm goes out to sea, everyone's happy. No one got hurt and it didn't stop any shopping days.

"Look at men's and women's boots. The first chill in the air in September or October, women's boot sales go right through the roof. Now, the weather's still nice at that time of year in a lot of the U.S. Men's boot sales don't budge. Men's boot sales move much later in the season, in late October or November when it's really cold and really wet and men's socks are getting wet.

"If you look at demographics, let's say, between Tampa and Miami: you have a nice-weather day down in that part of Florida—it's clear, it's warm. Tampa sales go up; Miami sales go down. If it's rainy, Miami sales go up, Tampa sales go down. What's going on? This can't be right, these [cities] are a few hours apart, it's the same weather system, and basically the same temperature, so how could it be that different? Miami is a younger demographic. They're outside when it's nice. They're shopping when it's not nice. In an older-skewed demographic [like Tampa], they're not going shopping when it's raining. In Seattle, people shop when it's raining because there are so few days of clear weather. When it's clear weather, that's not normal, so they're outside having fun—not shopping.

"Last year [2010] in New York, when did spring sales start for most retailers? April. When did it start this year? Late February. Almost a five- to six-week difference. Huge. I mean, that is an earthquake in change in terms of the selling season. We had an early spring in the eastern half of the nation. February got warm. That's going to drive sales much earlier. It's going to make the numbers look great. The impact is a hundred times larger than any one hurricane that could hit. Economists and people start to say, 'Oh, the economy's coming back.' And yet we saw that the weather later in the season was not going to be favorable. Sales are depressed now, after they were up earlier, you hear people say, 'The economy was up, now it's down.' We're sitting back and saying, 'Well, a lot of it is also the weather that's driving this. What's driving economics at the store level is the subtle week-to-week changes of warm and cold.'

"Why did the Israelites end up in Egypt? It wasn't because they liked the palm trees, it was because of drought. Jacob and his kids were starving, so they moved to Egypt. Joseph was a forecaster of drought. Pharaoh actually listened to him and bet the whole ranch on him. They built granaries, which gave them great power.

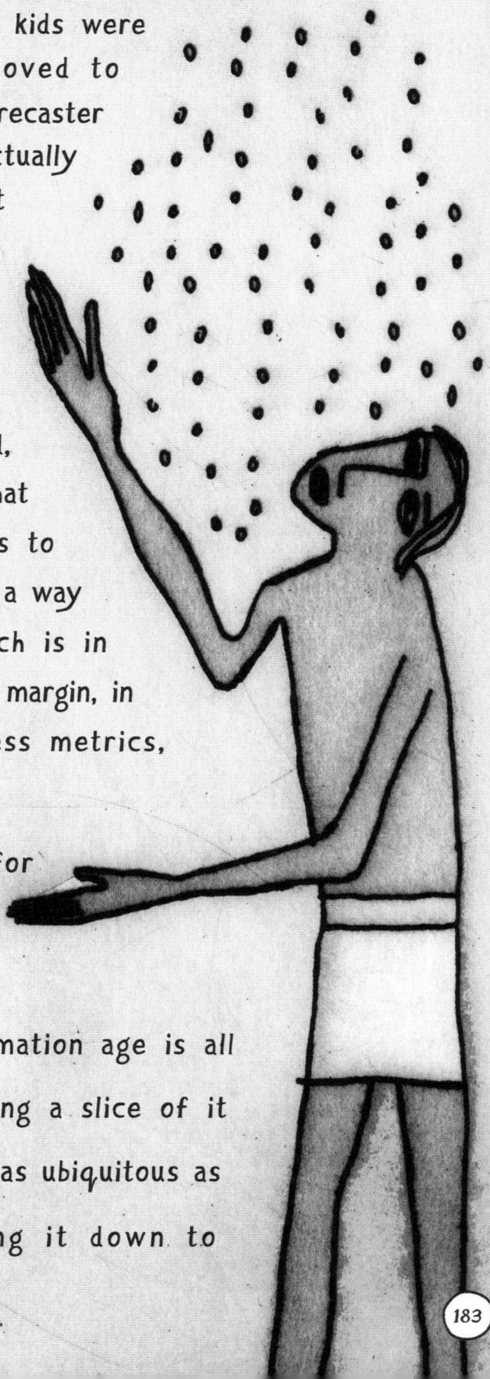

"Forewarned is forearmed, as the saying goes. What we try to do today is to forewarn our clients in a way they understand, which is in units sold, in dollars, in margin, in inventory—all business metrics, which is what we track.

"It was great power for Joseph. It's great power today. And that's what the information age is all about. We're just taking a slice of it from something that's as ubiquitous as weather and bringing it down to something useable."

At Walden,
Thoreau lived on land owned by Ralph Waldo Emerson.
In a letter to Thoreau in March of 1847,
Emerson considered the ice trade's effect
on his property's worth:

"I am not without a prospect that my woodlot by Walden Pond will get an increased value soon; as Mr. Tudor has invaded us with a gang of Irishmen and taken 10,000 tons of ice from the Pond in the last weeks. If this continues, he will spoil my lot for purposes which I chiefly value it,
and I shall be glad to sell it."

Fickle weather and technological progress spared Emerson's investment. Mechanical refrigeration made creeping gains beginning in the 1860s. By the 1890s, distillation, purification, and freezing processes had improved enough to threaten the natural ice industry. Ice manufacturers launched a propaganda campaign saying that natural ice contained "intestinal germs" which could cause typhoid and other diseases. Variations in winter weather made natural ice supplies unreliable. In 1906, *The New York Times* reported,

"Unless there is cold weather and plenty of it
within the next six weeks New York will enter upon
the coming Summer face to face with an ice famine."

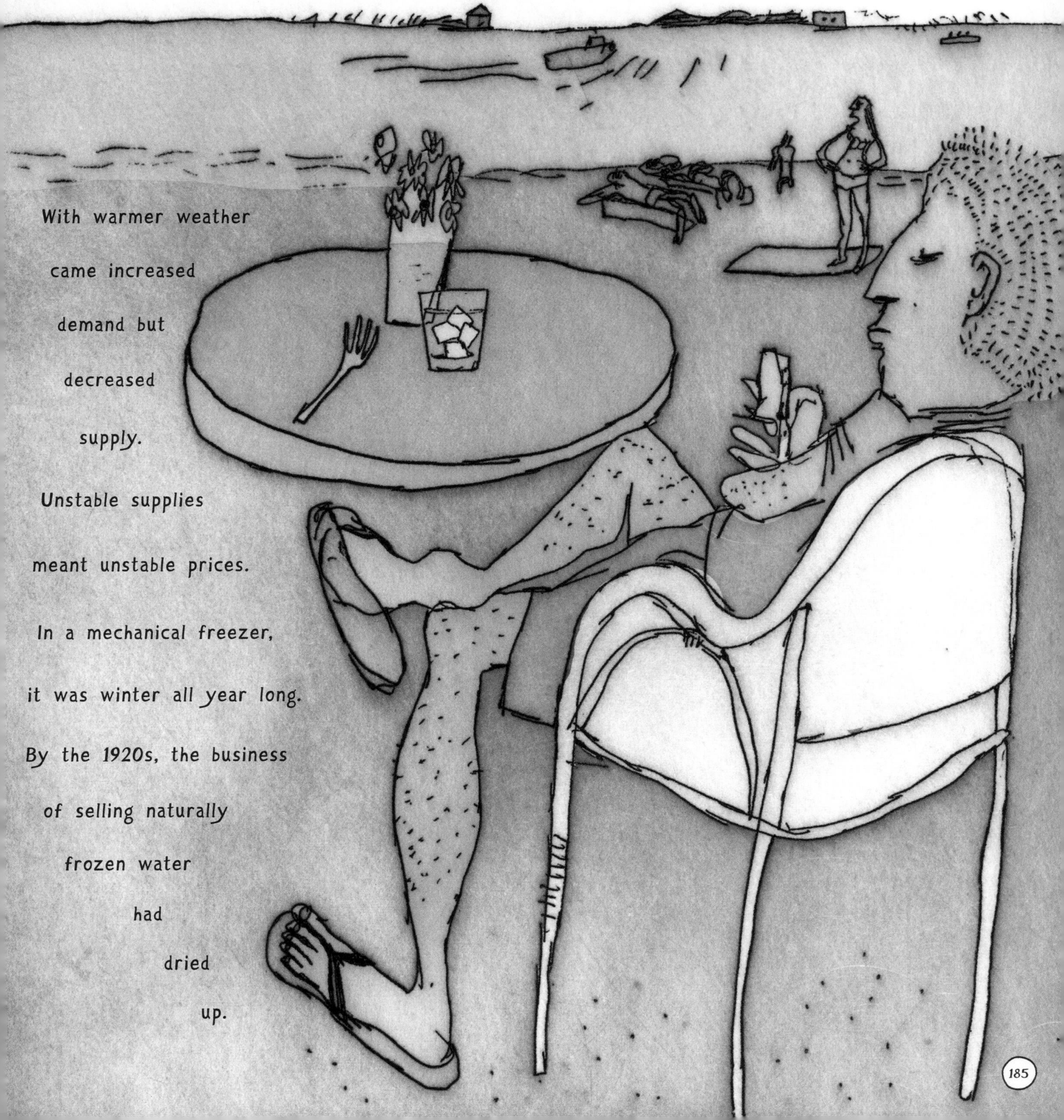

With warmer weather came increased demand but decreased supply.

Unstable supplies meant unstable prices.

In a mechanical freezer, it was winter all year long.

By the 1920s, the business of selling naturally frozen water had dried up.

CHAPTER 11:

PLEASURE

"There is really no such thing as bad weather, only different kinds of good weather."

— John Ruskin

During hurricanes Irene and Sandy, New Yorkers posted personal ads on Craigslist.com, seeking shelter from the storm in romance and sex.

"If the hurricane gonna destroy nyc, let's watch it together ;)-m4w-30 (queens, Brooklyn, manhattan, bronx, etc): if weather forecasters are right and Irene gonna make a mess in new york...I'm gonna find the best place where I can look on this sunday's morning hurricane and get some coffee with a beautiful girl let me know if it's you ;-)"

"How hot would it be to fuck while a hurricane is raging outside? I say very...Hurry, Sandy's coming!"

"Just met at Soho Evacuation Center-m4w-38: I realize you'll never see this but I don't care because being a fucking romantic is my thing especially during hurricane evacuations. You are the most amazing girl I've ever seen...if we ever get out of this alive I will remember the tears we shared, clutching our chunks of Government cheese snacks they just gave us."

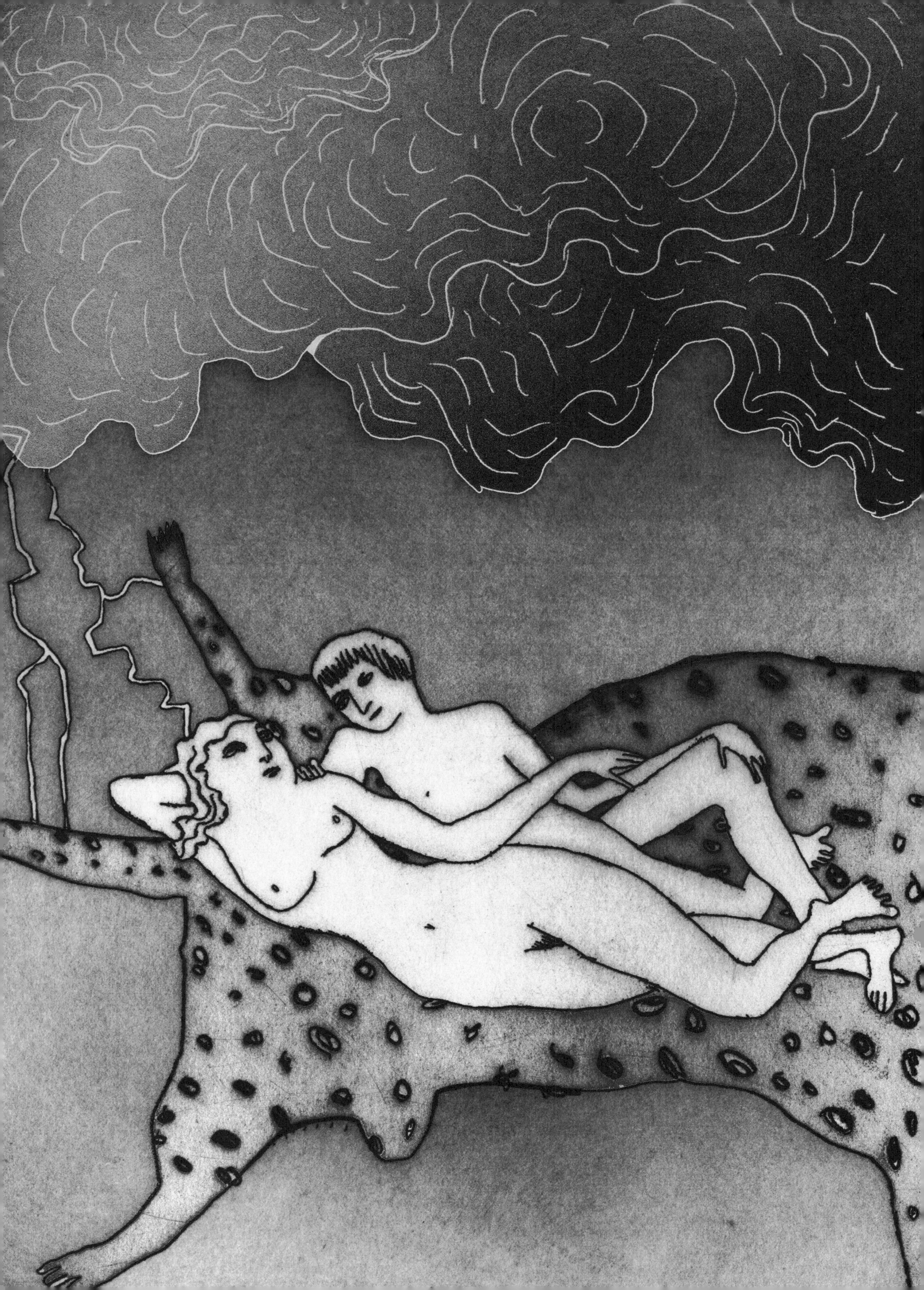

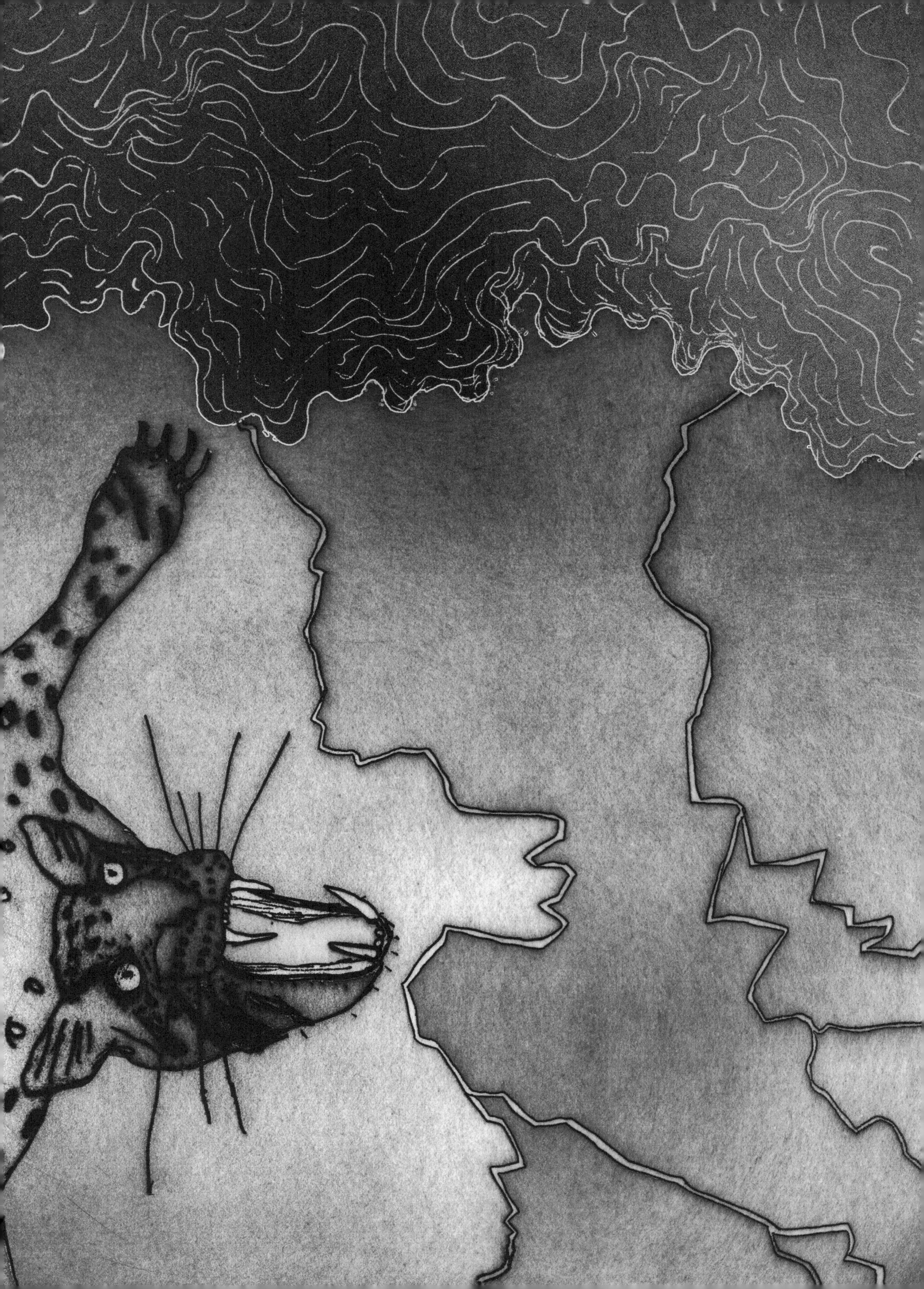

Benjamin Franklin was a proponent of air baths, the practice of sitting naked by an open window. "I rise early almost every morning and sit in my chamber, without any clothes whatever, half an hour or an hour, according to the season, either reading or writing." In the sandstorms of the Mongolian desert, Mark Norell, chief paleontologist at the American Museum of Natural History, has enjoyed a version of Franklin's bracing ritual. "You take off all your clothes and you stand there. When the sand runs over you it generates static electricity. You are just being pelted by the sand. All your hair stands on end."

When rivers and canals froze over during Europe's Little Ice Age, temporary carnival cities sprung up on the ice. There were "frost fairs" in Venice, Amsterdam, and London. Attractions included bull-baiting, bear-baiting, horse races, archery, puppet shows, music, football, feasting, boozing, and brothels. A 17th century poem described a London frost fair: "There is such whimsies on the frozen ice/Make some believe the Thames a paradice [sic]."

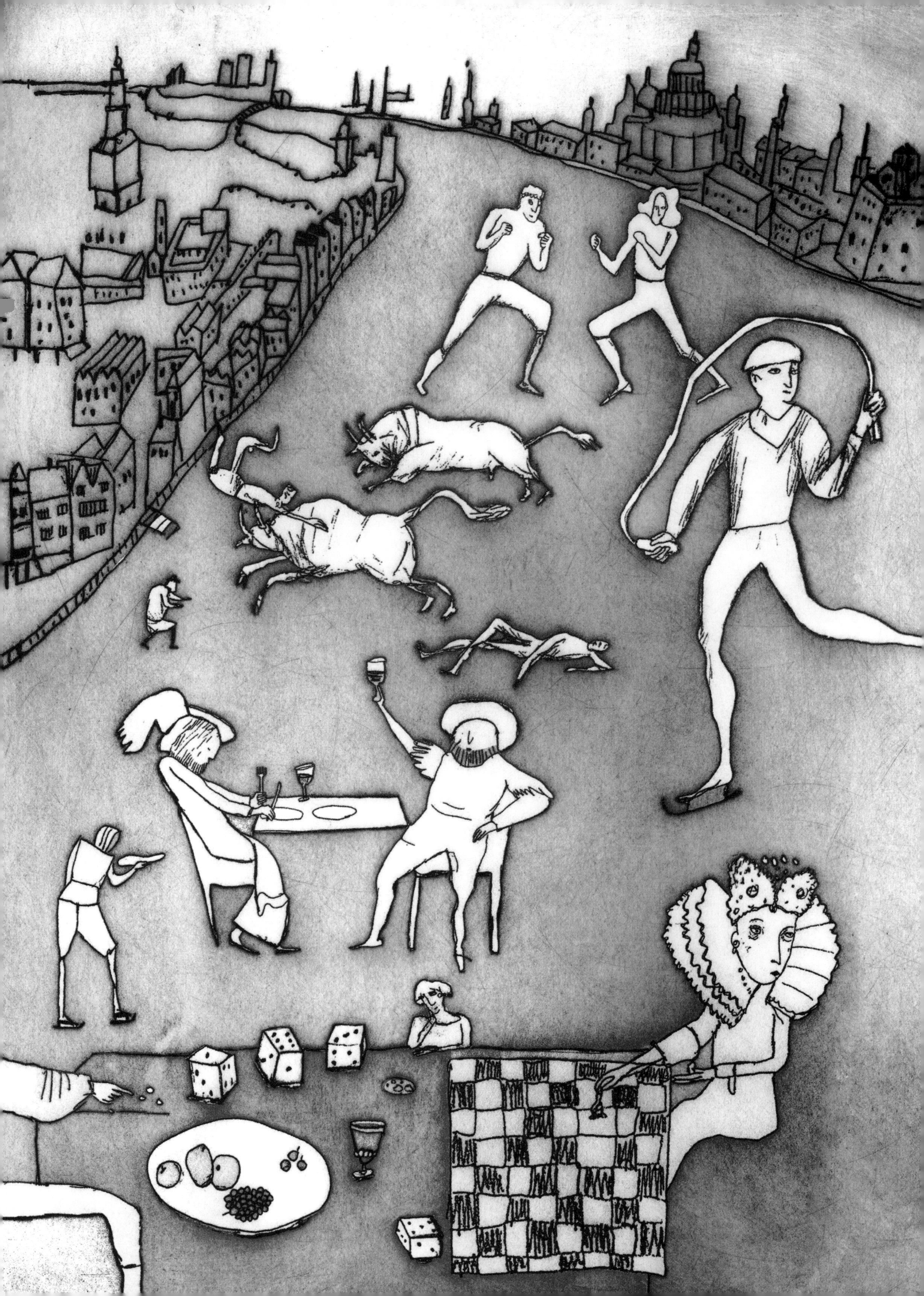

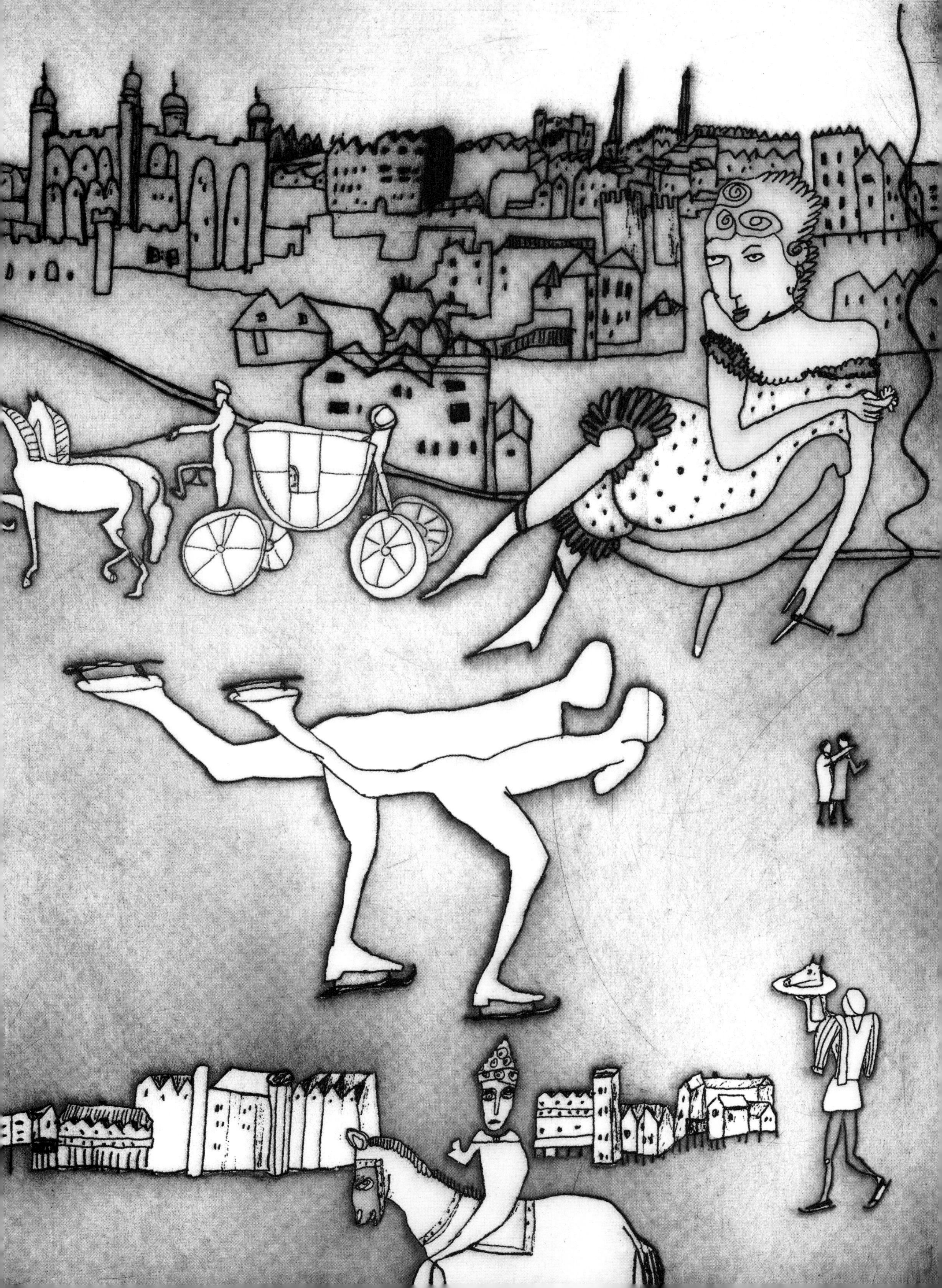

Weather reports can provide more than information. Generations of Britons have found comfort in the lulling rhythms of the Shipping Forecast, a maritime weather update that airs four times daily on BBC Radio 4. "Can there be anything in any language to match the poetry of the shipping forecast?" asks a writer in *The Guardian*. "The voice of the announcer...[is] the voice of God, all-knowing, untroubled...concerned to protect you from the violence of the world.... 'Rockall, Hebrides. Southwest gale 8 to storm 10, backing southerly, severe gale 9 to violent storm 11. Rain, then squally showers...Faeroes, Southeast Iceland. North 7 to severe gale 9, occasionally storm 10 later. Heavy snow showers.' That is the poetry: vastness and violence described in tranquility."

Falling snowflakes interfere with sound waves, limiting the distance they travel and contributing to the muffled quietude that accompanies a snowstorm. Fresh snow on the ground is airy and porous, and sound is further absorbed into these air pockets. Temperature adds to the effect: sound moves faster in warm air. When it snows, air near the ground is typically warmer than air above, curving sound waves upward, out of earshot and into the atmosphere. "It snowed all week," wrote Truman Capote in "Miriam" (1945). "Wheels and footsteps moved soundlessly on the street, as if the business of living continued secretly behind a pale but impenetrable curtain."

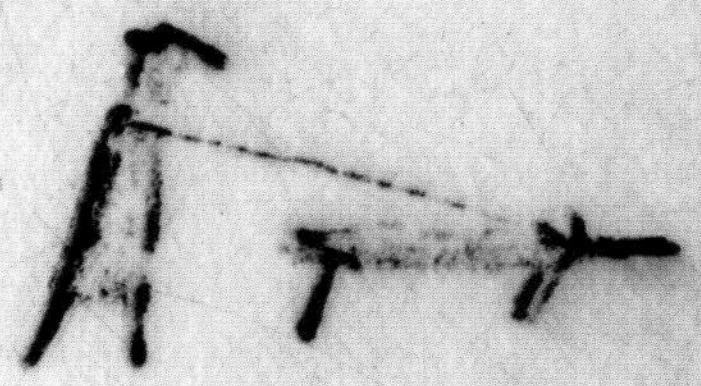

"They now had to run for it, and did not reach home till they were nearly soaked through. The lightning and thunder still continued, and the rain seemed to smoke along the ground and upon the thatched roof of a shed opposite to their house. Sometimes the thunder sounded very high in the air, as if above the clouds; at others, as if it were down in the road. That which but a few minutes before had been a lovely day, with a blue sky, and stately clouds like snowy rocks that scarcely moved at all, was now one dull, lead-coloured covering. In about an hour it became lighter, and in another hour they had the pleasure to see that stormy cloud sailing away from them still looking black with its edges touched by the light of the golden sun. From time to time they heard that the storm had not ceased, though it was not so loud; at length it was so far off that the thunder made only a low, surly rumbling; and the cloud which had before looked so angry, when over and near them now shone like a snow-covered mountain, with crags and precipices, and deep hollows and caverns. The family all remarked how pleasantly cool the air had become, and how calm; and admired the fresh and glittering appearance of the grass, and the leaves of the trees, and flowers in the sunshine; and they snuffed up with delight the smell of the earth after the rain."

— *Adam the Gardener,* by Charles Cowden Clarke, 1834

In dry weather, oils build up on rocks and vegetation. Rain releases the oils, and with them, a fresh, earthy scent that perfumes the air. In a 1964 article in *Nature,* mineralogists Isabel Joy Bear and R. G. Thomas coined a word for the smell, "petrichor": *petr-* for rock or stone, and *ichor,* a reference to ambrosial liquid said to run through the veins of Greek gods.

"I'd like to just get one of those pink clouds and put you in it and push you around."

— F. Scott Fitzgerald

CHAPTER 12:

FORECASTING

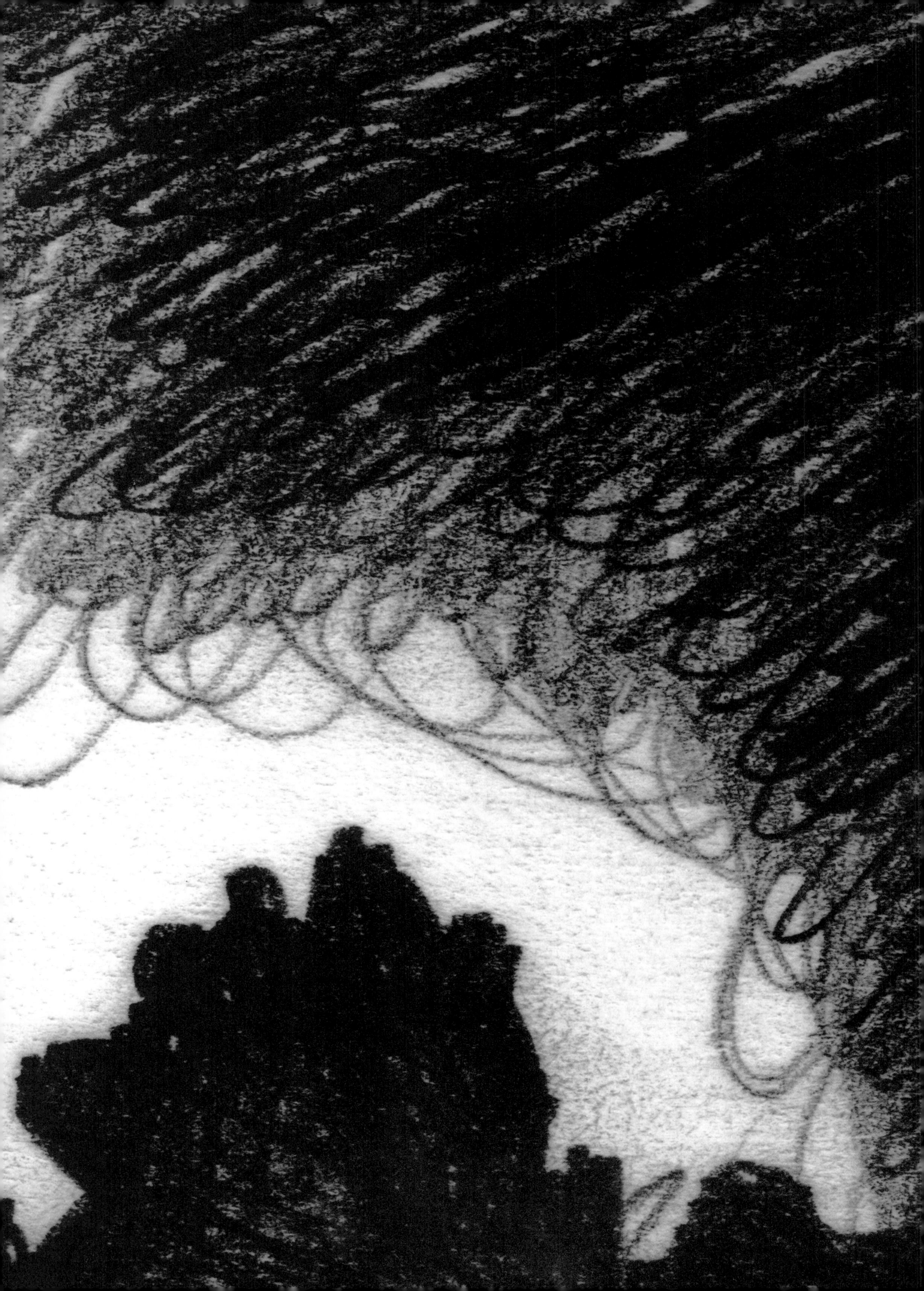

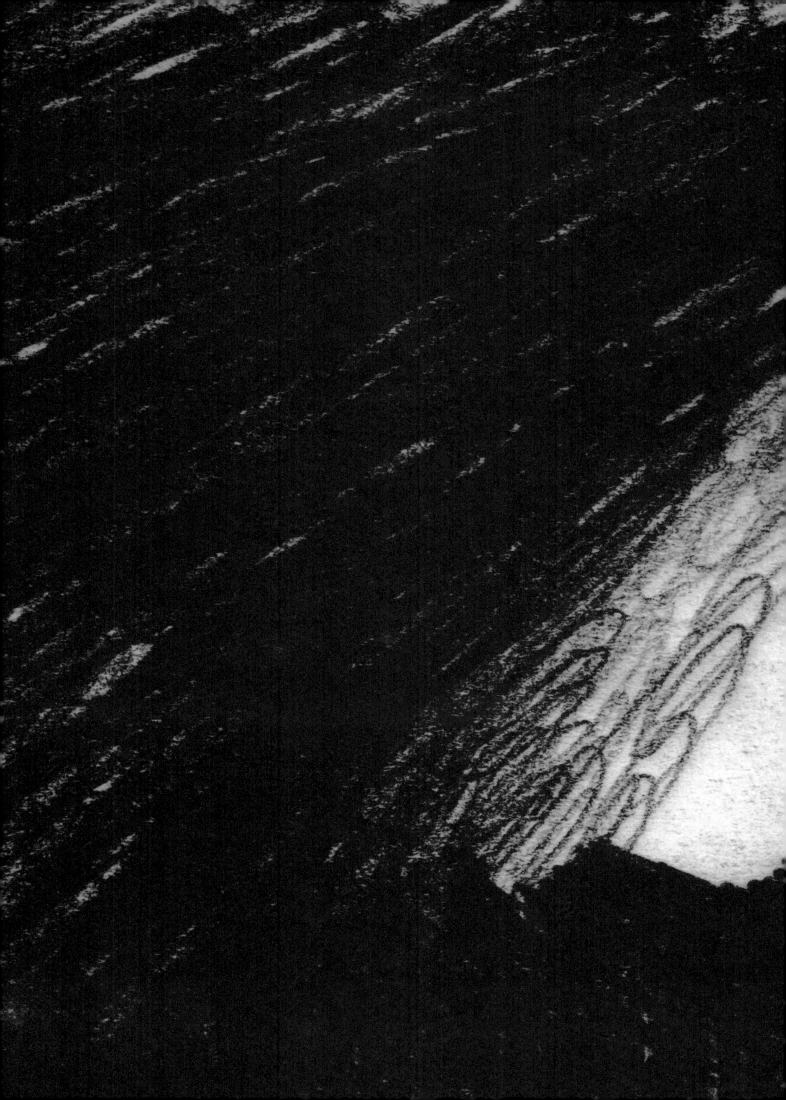

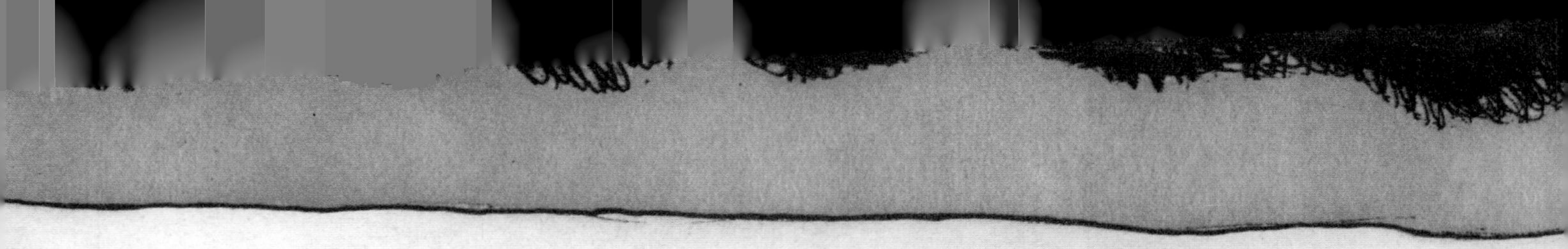

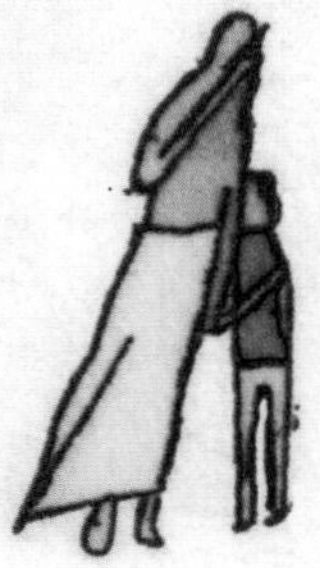

On the morning of June 10, 1953, Charles Golub, a grocer, took his four-year-old daughter Robin for a drive. He wanted to see the damage caused by a tornado that had hit their city, Worcester,

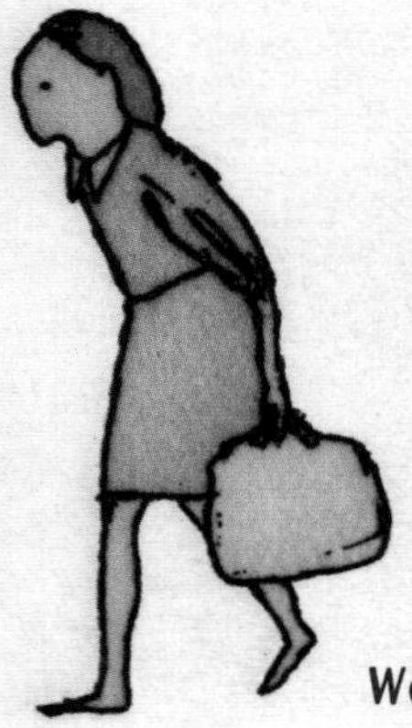

Massachusetts, the previous day. The tornado had spun through Worcester County for an hour and a half, at times reaching a mile in width. It killed 94 people and left

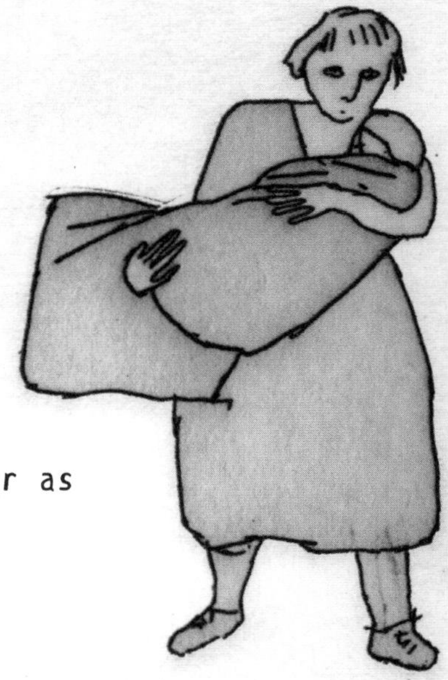

15,000 homeless. Debris was flung as far as Eastham, on Cape Cod, more than 100 miles away.

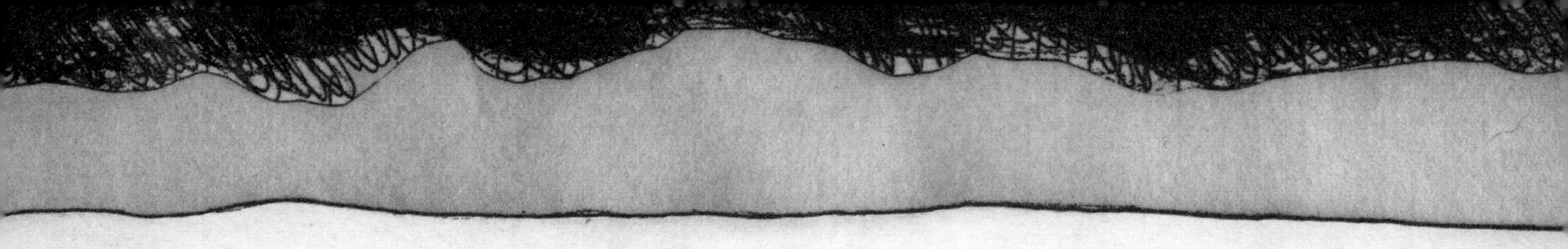

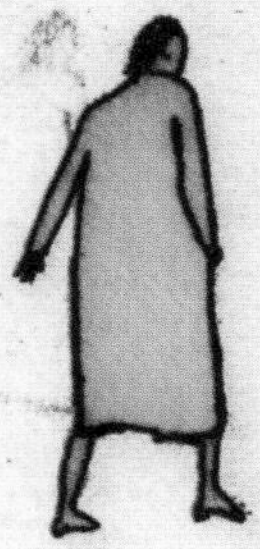

The little girl, grown now, has scattered memories of what she and her father saw from the car window that next morning. "I remember jagged wood, and roofs blown open, the side of a house open, looking in and seeing a bedroom,

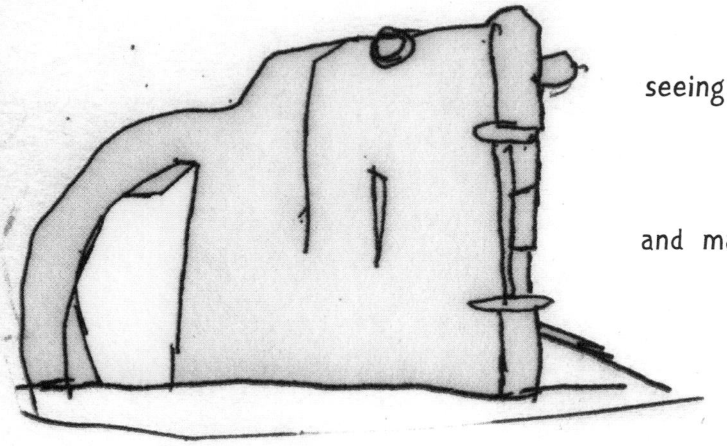

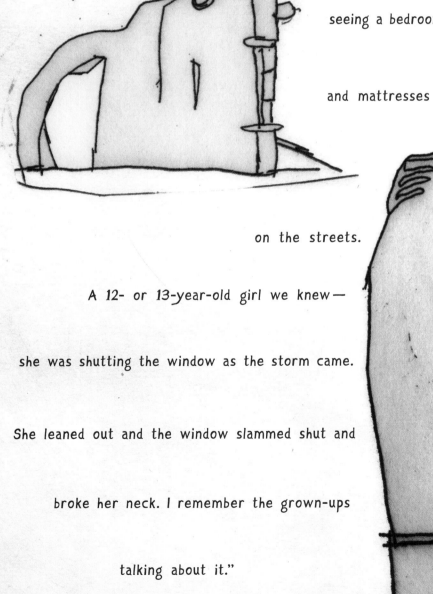

and mattresses on the streets.

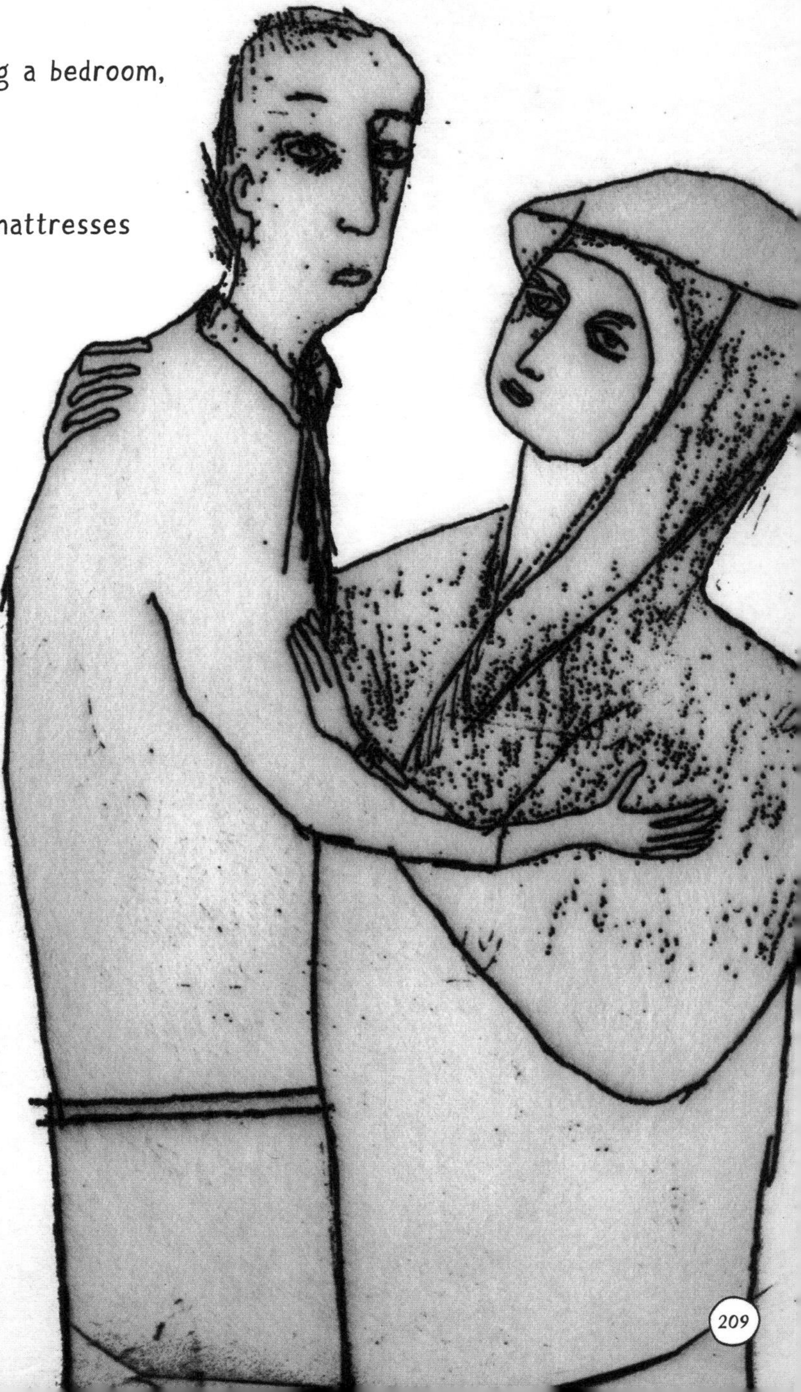

A 12- or 13-year-old girl we knew—she was shutting the window as the storm came. She leaned out and the window slammed shut and broke her neck. I remember the grown-ups talking about it."

Temperatures in Worcester had reached 90°F on June 6, unusual before summer had even begun, before dropping steeply to the mid-seventies on the following days. The Midwest was experiencing thunderstorms; a series of tornadoes hit Michigan and *Ohio*. As the storm system pushed eastward, the Weather Bureau at Boston's Logan Airport saw the possibility of tornadic activity developing in Massachusetts, too. But the word "tornado" had never been used in a New England forecast. Officials deliberated and, fearing panic, decided against issuing a warning. When the twister wound into Worcester at the end of the Tuesday workday, the public had been given no time to prepare.

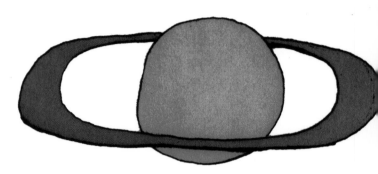

One publication, though, claimed it had predicted the tornado. The *Old Farmer's Almanac*, published each September, contains a year's worth of weather forecasts, covering the entire United States. The 1953 edition included a forecast for the first week in June, written in the *Almanac*'s signature rhyming couplet form: "A heavy squall and that's not all." Then, according to the *Almanac*, the weather would turn, in a word, "nasty."

After the tornado, readers wrote in to praise the *Almanac*. Sixty years later, its editors still cite "A heavy squall and that's not all" as an example of the uncanny accuracy of the magazine's weather forecasts.

The *Old Farmer's Almanac* has predicted the weather for more than 220 years. The *Almanac* predates railroads and electric light. When the first issue appeared in 1792, there were fifteen United States and George Washington was president.

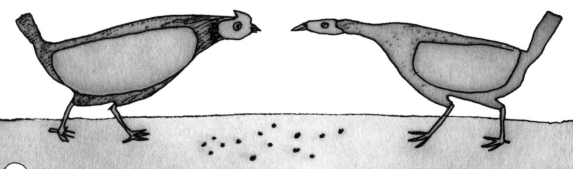

Since the Middle Ages, men have published almanacs—"calendars of the heavens"—charting the movements of the moon, the sun, and the planets. Typically these books have included tables of high and low tide, times of sunrise and sunset, and weather predictions for the coming year. Years before his Bible, Gutenberg printed an almanac. A Bible and an almanac were often the only books found in the homes of American colonists. Almanacs provided essential guidance for planting and harvesting, and the care of livestock. They prescribed home remedies, listed stagecoach schedules, important roads, and the names of innkeepers along those roads. According to the American Antiquarian Society's Richard Anders, cataloger of the Society's vast almanac collection, "If the almanac had a comprehensive subject, it was: How to get through life."

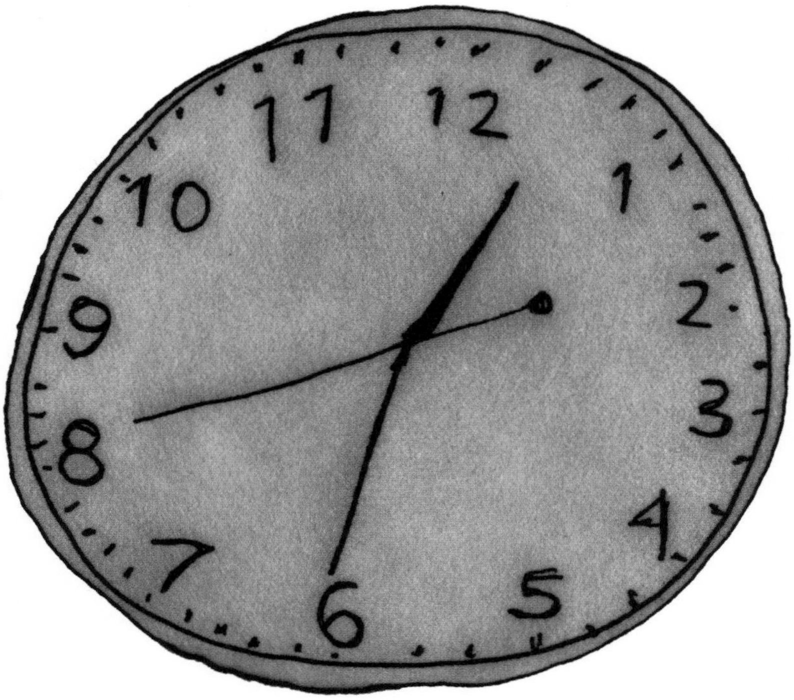

11
12
1
10
2
9
3
8
4
7
6
5

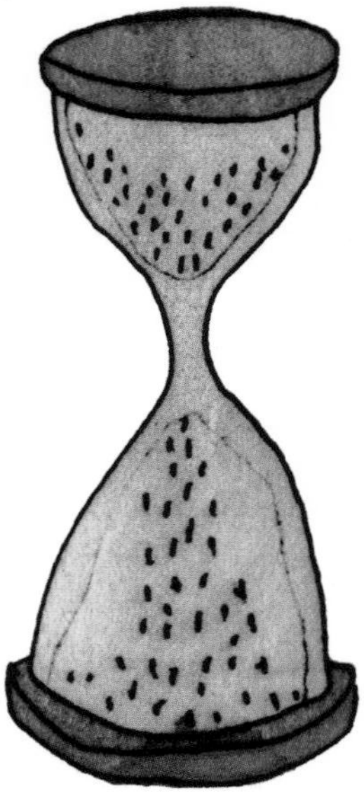

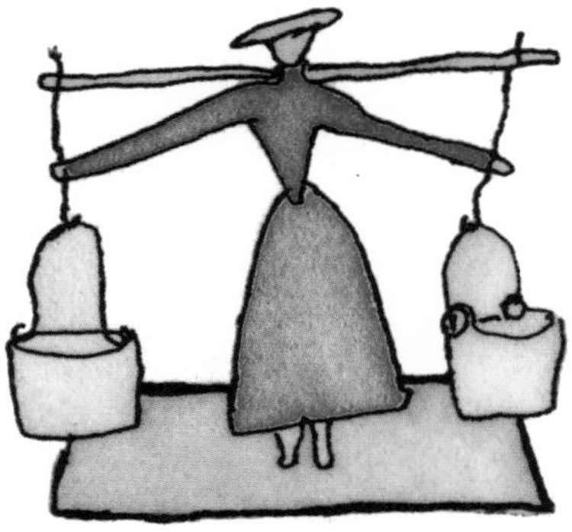

The *Old Farmer's Almanac*'s calendar pages included all the familiar information. It also promised novelty. Even the debut issue (published in 1792 for the year 1793) proclaimed on its title page that it was "New and Improved." Peppered with aphorisms and one-liners, the *OFA* channeled the voice—the knowing, dry High Yankee tone—of Benjamin Franklin's *Poor Richard's Almanack*, which had ceased publication 34 years earlier. *Poor Richard's* advised readers that "Fish and Visitors stink after three days" and "Haste makes Waste." In the *OFA*, "Poor Ned" warned, "When poverty comes in at the door, love creeps out the window." Both almanacs extolled the virtues of thrift, marriage, and discretion. The *OFA* promised to be "useful, with a pleasant degree of humor." The upper left-hand corner of the *OFA* to this day is punched through with a hole, so it can be hung and referenced. According to former editor Judson Hale, "It's not a shelf book." In recent years, annual circulation has hovered at over 3 million print copies. The *OFA* today maintains a Facebook page, a Twitter account, and a number of mobile apps.

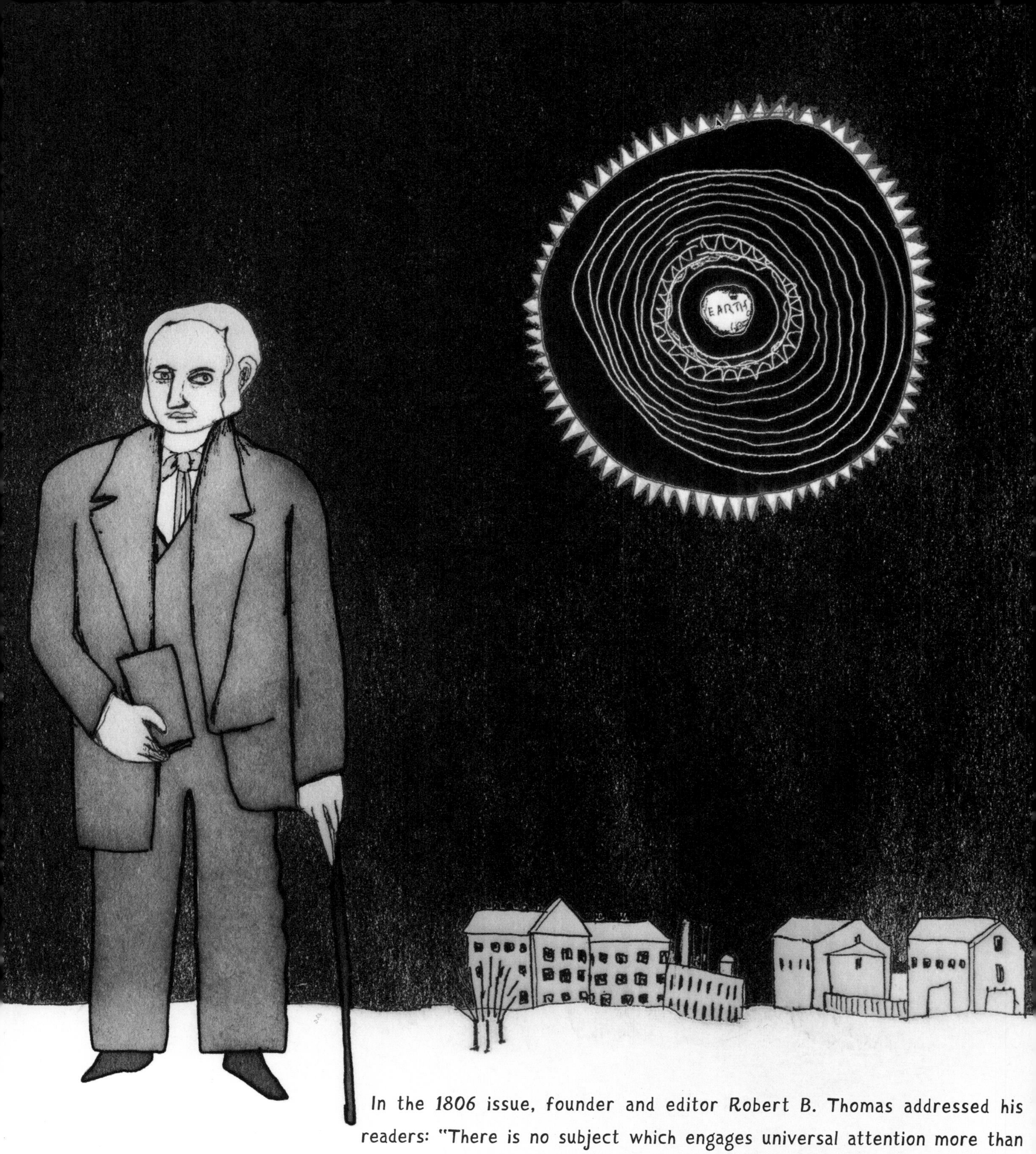
EARTH

In the 1806 issue, founder and editor Robert B. Thomas addressed his readers: "There is no subject which engages universal attention more than the Weather." He set out seven "principal signs" essential for weather forecasting: "The preceding state of the weather," "The undulations of the atmosphere," "The apparent colour of the sky," "The appearance of the clouds," "The wind," "The changes in temperature," and "The apparent colours of the sun and moon, etc." Thomas contrived a "secret forecasting formula," which, to this day, the *Almanac* holds in a black tin box in its Dublin, New Hampshire, headquarters, and claims to still rely upon for weather prediction. With his formula, Thomas was able to tell readers to expect weather that was, for instance, "Very fine for the season."

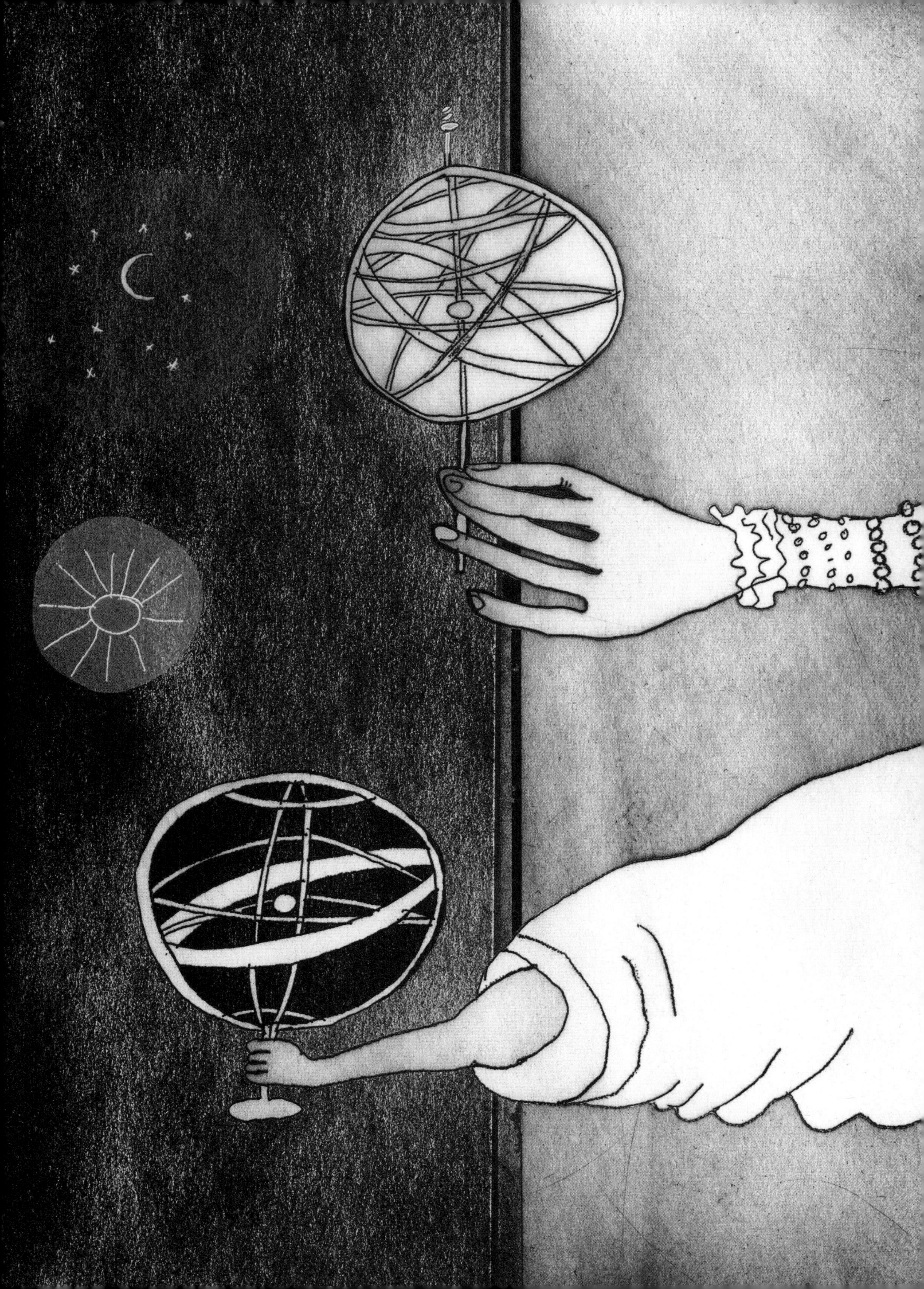

Later editors were even more succinct. Robb Sagendorph bought the *Old Farmer's Almanac* in 1939. A typical prediction during Sagendorph's stewardship would declare "mild" or "wet" or "frosty" weather—for an entire season, for the whole country. When one reader wrote in appealing for greater specificity, Sagendorph replied:

> "You request the actual number of snowflakes that fell in New England during the months of December 1947. Our staff reports that the actual count they made, which came to quite a figure, remains inaccurate inasmuch as several of the flakes that fell on the eastern side of Mount Mansfield, near Stowe, Vermont, became mixed up with some that had blown up from the ground (already counted). Sorry."

Sagendorph claimed an accuracy rate of 80 percent for his forecasts. In a 1966 profile, *Life* magazine described his technique:

> "He begins with a series of cycles which he keeps a weather eye on, including cycles of sunspots, hurricanes and storms, the Bruckner (35-year) weather cycle, 40-day biblical cycles, and a few culled from venerable axioms ('Heavy winters go by the decades'). Then he breaks the year into divisions: spring, summer, fall, hurricanes, northeast storms, cold, snowstorms, blizzards and tornados, in that order. He also checks ocean temperatures, the course of storm tracks and weather averages. Finally he consults the mysterious data kept in the Book of Days, a manual which has passed from one *Almanac* editor to the next since 1792—this is the classified formula."

Sagendorph was self-effacing about his efforts: "This isn't science. Frankly, I don't know what it is." Still, he collaborated with astronomers from Harvard, and eventually hired a NASA scientist as a full-time forecaster. Under Sagendorph and the two editors who have followed him, the *Almanac* has "refined and enhanced that formula with state-of-the-art technology and modern scientific calculations." Yet Sagendorph was aware that a whiff of fortune-telling hovered over *Almanac* predictions. "There is, I am almost convinced, some mystic quality in anything as hoary with tradition as the *Old Farmer's Almanac*, which now and then brings to it an aura of prophecy in spite of all attempts to avoid it."

On the *Almanac*'s page for November 1963 (written more than a year earlier), the month's "calendar essay"—a thin column to the right of the tide listings and moon cycles—tells a cryptic story

that reads like an allegory.

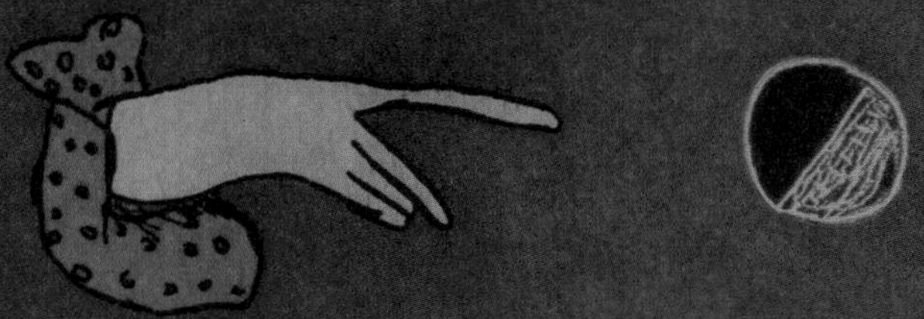

A squire smokes a pipe

and talks

with his son.

A blue jay caws the sound

"havoc, havoc."

The squire relishes the beautiful autumn day and the leaves gently falling.

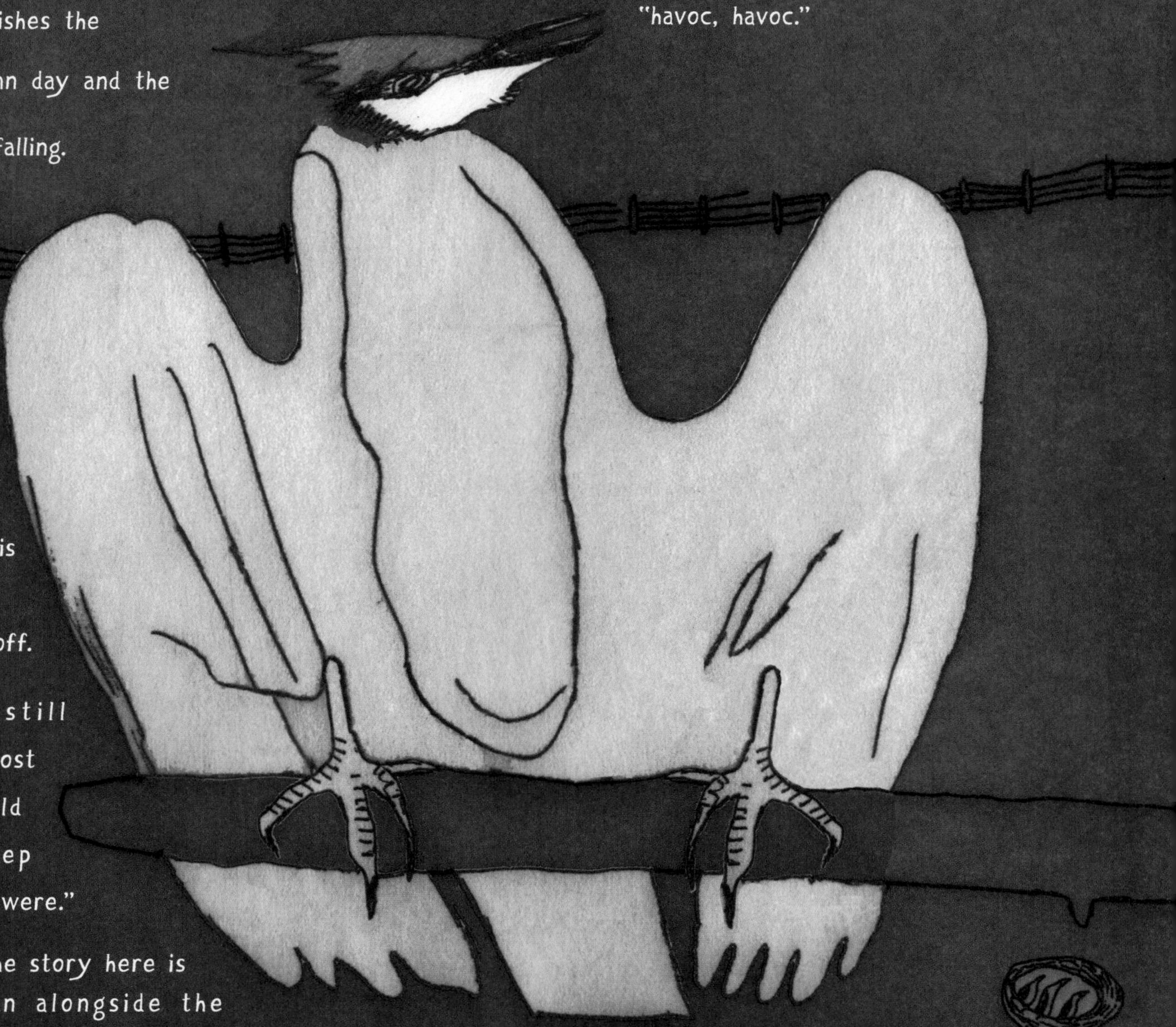

"It was not yet a dead world."

As the squire speaks about his good fortune, the bird flies off.

"It was so still you could almost hear the world trying to keep things as they were."

The text of the story here is running down alongside the calendar dates, approaching the third week in November. The son laments the discomfort of feeling blessed while living in a tumultuous world. He tells his father, "Night is coming on"—and here the line reaches November 22, the day of President John F. Kennedy's assassination—"and murder perhaps." For the remaining eight days of the month the *Almanac* foresees chaotic weather: a storm, rain, snow, wind, and mist. It also notes the third birthday, on November 25, of John F. Kennedy, Jr. At the very bottom of the page is an added note: "Two full moons this month—Guard against crime."

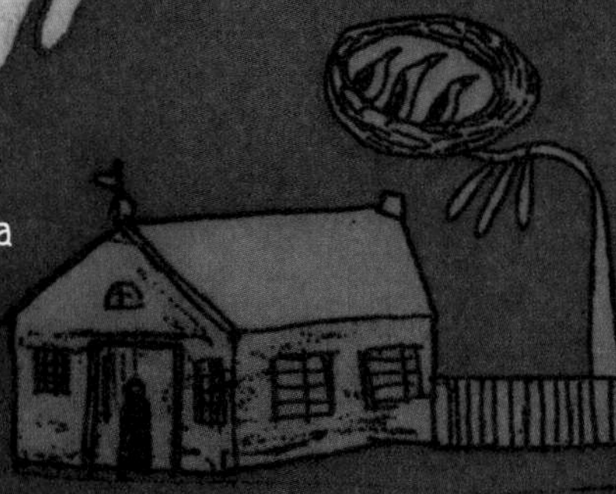

When Robb Sagendorph died in 1970, his nephew, Judson Hale, became editor — the 12th editor of the *Old Farmer's Almanac* in 182 years. Hale had begun working at the *Almanac* twelve years earlier, writing readers' letters. ("Most readers' letters were kind of boring, so I was to write interesting letters. I sent them on to the person entering them, who thought they were legit.") Since 2000, Hale has occupied a kind of emeritus position at the *Almanac* and he still comes to the office. On a November day in 2011, Hale, then in his early 80s, arrived in a colorful plaid shirt, corduroys, and a tweed blazer, carrying his things in a wicker basket.

Jud Hale: "We didn't mean to predict Kennedy's assassination, but a lot of people interpreted it that way. People wrote us from all over the world. He was shot on a Friday, and it says, 'Night is coming on, and murder perhaps.' Yeah. 'Night is coming on, and murder perhaps.' I asked Ben Rice, who wrote this, about it. And he said, 'Oh, I just felt funny about November, and I just wrote it. I don't know why.' Some people have said, 'It was in the air and he just tuned into that.' Who knows?"

In 1858, a young trial lawyer named Abraham Lincoln represented William "Duff" Armstrong, who was accused of murder. A witness swore to having seen Armstrong kill the victim by the light of the full moon on the 29th of August the previous year. In court, Lincoln asked the witness to read an almanac entry for August 29, 1857, and showed the jury the page. "The moon rides low," the *OFA* had written. Here was scientific evidence: there could not have been enough light for the witness to have seen a crime, Lincoln said. The defendant was acquitted.

During World War II, when a German spy was caught by the FBI at New York City's Penn Station with a 1942 *Old Farmer's Almanac* in his pocket, the U.S. Office of Censorship apparently became concerned the Germans were accessing critical intelligence. They asked, under the Code of Wartime Practices for the American Press, that the *Almanac* substitute weather "indications" for weather "forecasts."

Almanac editors repeat these stories, proud of the magazine's history and reputation for reliability. But they do it with a wink. When it was suggested that the Nazis were indeed using *OFA* forecasts, Robb Sagendorph was known to say, "Maybe they were. They did lose the war." When he tells the Lincoln story, Jud Hale adds that the defendant (found innocent thanks to the *Almanac*) confessed to committing the murder on his deathbed.

Jud Hale: "The question I get more than any other is, 'What's the winter going to be like?' The second question is, 'How come the *Almanac* is still being published and is still successful?'

"I don't know the answer to either one. I like to say it's going to be wintry, followed by spring. And I've always been pretty much on the button. Then I get serious and give them what our forecast is. As far as why it's lasted so long, I think people, over the years, began to think of it as an old friend in the changing world. I try to point out everything in the *Almanac* is brand-new every year. But the format, the look, the way it's put together, the cover — it's an unchanging friend. It kind of reassures people.

"I have one at my breakfast table. I'm always curious: 'Where are we today?' And I look at it and get oriented. Yeah, the full moon is tomorrow. It'll tell you what the evening star is and when the sun sets, when it rose that day, and you'll think, 'My gosh, we sure live in an orderly universe. Maybe life isn't as confusing as my day-to-day would indicate it is.'

"When it first came out it was like *Playboy* — I mean, like any other magazine, like *The New York Times*. It was just something that people bought when they bought their notions, and the other things from the wagons that went by once a week or whatever. Can you imagine back when there were no lights? When it got dark this time of year, it got really dark. You had to light some candles or some lanterns. Well, the *Almanac* would tell you when it was going to get dark."

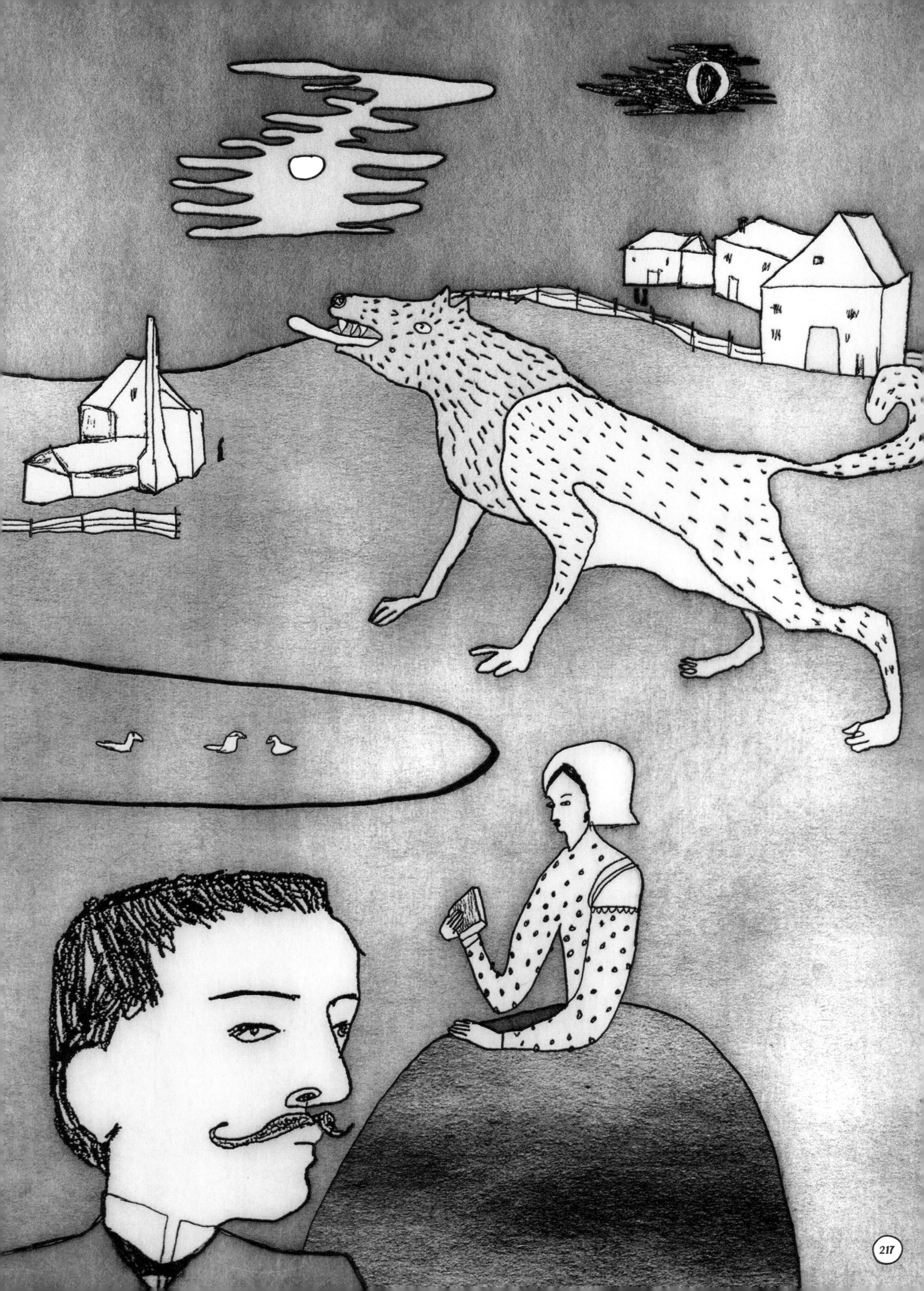

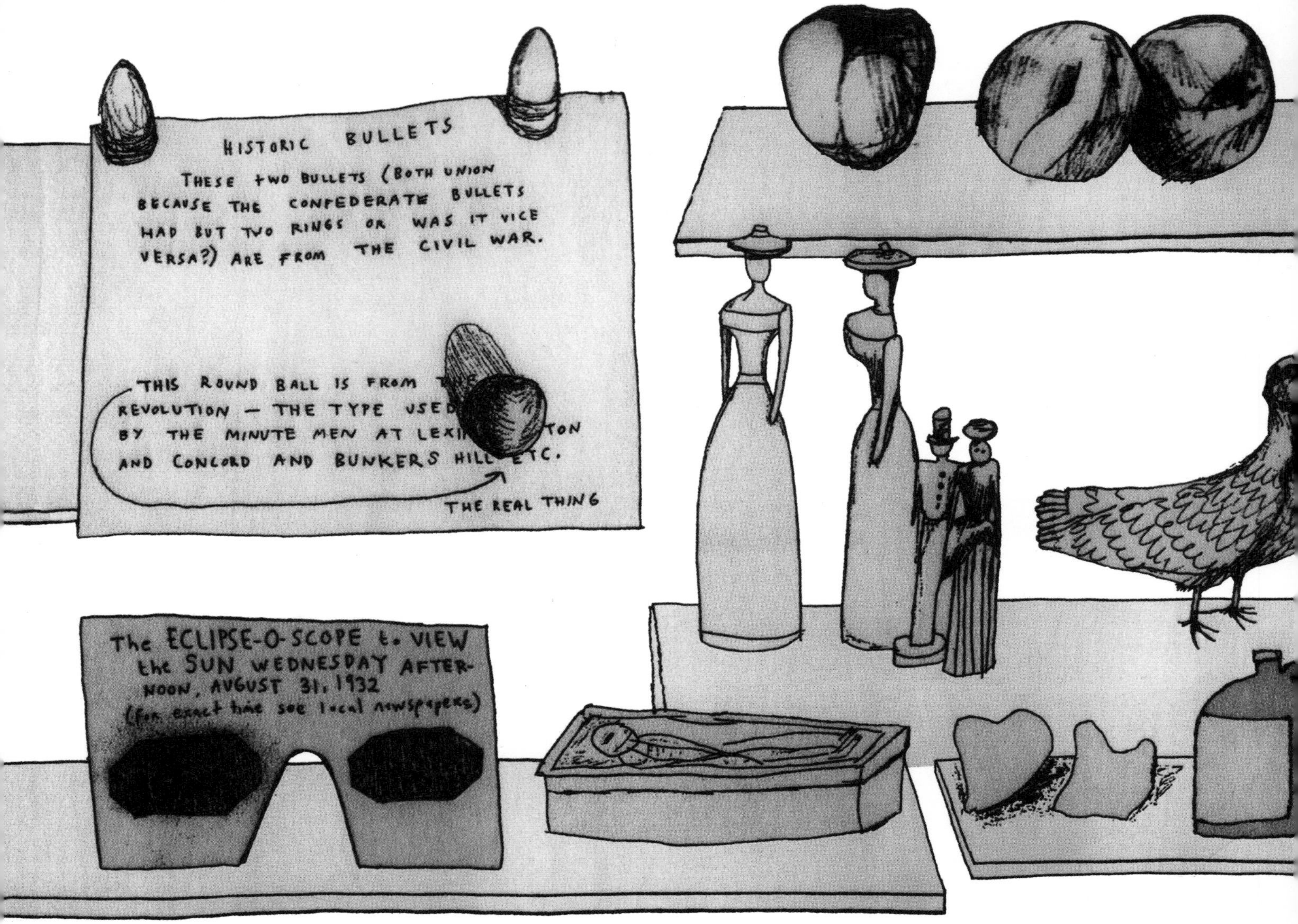
HISTORIC BULLETS
THESE two BULLETS (BOTH UNION BECAUSE THE CONFEDERATE BULLETS HAD BUT TWO RINGS OR WAS IT VICE VERSA?) ARE FROM THE CIVIL WAR.
THIS ROUND BALL IS FROM
REVOLUTION — THE TYPE USED
BY THE MINUTE MEN AT LEXI TON
AND CONCORD AND BUNKERS HILL ETC.
THE REAL THING
The ECLIPSE-O-SCOPE to VIEW the SUN WEDNESDAY AFTER-NOON, AUGUST 31, 1932
(for exact time see local newspapers)

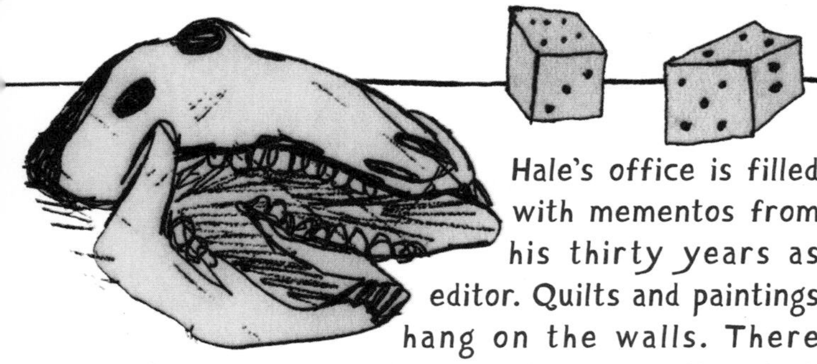

Hale's office is filled with mementos from his thirty years as editor. Quilts and paintings hang on the walls. There are diplomas from Choate and Dartmouth, which expelled Hale in 1955 for throwing up on the dean and his wife. (He was readmitted after serving in the U.S. Army and eventually graduated.) There is a flyer for a 1984 *Walter Mondale for President* rally in Detroit, a rubber chicken, an old-fashioned telephone, and a Boston Celtics t-shirt. There are photographs of Hale's family, of Ted Williams and Dom DiMaggio, of snowy New England scenes.

One side of the room is dominated by Hale's "Witness to History Museum." The Museum consists of four shelves laden with Ziploc bags and open jewelry boxes containing artifacts. Next to each item is a hand-lettered description on a white notecard. There is a piece of wood with the label "From an orchard at the home of Johnny Appleseed." There are "gizzard stones from a dinosaur," stones from "the ruins of King Arthur's castle," stones from "the legendary city of Troy," stones from the Alamo, stones from Stonehenge. There is a torn swatch of fabric from Charles Lindbergh's plane *The Spirit of St. Louis* and a box displaying Napoleon's embroidered handkerchief. (**Hale:** "I don't use that handkerchief as a handkerchief. If I had to sneeze and it was right there, I might grab it, but other than that, it's strictly for display.") There is a copper spike made by Paul Revere and two sheets of yellowed, water-logged stationery with no visible text, one marked "Letter from J. P. Morgan," the other "Letter from the *Titanic*."

When asked about the authenticity of the objects, Hale says, "No one knows if they're real or not. Just like the *Almanac*."

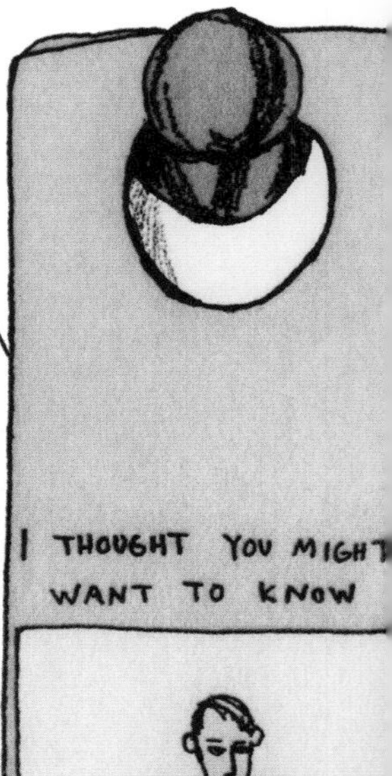
I THOUGHT YOU MIGHT WANT TO KNOW
I'M NAKED FROM THE WAIST DOWN.

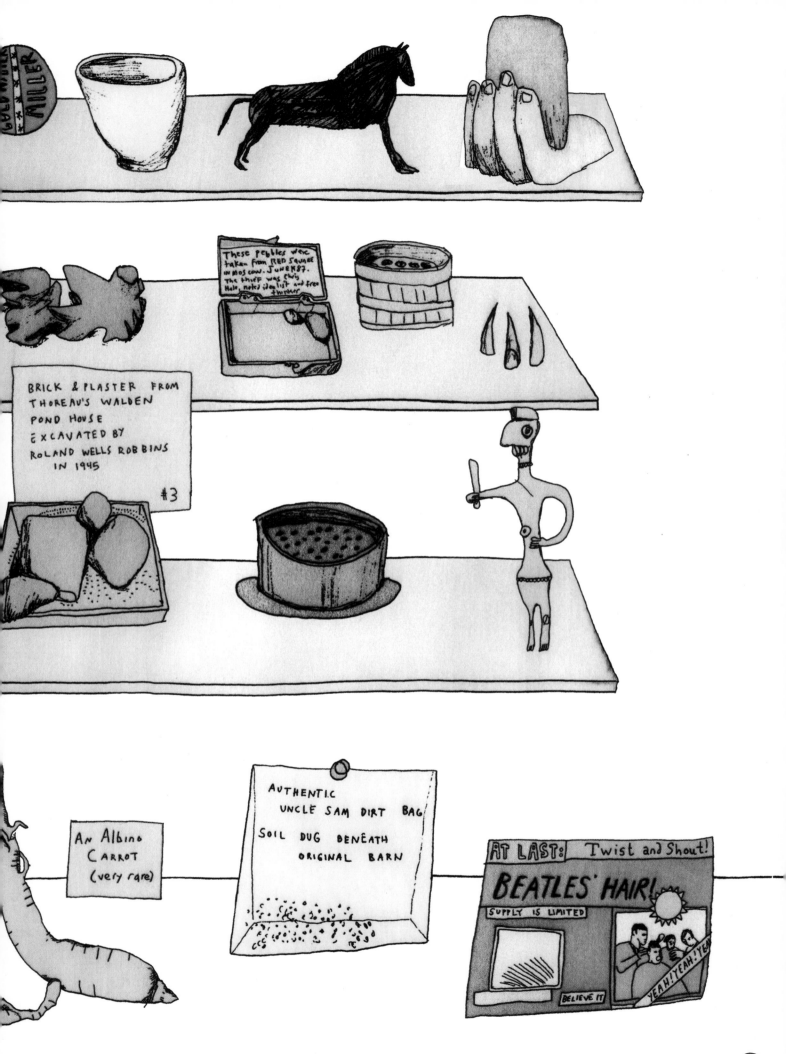

These pebbles were taken from RED SQUARE in Moscow June 1987. The thief was Chris Hale, noted idealist and free thinker.

BRICK & PLASTER FROM THOREAU'S WALDEN POND HOUSE EXCAVATED BY ROLAND WELLS ROBBINS IN 1945

#3

An Albino Carrot (very rare)

AUTHENTIC UNCLE SAM DIRT BAG SOIL DUG BENEATH ORIGINAL BARN

AT LAST: Twist and Shout!

BEATLES' HAIR!

SUPPLY IS LIMITED

BELIEVE IT

YEAH! YEAH! YEAH!

Jud Hale: "If there's a huge storm and people call us and say, 'Did you get that blizzard?' You know, maybe we did, maybe we didn't. There's three places in the *Almanac* you can look. We look it up and we think, oh, we missed it on the local forecast, the New England Region, we missed it there. Well, let's look on the national. Well, in the national we missed it, too. So we go to the calendar page. Oh yeah, it says 'wintry' or something. And we go with that! It is—well, we say—80 percent accurate. Tradition is very important in the *Old Farmer's Almanac*. And it's a tradition to claim 80 percent.

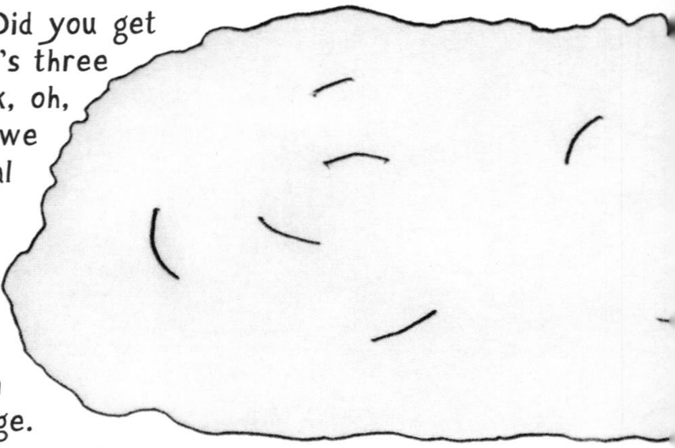

"Every month we make a temperature forecast and a precipitation forecast, and it's either above average, below average, or average. If we say it'll be above average, and it turns out to be above average, we count that as correct, whether it's one degree above or ten above. It's still correct in our book. Looking at it that way, we're often 85, 90 percent accurate for the year. Very accurate.

"The National Weather Service, I think they do it two or three months in advance. We do it eight or nine months in advance. I think they copy us."

On an exterior wall of the National Weather Center on the University of Oklahoma's Norman campus there is a plaque with a Latin inscription. "TOTUM ANIMO COMPRENDERE CAELUM." A translation is provided: "To embrace the whole sky with the mind." The building houses a weather brain trust. The National Oceanic and Atmospheric Administration's Severe Storms Lab is on the second floor, down the hall from the National Weather Service Forecast Office and the Storm Prediction Center. The Radar Operations Center and the Warning Decision Training Branch are on the southeast side of an airy atrium. On the fifth floor is the University's School of Meteorology. In the parking lot are stormchasing SUVs, identifiable by the hail dents that pockmark their bodies and windshield glass shattered into spiderweb patterns.

Harold Brooks is a meteorologist at the National Severe Storms Lab.

Harold Brooks: "The *Old Farmer's Almanac* claims to be right, like, 80 percent of the time. Every forecast system ever has been right that often. When you look at the mid-19th century, the forerunner of the Met Office, Britain's national weather service, used to claim their forecasts were right 80 percent of the time. And the Met Office still claims their forecasts are right 80 percent of the time. They forecast different elements, different things, but clearly the standard of what we call right has changed in order to preserve the magic 80 percent number."

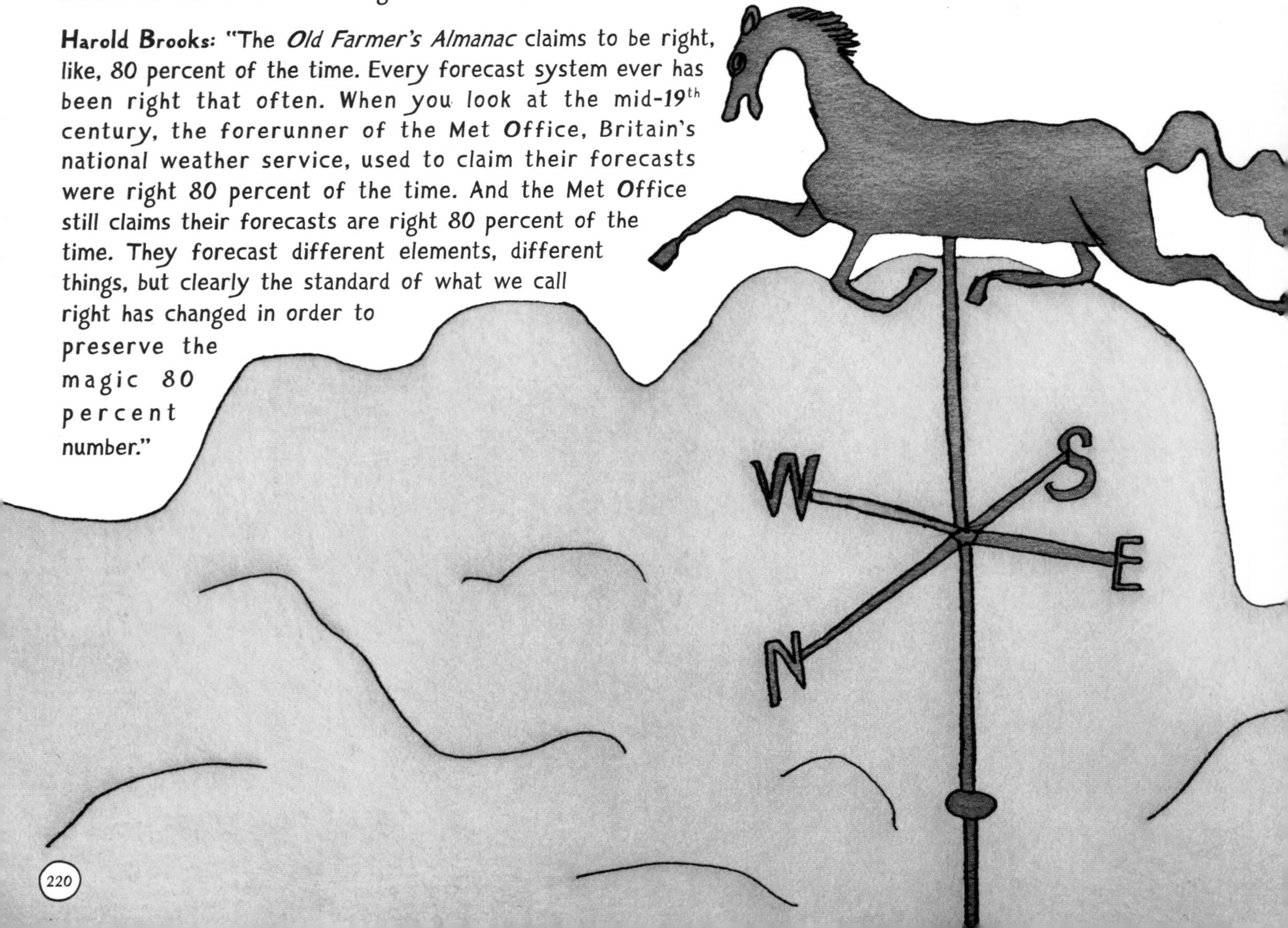
W
S
E
N

In 1972, meteorologist Edward Lorenz gave a paper at a meeting of the American Association for the Advancement of Science entitled "Predictability: Does the Flap of a Butterfly's Wings in Brazil Set *Off* a Tornado in Texas?" In it, Lorenz laid out his theory describing unpredictability in complex systems. A decade earlier, Lorenz had been crunching numbers on a computer. The initial input data represented atmospheric conditions and the results were meant to simulate weather predictions for a few months in the future. At one point, Lorenz decided to replicate a calculation. When he re-entered the data, he introduced a rounding error—the change from .506127 to .506. This seemingly insignificant adjustment generated dramatically different results. Lorenz said the system was sensitive to changes in initial conditions: the few decimal points stood in for tiny perturbations in the atmosphere, like the flutter of a butterfly's wing. To Lorenz, the divergent results meant "long-range weather forecasting must be doomed."

Lorenz: "Since we do not know exactly how many butterflies there are, nor where they are all located, let alone which ones are flapping their wings at any instant, we cannot...accurately predict the occurrence of tornados at a sufficiently distant future time."

The "butterfly effect" would become the signature metaphor for sensitivity to changes in initial conditions, known now as chaos theory. As *The New York Times* put it later, "A perfect forecast would require not only a perfect model, but also perfect knowledge of wind, temperature, humidity and other conditions everywhere around the world at one moment of time. Even a small discrepancy could lead to completely different weather."

Harold Brooks: "We don't know the current state of the atmosphere perfectly. Are you really going to measure the temperature of every molecule in the air? But okay, fine, the errors in our understanding are, on some occasions, small enough that they don't really matter, in a real practical sense.

"You're crossing the street in New York and you see that there are cars coming, you have this mental model: that car looks like it's coming but I've got time to cross the street. If you're off on your estimate of how fast it's moving—you thought it was moving 30 miles an hour and it's actually moving 31 miles an hour, unless you were cutting it real close, that difference doesn't matter.

But if that car in some bizarre way suddenly accelerated to 90 miles an hour, your model would've been bad. Or, if you were taking a cab ride from Morningside Heights down to the Village, now my errors matter, because if I was off by two miles an hour on the cab's average speed, that's a long enough cab ride that I missed by 10 minutes when I'm going to arrive. Even worse if you were driving from New York to L.A. Now all those little things really compound. That's your long-range forecast."

While we expect a weather forecast to help us get dressed or prepare for a storm, we intuitively accept certain limitations in a prediction's scope and precision. We don't, for instance, wait for a forecaster to give us the size and shape of a particular cloud at an exact moment or location. But there remains confusion about what we can demand of our forecasters and their predictions. In 1993, meteorology professor and statistician Allan H. Murphy published a paper titled "What Is a Good Forecast?" Murphy's article aimed to clarify where on a spectrum between guesswork and clairvoyance we should place our expectations.

Murphy sketched three types of "goodness," as he put it, essential to a forecast. First, consistency: a good forecast directly reflects a forecaster's truest and best judgment of coming conditions. Second, quality: a good forecast shows close similarity between the forecast and subsequently observed conditions. Finally, value: a good forecast helps its users make decisions that benefit them economically or otherwise: a family listens to the weather forecast, evacuates from a hurricane-threatened area, and therefore survives the storm.

Included in Murphy's understanding of a good forecast is the importance of communicating uncertainty—the inherent unpredictability Lorenz described.

But people don't like uncertainty. Because of uncertainty, people mock weather forecasters for being lousy at their jobs.

Greg Carbin is a warning coordination meteorologist at the National Weather Service's Storm Prediction Center: "There's a fuzziness you have to live with. There's a coarseness to the observational network that doesn't allow you to truly know what certain variables of the atmosphere are at certain locations. There are large areas on the surface of the earth that have no good information about the atmosphere and you have to just kind of interpolate between those. It's frustrating at times, not really knowing, but having to make decisions with limited information."

A forecaster, for instance, may feel pressure to err on the side of issuing a severe weather warning when lives could be at stake.

Greg Carbin: "The penalty function for missing, in most cases, is greater than that for false alarms. But you want a balance. You don't want too many false alarms, because then nobody's going to pay attention."

Some forecasting practices may reinforce skepticism. In certain circumstances forecasters deliver their predictions with deliberately impoverished consistency. In his book *The Signal and the Noise*, Nate Silver cites the example known as "wet bias": commercial weather providers are known to overstate the chance of rain because, while people may be pleased if a predicted shower does not occur, they are likely to be distraught if caught unprepared.

Rick Smith is a meteorologist at NOAA who serves as liaison between the agency, the media, and emergency managers.

"People want a yes or a no. It's hard; we don't know yes or no. 'There's a 30 percent chance of thunderstorms.' Probabilities don't always translate well to decisions, whether it's a community trying to decide whether to sound a tornado siren, or whether to close schools because of a snowstorm. On the day of an event, people will call up and want to know what to do. It always comes down to the question: what would you do? I'll say, 'I would sure be calling my wife and kids and telling them to take cover' or 'My fifth-grade son is supposed to have an elementary school graduation ceremony tonight, and we're not going.' You know, trying to put it in different terms.

"We provide information. People have to take that information and use it to make their own decisions. The main reason we have a job is to tell people that bad weather's coming so they can take action. The final step of whether you shut the steel door on that shelter is up to you."

Allan H. Murphy: "Forecasts...acquire value through their ability to influence decisions made by the users of the forecasts....Forecasts have no intrinsic value."

Since 1996, Michael Steinberg has been the meteorologist for the *Old Farmer's Almanac*. Each year he single-handedly creates the long-range forecasts for 16 regions in the U.S. and five in Canada. Steinberg has a degree in atmospheric science from Cornell University and a master's degree in meteorology from Pennsylvania State University. He is a senior vice president at AccuWeather, which he joined in 1978 as forecast meteorologist specializing in snow and ice prediction. He describes the *OFA*'s secret formula as "a methodology that looks at relationships and utilizes them to predict what will happen."

Michael Steinberg: "I saw something online a number of years ago which listed the top five trade secrets in the U.S. They listed the secret formula for Coke and the secret spices for Kentucky Fried Chicken. The secret formula used by the *Old Farmer's Almanac* for its long-range weather forecasts was number three.

"It would be nice if it were as simple as 'take the solar output and multiply by three and subtract from that the normal temperature and that'll tell you what it will be.' But it's much more complex than that. Even if you were able to look in the black box, you wouldn't be able to just sit down and create the forecast right away."

Given the *OFA*'s publication cycle, Michael Steinberg needs to produce his forecasts almost two years in advance. When asked to comment on the *Almanac*'s traditional claim of 80 percent accuracy, Steinberg says, "Who am I to argue with tradition?"

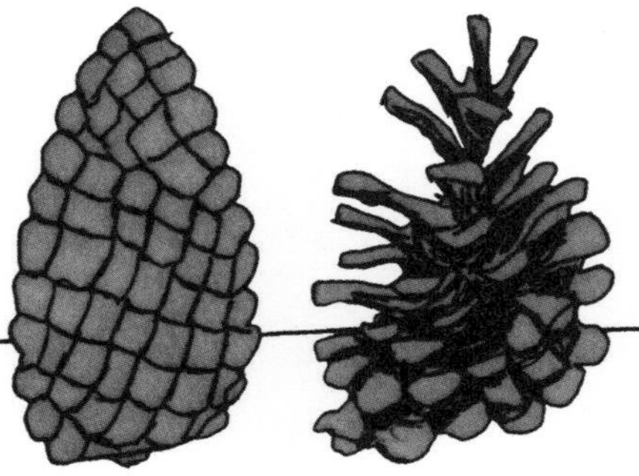

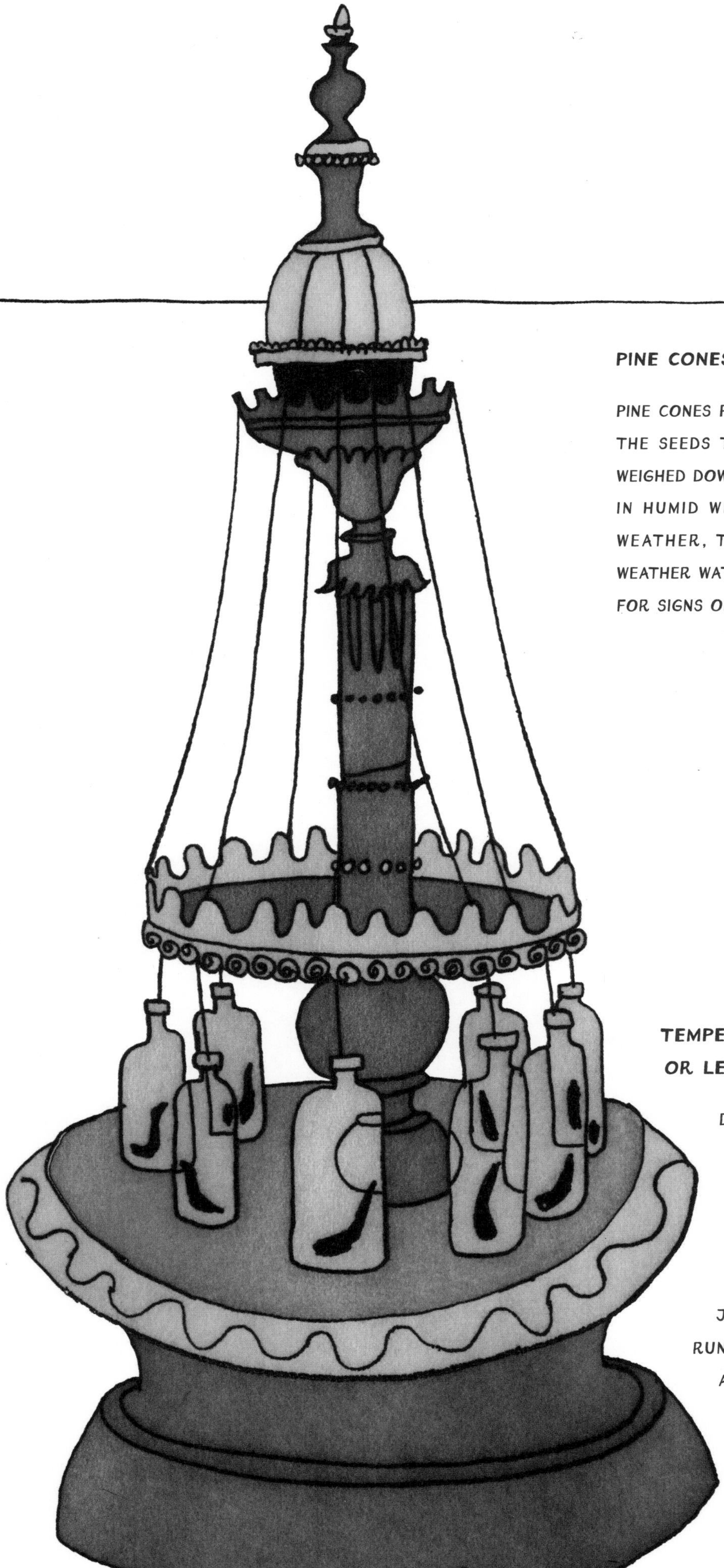

PINE CONES

PINE CONES PRODUCE SEEDS THAT ARE DISSEMINATED BY WIND. THE SEEDS TRAVEL FARTHER IN DRY, WARM AIR WHEN NOT WEIGHED DOWN BY MOISTURE. PINE CONES CLOSE THEIR SCALES IN HUMID WEATHER, LOCKING THE SEEDS INSIDE. IN DRY WEATHER, THE SCALES OPEN AND SEEDS ARE RELEASED. WEATHER WATCHERS THEREFORE LOOK TO CLOSED PINE CONES FOR SIGNS OF INCREASED HUMIDITY, AND THE CHANCE OF RAIN.

TEMPEST PROGNOSTICATOR, OR LEECH BAROMETER

DR. GEORGE MERRYWEATHER'S DEVICE, DISPLAYED AT THE 1851 CRYSTAL PALACE EXHIBITION IN LONDON, USED LEECHES TO FORECAST WEATHER. IN ADVANCE OF AN APPROACHING STORM, THE LEECHES, MERRYWEATHER SAID, WOULD CLIMB THE WALLS OF THEIR GLASS JARS, TRIGGERING A BELL: THE MORE BELLS RUNG, THE HIGHER THE PROBABILITY OF A STORM. ACCORDING TO LEECH EXPERT MARK SIDDALL, CURATOR OF ANNELIDA AND PROTOZOA AT THE AMERICAN MUSEUM OF NATURAL HISTORY, THE TEMPEST PROGNOSTICATOR IS "TOTAL BULLSHIT."

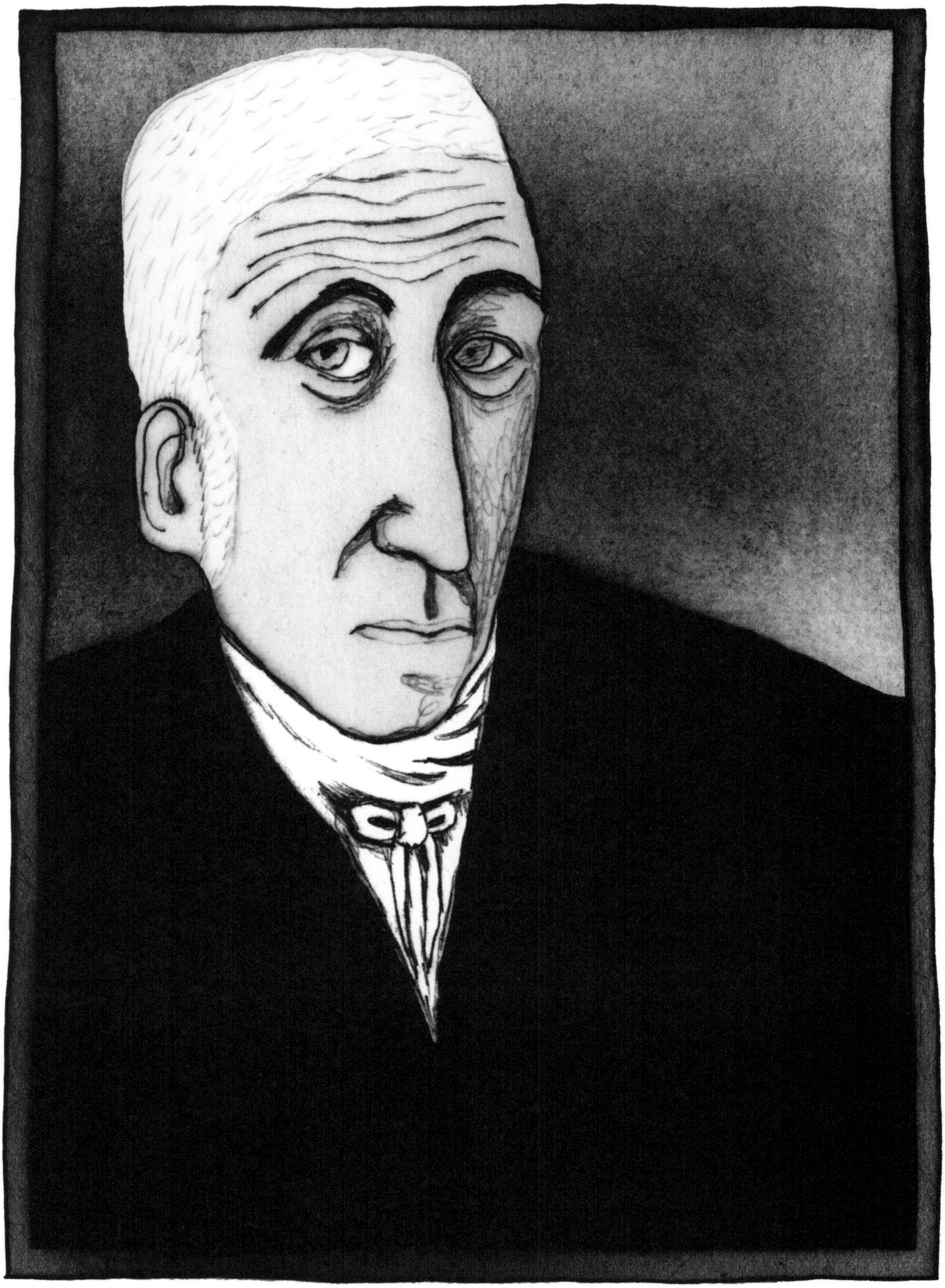

In November 2011, I visited the *Old Farmer's Almanac* on Main Street in Dublin, New Hampshire. The clapboard building, originally built in 1805 and extended more recently to add office space, is low slung and painted barnhouse red. It sits across the road from the Dublin Town Tax Collector, the Town Hall, and the Public Library. The white steeple of the Dublin Community Church next door rises over the parking lot.

Receptionist Linda Clukay, who has worked at the Dublin headquarters since 1966, greets visitors. Behind her desk hang portraits of *Old Farmer's Almanac* founder Robert B. Thomas and his wife, Hannah.

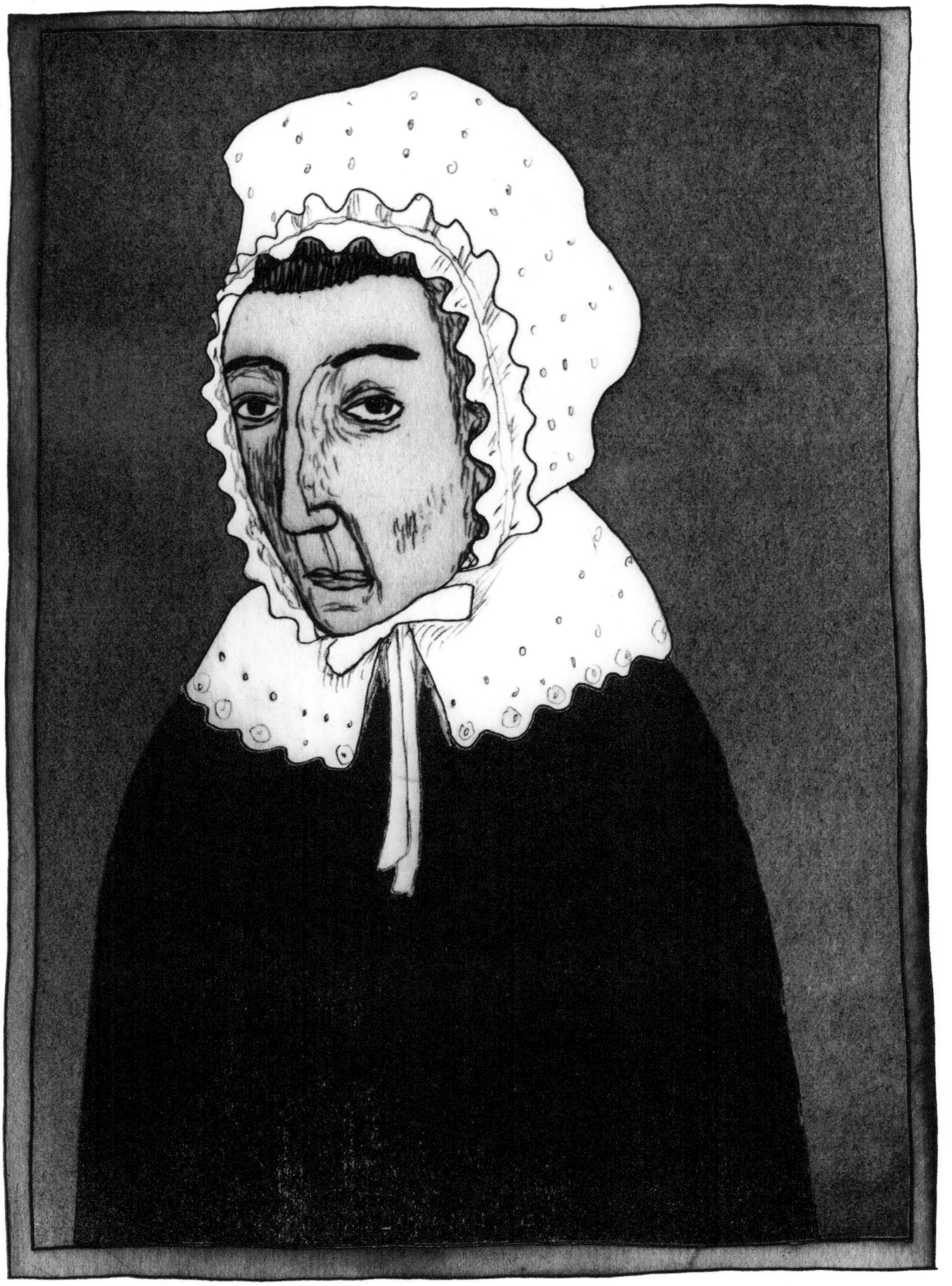

Robert has white hair, mutton chop sideburns, and raised eyebrows. Hannah wears a white bonnet, a delicate lace collar, and a pained expression. The placard labeling her picture reads: "At some point a smiling face was painted over this portrait. The original face was discovered during restoration in 1961."

Jud Hale's office, with the Witness to History Museum, is on the second floor, down the hall from a modest library containing every back issue of the *Almanac* and other antique reference books. I asked Hale about the famous black box that holds the *Almanac*'s secret formula. He picked it up off the floor of his office and handed it to me.

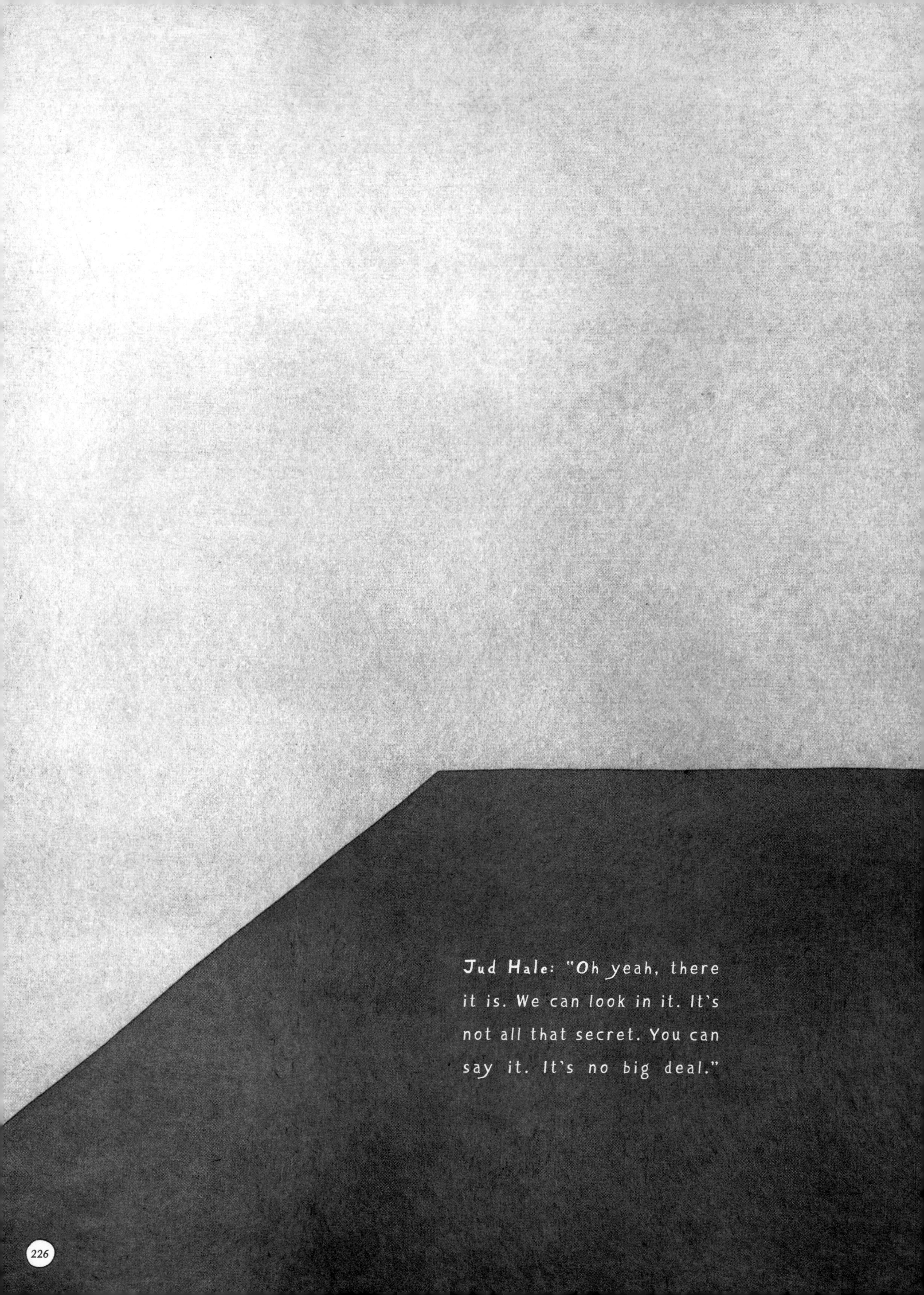
Jud Hale: "Oh yeah, there it is. We can look in it. It's not all that secret. You can say it. It's no big deal."

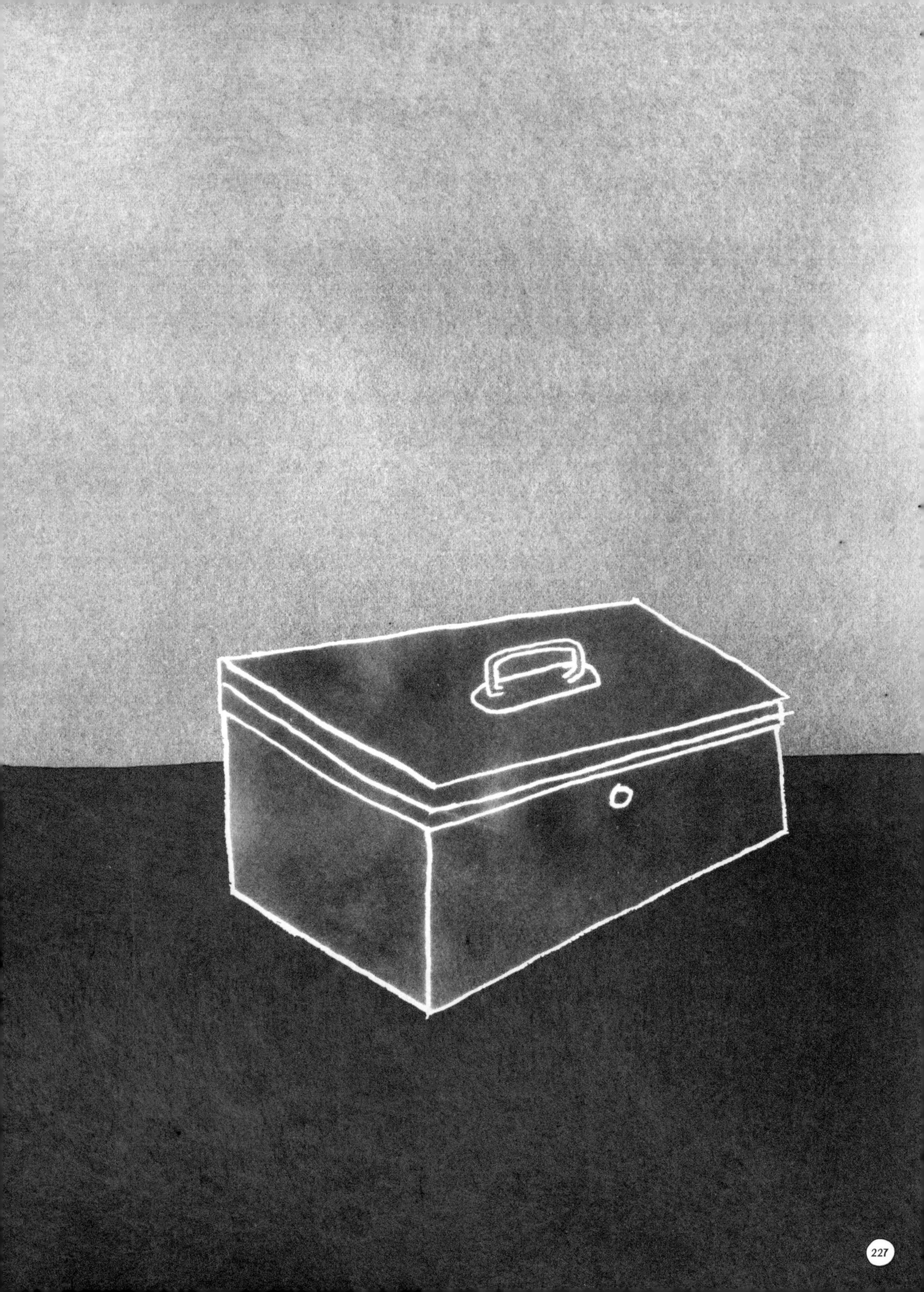

The box was dusty. It was black with gold trim, about the size of a tackle box or a makeshift coffer that might be used for cash at a bake sale. The lock was open. Inside were a few leathery spiral notebooks, two keys with clover-shaped heads held together on a jumbo paper clip. There were a number of loose documents—some typed, some handwritten—and one envelope stamped twice in red: "CONFIDENTIAL." "CONFIDENTIAL."

I was left alone to sort through the contents.

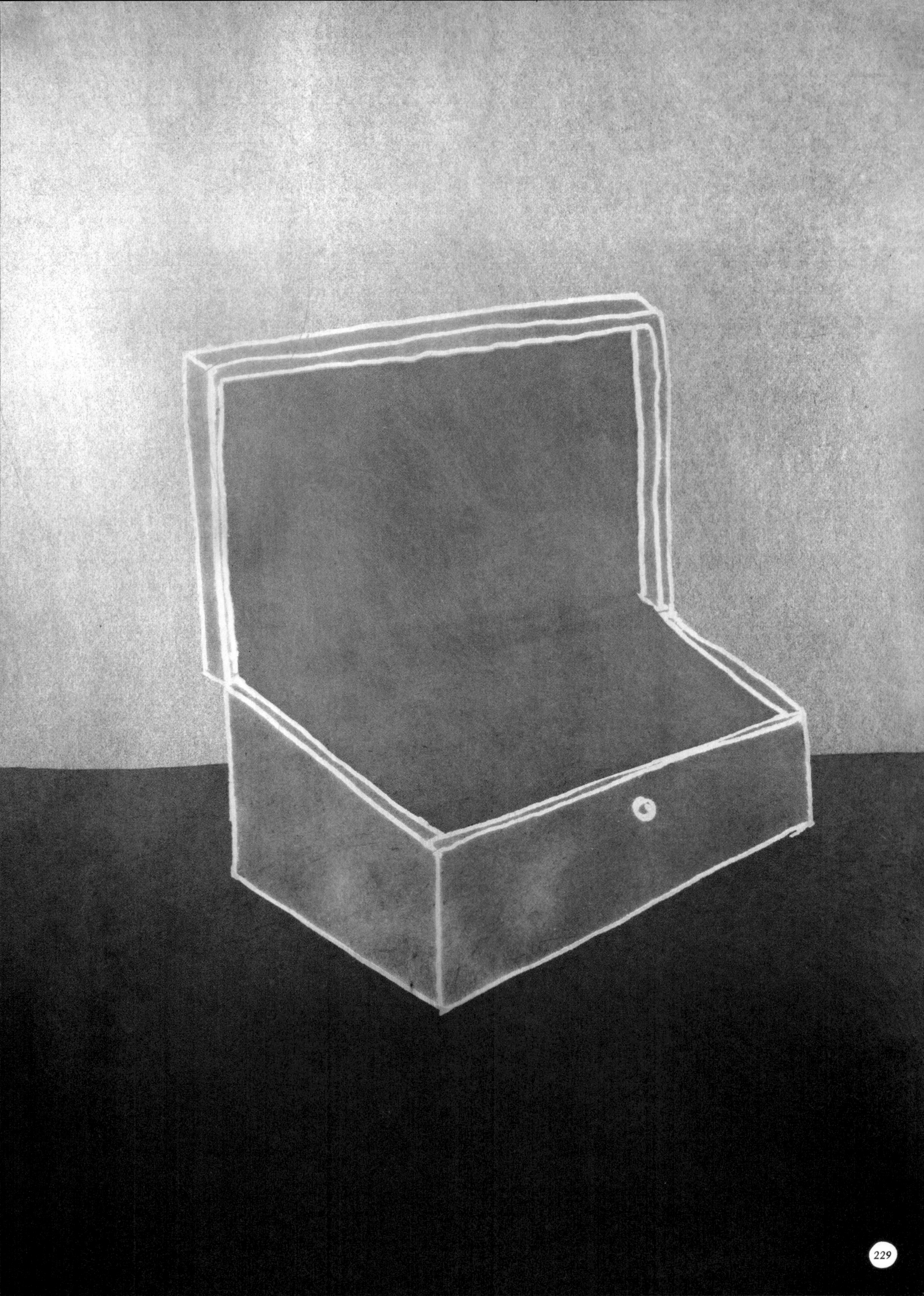

Most of the notebooks were filled with facts—lists of interesting anniversaries and bits of trivia collected to fill tiny spaces between tidal information and astronomical data on the *Almanac's* calendar pages. "Jan 9: 1ˢᵗ shot fired in Civil War, 1860," "July 2: Amelia Earhart lost, 1937," "April 8: Vermont petitions Congress for admission to the Union, 1777—Denied," "April 1: All Fools Day (put in a joke)."

Inside the double-stamped CONFIDENTIAL envelope were three typewritten pages, stapled together, undated and unsigned. The header on the first page was in capital letters:

"WEATHER FORECASTING—THE OLD FARMER'S ALMANAC"

and underneath, in larger type:

"*FOR INTERNAL USE ONLY*."

And then:

"The procedure used for forecasting the weather for the 48 contiguous states for periods up to 1 to 2 years in advance currently involves the following steps."

The seven steps went like this: You begin by making a forecast of solar activity. Next, you determine the "orientation of the earth and its magnetosphere." The third step involves "the location of the earth relative to the Sun's equator and the tilt of the earth's geomagnetic axis relative to the direction of the solar wind"; "The latter determines how much of the energy contained in the particles and magnetic fields that are impinging upon the magnetosphere actually gets transferred unto the earth's atmosphere (primarily via the earth's magnetotail to end up ultimately, through a complex process, in the auroral zones)."

Next, you study variations in cosmic rays. Fifth, you examine past solar activity to gauge future conditions.

For the sixth step, you analyze prior steps 3, 4, and 5 and use that data "to forecast those times of significant deepening of the upper-air troughs and the attendant outbreak of cold waves in winter and the generation of storm systems on their lee side; of intensification of high pressure systems with their generally clear and stable air and of deepening of low pressure systems and their frequency of outbreak."

Finally, "a lunar effect is then incorporated into the forecast." The lunar effect "is associated with the moon crossing through the magnetosheath in the earth's magnetotail at Full Moon and with the moon disrupting the flow of the solar wind at the New Moon (in turn affecting the particle energy that gets transferred into the earth's atmosphere), then the times at which the moon is close to the ecliptic plane at the Full or New Moon become important."

Obscure from the beginning, the longer the description goes on, the more confusing it gets. Each detail added to the formula further frustrates any possible understanding. It reads like a riddle. A list of disclaimers comes at the end: the formula as currently configured does not account for human influence on the weather, or localized effects like urban heat islands, or the impact of natural phenomena like volcanoes or forest fires. Nevertheless, the list is followed by a reiteration of confidence in the system. The anonymous writer concludes: "From this follows increasingly reliable forecasts for the benefit of mankind."

Asked to assess the *OFA* forecasting formula, NOAA meteorologist Greg Carbin said, "It was rather hard to follow."

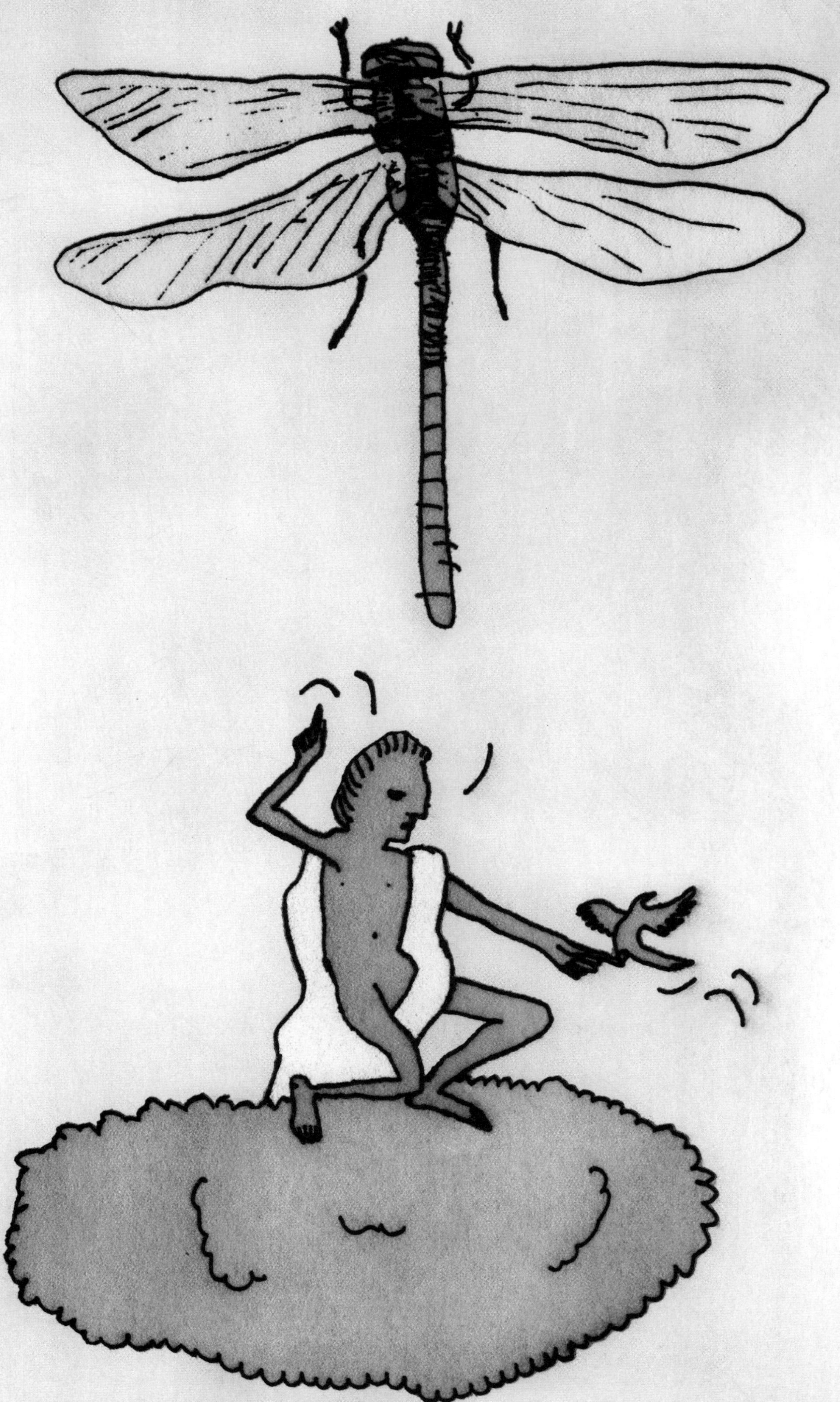

A weather forecast takes in our knowledge of history, our present understanding of science,

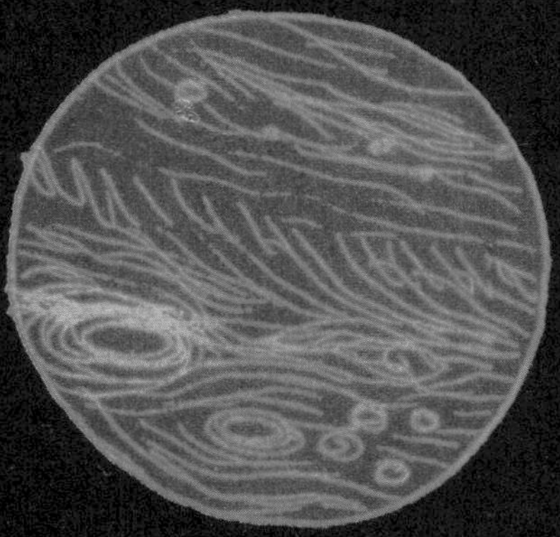

our best guesses about the future.

Each factor is imperfect;

a forecast is a summary of human accomplishment

and human limitation.

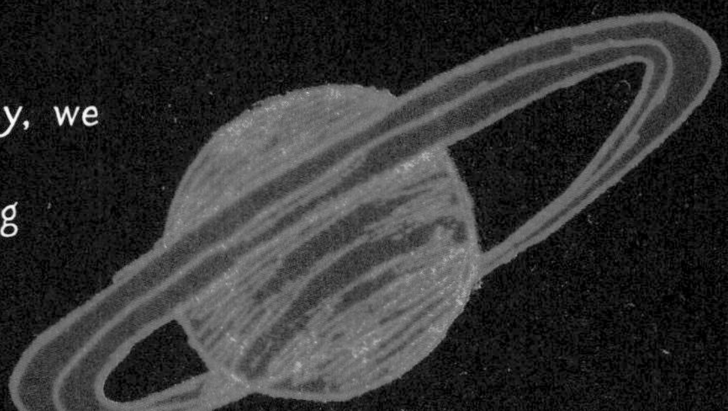

Through modern meteorology, we

have gained an understanding

of the weather, and an

ability to predict it, that

would astonish our ancestors.

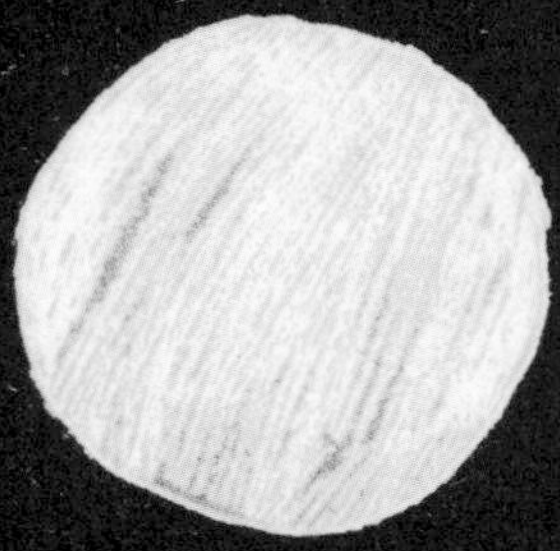

But science gets us only so far:

what the weather will be on Tuesday

remains, on some level, a mystery.

We look up at the sky,

we study Doppler radar screens,

we peer into a black box.

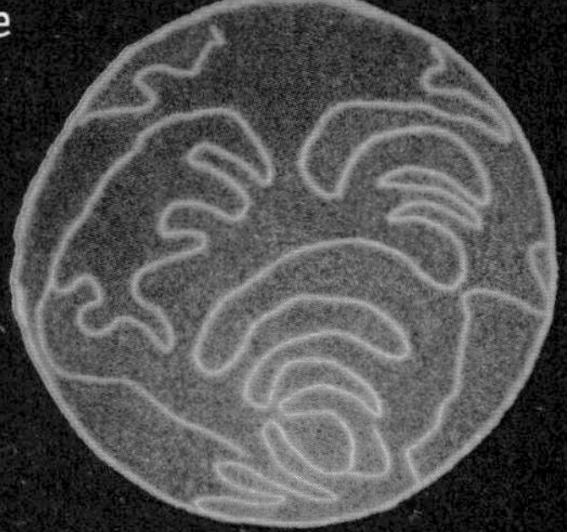

Greg Carbin: "We can see that the motions of the

sun and earth create the seasons. We know

that we go into a period where in the

northern hemisphere,

it's winter,

and

in the southern hemisphere, it's warm and summer.

There's a periodic nature to that.

If you back out on a large scale:

it will be cold in winter and it will be warm in summer.

If only it were that simple, you know. The devil is in the details."

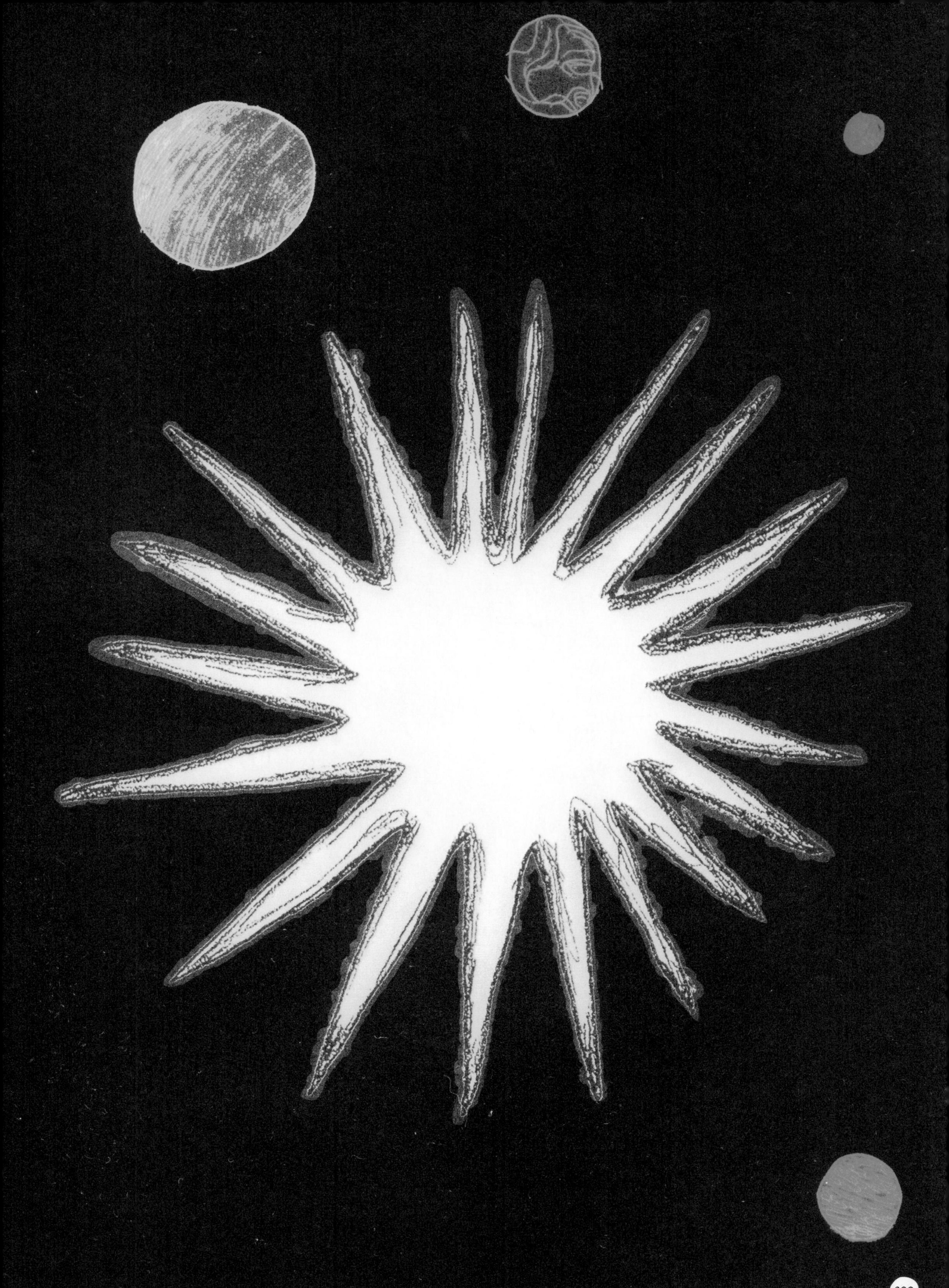

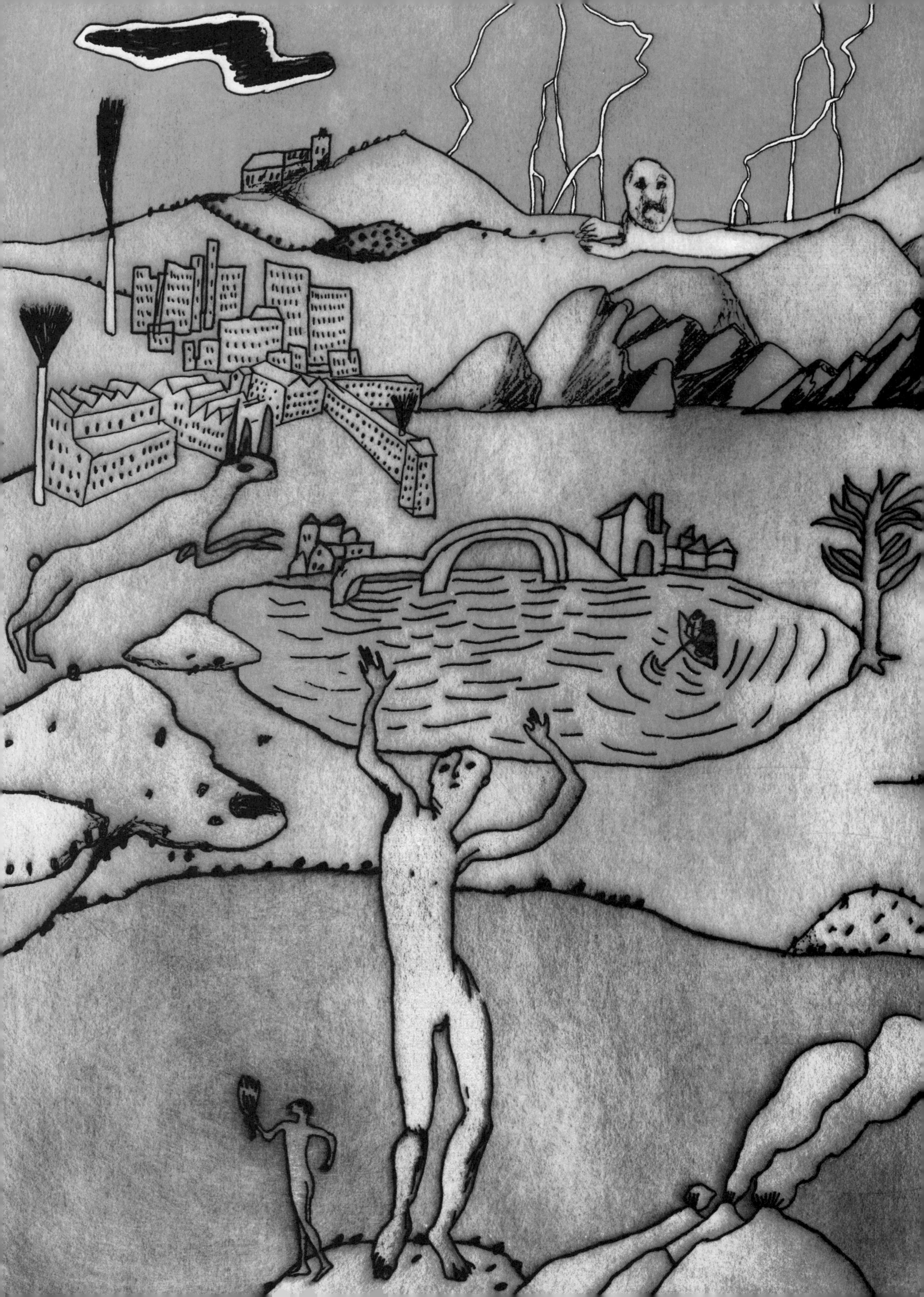

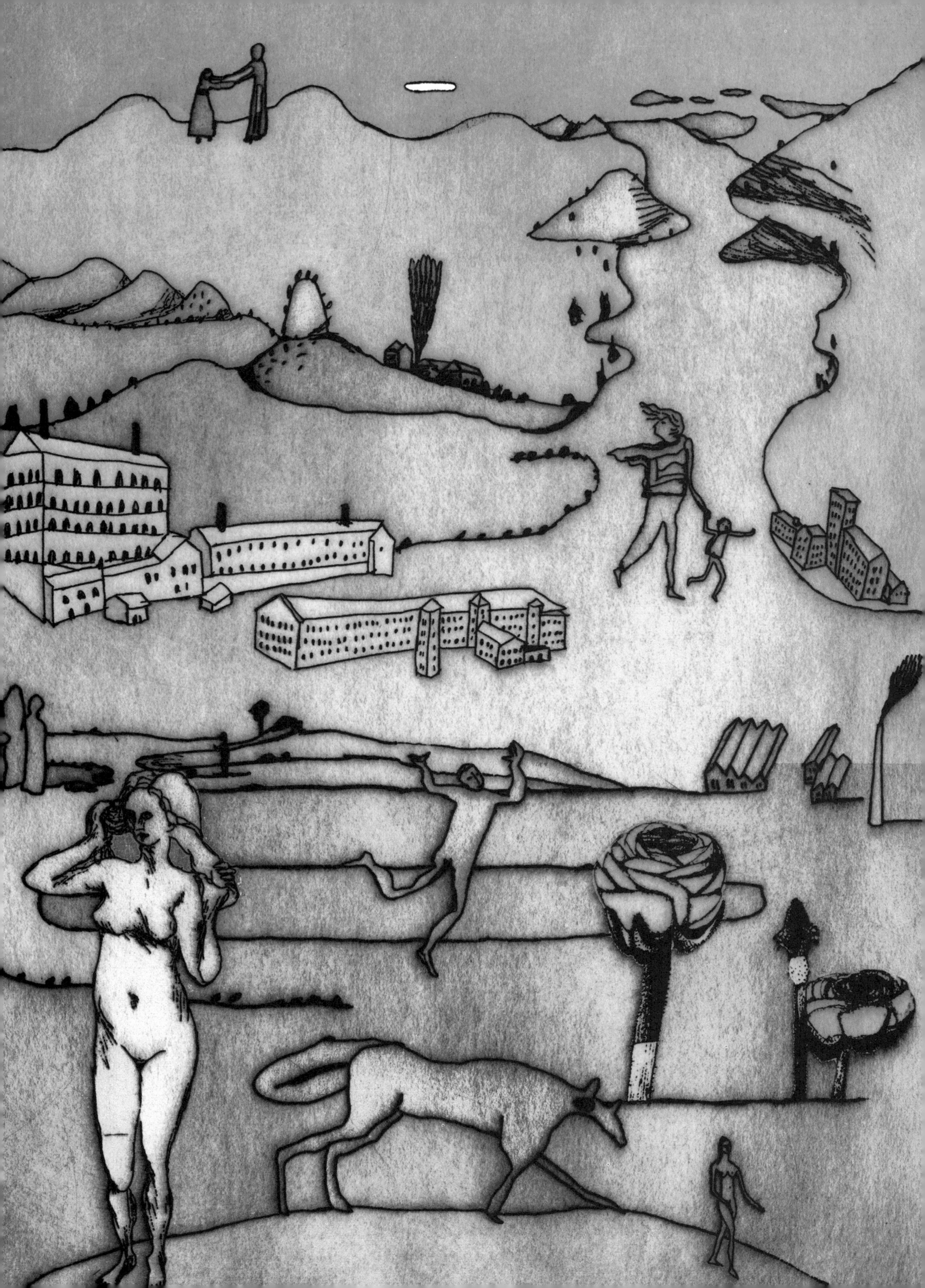

NOTES

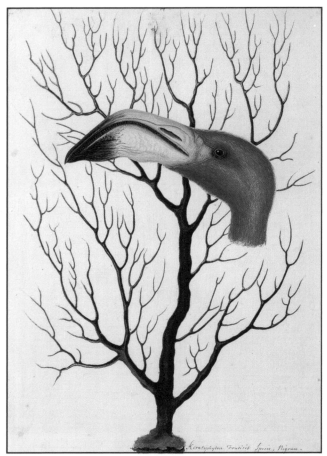

MARK CATESBY, *HEAD OF THE FLAMINGO* (PHOENICOPTERUS
RUBER) *AND GORGONIAN* (PLEXAURA FLEXUOSA), 1725.

P. HENDERSON, *THE AMERICAN COWSLIP* (DODECATHEON
MEADIA), 1801.

NOTE ON THE ART

Many images in this book were drawn on location: in the Atacama Desert, in the Arctic, in Newfoundland, at the offices of the Old Farmer's Almanac in Dublin, New Hampshire, among other places. I also looked at a variety of reference material, including scrimshaw, Greek pottery, archival photographs and prints, and Japanese Byōbu screens. In addition:

Gerrit de Veer's 16th century diary of the Dutch explorations to Svalbard, published as *The Three Voyages of William Barents to the Arctic Regions: 1594, 1595, and 1596* (Hakluyt Society: 1853), provided reference for Chapter 2, *Cold*.

Cunradus Schlapperitzi's 1445 "Picture Bible" was a particular inspiration for Chapter 8, *Dominion*.

The drawings of Ben Livingston in Chapter 9, *War*, were done from personal photographs lent to me by Mr. Livingston.

The images of ice harvesters in Chapter 10, *Profit*, were drawn from 19th-century photographs and etchings reproduced in *Refrigeration in America* by Oscar Edward Anderson, Jr. (Princeton University Press, 1953), *The Frozen Water Trade* by Gavin Weightman (Hyperion, 2003), and Richard O. Cummings's *The American Ice Harvests* (University of California Press, 1949).

The foghorn in Chapter 4, *Fog*, was drawn from a photograph of a foghorn built by British glassworkers Chance Brothers and Company, who also contributed glasswork to London's Crystal Palace for the Great Exhibition in 1851, the Houses of Parliament, and the White House in Washington, D.C.

The worm on page 67 of Chapter 4, *Fog*, first appeared in the beak of John J. Audubon's Willet (1824), an alternative study that Audubon did not include in *Birds of America*.

***

I chose to create the images in *Thunder & Lightning* using two printmaking techniques: copper plate photogravure etchings and photopolymer process.

In photogravure, acid etches a copper plate with the lines and tones of an image. When the plate is inked, pigment fills the etched grooves in the copper. The plate is run through a printing press where dampened paper picks up the inked areas, revealing the image. Photopolymer process is a contemporary variation on traditional gravure which substitutes polymer plates for copper.

For centuries, artists and scientists have used printmaking to represent their observations and transmit their ideas. During the Renaissance, printed images functioned, according to art historian and curator Susan Dackerman, as "instruments in the processes of inquiry into the natural world." Artists working in the name of science veered from visual conventions of scale, of perspective, of color and light. The need to convey specific information — about the anatomy of an animal or the structure of a plant — produced in some cases a kind of proto-surrealism.

One of my favorite images is British naturalist Mark Catesby's 1725 watercolor painting of a flamingo. The bird's head is exquisitely detailed and perfectly proportioned; fine lines along the beak delineate the threads of the *lamellae*, indicating how the bird feeds, filtering algae and tiny crustaceans from gulps of water. Every feather on the bird's head is meticulously drawn. But the head hovers in the air — disembodied, enormous, almost psychedelic against a spray of coral.

P. Henderson's 1801 *American Cowslip* (Dodecatheon meadia) is a hand-colored botanical engraving. In the foreground, the flower rises on a curving stem, massive and solitary against seaside cliffs. Its spiky purple petals jab the air. At its base, eight leaves unfurl like tentacles. In the distance, ocean waters foam, the sky is darkened by storm clouds, and two tiny ships tilt in the wind. In the murk, the flower glows, a mutant specimen looming ominously.

It was to this tradition that I hoped to pay homage with my choice of medium. Catesby's flamingo and Henderson's cowslip capture a certain feeling — a sensation of strangeness, wonder, terror — that we experience in the presence of nature, most powerfully perhaps when encountering the forces of the elements: a howling wind, a thunderstorm, the beating sun.

All the prints in *Thunder & Lightning* were made in black and white. I colored each print individually.

I worked with two master printers to produce the prints. Paul Mullowney, assisted by Courtney Sennish, printed the copper plate photogravure etchings. Paul Taylor, assisted by Oliver Dewey-Gartner and Emil Gombos, produced the photopolymer prints.

I drew the images in Chapter 7, *Sky*, in oil pastels.

IMAGES IN THE FOLLOWING CHAPTERS WERE PRINTED USING THE TECHNIQUE OF PHOTOGRAVURE: CH 1. CHAOS; CH. 2 COLD (WITH THE EXCEPTION OF PP. 20–21); CH. 3 RAIN; CH 4. FOG (PP. 48–49, 51–52, 64–65, 66–67); CH. 5 WIND (PP. 70–71, 72–73, 76–77, 78–79, 80–81, 82–83, 86–87), CH. 8 DOMINION (PP. 130–131, 134–35, 142–43, 144–45, 148–149, 150–51).

IMAGES IN THE FOLLOWING CHAPTERS WERE MADE USING THE TECHNIQUE OF PHOTOPOLYMER PRINTING: CH. 2 COLD (PP. 20–21); CH. 5 WIND (PP. 74–75, 84–85); CH. 6 HEAT; CH. 8 DOMINION (PP. 132–33, 136–37, 138–39, 140–41, 146–47); CH. 9 WAR; CH. 10 PROFIT; CH. 11 PLEASURE; CH. 12 FORECASTING. THE FRONT MATTER AND COVER ARE ALSO PHOTOPOLYMER PRINTS.

NOTE ON THE TYPE

The typeface I created for this book is called Qaneq LR, for the Inuktitut word for "falling snow."

The claim that Eskimos have many words for snow is well known. It has also been attacked as folklore and is the subject of scholarly debate. In "Eskimo Words for Snow" (1986), Cleveland State University professor Laura Martin wrote that repetition of the "snow example" in academic circles demonstrated "trivialization of the complexity inherent in linguistic structures" and "careless disregard for the essential requirements of responsible scholarship." In his essay, "The Great Eskimo Vocabulary Hoax," linguist Geoffrey Pullum says, "Nine, forty-eight, a hundred, two hundred, who cares? It's a bunch, right?...The truth is that the Eskimos do not have lots of different words for snow, and no one who knows anything about Eskimo (or more accurately, about the Inuit and Yupik families of related languages spoken by Eskimos from Siberia to Greenland) has ever said they do."

Cultural anthropologist Igor Krupnik and Ludger Müller-Wille, a professor of human geography, disagree. Krupnik and Müller-Wille have identified dozens of words for snow forms and snow conditions from various dialects including *mannguumaaq* (snow softened by warm weather), *katakartanaq* (snow with a hard crust that gives way under footsteps), and *kersokpok* (frozen snow, in which there are tracks). Krupnik and Müller-Wille maintain that the languages of peoples historically identified as "Eskimo" do in fact have a rich vocabulary of snow words — and, incidentally, an even richer one for ice-related terms.

CHAPTER 1: CHAOS

3. Hurricane Irene began...that evening: This description is pieced together from multiple sources, including National Weather Service, "Service Assessment: Hurricane Irene, August 21–30, 2011" (Silver Spring, MD: U.S. Department of Commerce, 2012) and Lixion A. Avila and John Cangialosi, "Tropical Cyclone Report: Hurricane Irene" (Miami: National Hurricane Center, December 14, 2011).

3. 49 people dead...$16 billion in damages: Ibid ("Tropical Cyclone Report: Hurricane Irene").

3. Quotes from Sue Flewelling and Elizabeth Bundock: I interviewed Elizabeth Bundock on September 23, 2011, and Sue Flewelling on March 20, 2012. Here, as elsewhere in *Thunder & Lightning*, I have edited interviews in the interest of narrative flow. My aim in this passage is not to suggest that Flewelling and Bundock were "in conversation," but rather to present a single account of the events at the Rochester, Vermont, cemetery in the voices of two different eyewitnesses.

CHAPTER 2: COLD

8. Eskimos believe that your eyeballs travel: Vilhjálmur Stefánsson, *The Friendly Arctic: The Story of Five Years in Polar Regions* (New York: The Macmillan Co., 1921), 409–10.

9. "Some sleepers have their eyes partly open": Stefánsson continues: "In this connection I asked whether the fact that we dream about hearing things did not imply that the ears traveled also. They both agreed that seemed reasonable but that they had never heard it so stated; privately they considered that in all probability the ears as well as the eyes travel. Still, that would not be the outer ear for they had frequently observed that those remain while persons are asleep."

10. "The daylight is negligible": Stefánsson, *The Friendly Arctic*, 288.

10. "It looks as if there were nothing there": Vilhjálmur Stefánsson, *Hunters of the Great North* (New York: Harcourt, Brace, and Company, 1922), 179–80.

10. "After throwing one of them": Stefánsson, *The Friendly Arctic*, 288.

10. "All this would not be so bad if": Stefánsson, *Hunters of the Great North*, 180.

10. "It might be inferred...can be seen anywhere": Stefánsson, *The Friendly Arctic*, 200.

10. "On the rough sea ice you may:" Stefánsson, *The Friendly Arctic*, 149.

10. "My first ambition": Stefánsson, *Hunters of the Great North*, 3.

14. Soil that does not thaw over the course of the year: This is beginning to change due to global warming.

14. "Blubbertown" in Dutch: Kristin Prestvold, "Smeerenburg Gravneset," (Longyearbyen: Governor of Svalbard, Environmental Section, 2001).

14. "Concerned about safeguarding": Helge Ingstad, *Landet med de kalde kyster* (Oslo: Gyldendal, 1948), 57–58, cited in Ingrid Urberg, "Svalbard's Daughters: Personal Accounts of Svalbard's Female Pioneers," *NORDLIT* 22 (Fall 2009), 167–91.

14. "First European woman": Christiane Ritter, *A Woman in the Polar Night* (1954), (Fairbanks: University of Alaska Press, 2010), 115.

14. "The ground is frozen hard as steel": Ibid, 127.

14. Little by little towards the surface: "For Some, No Rest, Even in Death," *The Milwaukee Journal* (August 28, 1985).

14. Decades ago the small local cemetery: Duncan Bartlett, "Why dying is forbidden in the Arctic," *BBC Radio 4* (July 12, 2008).

In the 1990s, bodies of Spanish flu victims buried in Longyearbyen were dug up to see if the virus had been preserved in frozen human bodies and, therefore, could be studied. See Malcolm Gladwell, "The Dead Zone," *The New Yorker* (September 29, 1997).

14. "In a kind of ironic way": Telephone interview with Liv Asta Ødegaard, March 2012.

14. "We can sink into poverty": "Portrait of an Artist 'Too Old,'" Mark Sabbatini, *Ice People*, Vol. 4, Issue 36 (September 11, 2012).

17. In 2012, Russian scientists reported: Vladimir Isachenkov, "Russians revive Ice Age flower from frozen burrow," Associated Press (February 21, 2012).

18. Food will "always be safe": "Is Frozen Food Safe? Freezing and Food Safety," Food Safety and Inspection Service, United States Department of Argiculture, fsis.usda.gov.

18. "We have seeds from more countries than actually exist": Interview with Cary Fowler, Svalbard, Norway, February 2012, and telephone interviews, 2014.

18. Income tax figures on Svalbard and mainland Norway: Correspondence with Jørn Erik Kvven, Svalbard Tax office, June 2014.

20. Thai immigrants make up the largest segment of the population: According to the Svalbard tax office, there were 102 Thai residents in Longyearbyen in 2012. Some Thai residents of Svalbard with whom I spoke suggested that the official figures fell short of an accurate total, noting that certain Thai newcomers fail to register with the government in order to avoid paying taxes.

22. "There is a lot of fruit in my garden": Interview with Tanyong Suwanboriboon, Longyearbyen, Svalbard, March 2012.

24. "Svalbard is a place where people come to work": Interview with Herdis Lien, Longyearbyen, Svalbard, February 2012.

25. "These light nights are strange": Ritter, 54.

CHAPTER 3: RAIN

30. Over a thousand journalists: Robert Mackey, "Latest Updates on the Rescue of the Chilean Miners," *The New York Times* (October 30, 2010).

30. "Absolute desert": According to scientist Julio Betancourt: "By 'absolute desert' we mean that it's without vascular plants, without regular plants. If it rains, it probably rains once every few decades...There are not a lot of weather stations in the absolute desert, because it'd be kind of a waste of time, right?" (Telephone interview with Julio Betancourt, April 2012.) See also: John Houston and Adrian J. Hartley, "The Central Andrea West-Slope Rainshadow and its Potential Contribution to the Origin of Hyper-Aridity in the Atacama Desert," *International Journal of Climatology*, 23 (2003).

30. NASA uses the Atacama as a proxy for the red planet: Like Mars, the Atacama is dry, with high levels of ultraviolet radiation. In addition, the Atacama's soil composition is similar to Martian sand. (Telephone interview with Carnegie Mellon University professor David Wettergreen, June 2012; Dr. Wettergreen leads robotic exploration projects in the Atacama.)

30. "Rain shadow" effect: Houston and Hartley, 1453–64.

31. "You're in the sweet spot": Telephone interview with Julio Betancourt, April 2012.

34. "It happens usually around every seven or eight years": Telephone interview with Pilar Cereceda, January 2012.

36. "That combination of rain and heat": Interview with Christopher Raxworthy, New York, NY, March 2012.

   Raxworthy and I also spoke about potential changes in Madagascar due to climate change: "We're just at the early stages of [climate change] research in Madagascar, but one of the predictions at the moment for higher temperatures is that you're going to get more cyclones hitting Madagascar. And also that the rainfall in Madagascar might actually increase a little bit. But unfortunately that means that the rain is going to be a lot more intense. When it falls it's going to be in much shorter pulses and that means with runoff that you have a much higher risk either from storm damage or erosion, so some serious consequences from that. People are very aware of cyclones in Madagascar, and particularly on the east coast, they suffer greatly when they get hit by cyclones."

39. As the warm air moves upwards, it reaches the subfreezing temperatures: My thanks to Mary Ann Cooper and Ronald Holle for their help describing the science of lightning.

39. One entire team was killed: Marcus Tanner, "Lightning kills an entire football team," *The Independent* (October 29, 1998).

A 2010 article in *Der Spiegel* quotes a healer in Cameroon: "All I have to do is cast a few shells and contact the spirit of the playing field, then our own goal will be nailed up and the opposition's goal will be wide open." The same article quotes retired Ghanaian defender and Confederation of African Football official Anthony Baffoe: "Every African team has a resident witch doctor." (Thilo Thielke, "They'll Put a Spell on You: The Witchdoctors of African Football," *Der Spiegel*, June 11, 2010.)

39. Thor ruled the heavens: Special thanks to Sasha Rosen for his expertise on Norse mythology.

39. "Eccentricities of its operation": Andrew Dickson White, *A History of the Warfare of Science with Theology in Christendom*, Volume 1 (New York: D. Appleton and Co., 1903), 332.

40. "If circumstances facilitating the original lightning strike": Mary Ann Cooper et al., Paul S. Auerbach, editor, "Lightning Injuries," Chapter 3, *Wilderness Medicine* (St. Louis: Mosby, 2007), 69.

40. Men, more often engaged in outdoor activities: Curran, E.B., R.L. Holle, and R.E. López, "Lightning casualties and damages in the United States from 1959 to 1994 (*Journal of Climate*, volume 13, 2000), 3448–64.

41. "In 1969, I was struck by lightning": Telephone interview with Steve Marshburn, October 2011.

41. "A person may have rainwater, or sweat, on them": Telephone interview with Mary Ann Cooper, October 26, 2011. See also: Cooper MA, Andrews CJ, Holle RL: "Lightning Injuries," *Wilderness Medicine*, 87–90.

41. "I felt lit up like a Christmas tree": *Life After Shock: 58 LS & ESVI Members Tell Their Stories* (Jacksonville, NC: Lightning Strike and Electric Shock Victims International, Inc. 1996), Introduction.

41. "Saw nothing, heard nothing": Ibid, 72.

41. Forensic pathologist Dr. Ryan Blumenthal: Email exchange with Ryan Blumenthal. See also: Ryan Blumenthal et al., "Does a Sixth Mechanism Exist to Explain Lightning Injuries?" Volume 33, Issue 3, *The American Journal of Forensic Medicine and Pathology* (September 2012), 222–26.

41. Victims may also be impaled by shrapnel-like debris: Ryan Blumenthal, "Secondary Missile Injury From Lightning Strike," Volume 33, Issue 1, *The American Journal of Forensic Medicine and Pathology* (March 2012), 83–85.

42. "I had become an exhibit at the zoo": *Life After Shock*, 81.

42. "Touching the face of death": Ibid, 97.

CHAPTER 4: FOG

48. Description of 1983 royal visit to Cape Spear: I viewed 1983 Canadian Broadcasting Corporation footage of Charles and Diana's visit to Newfoundland online: http://www.youtube.com/watch?v=H7qzSKYbpcY. The video appears to have been removed from YouTube.

48. The easternmost edge of North America: Cape Spear (47°31' N, 52°37' W) is the easternmost point of North America in a definition of North America that does not include Greenland.

49. "It happened in 1845": Interview with Gerry Cantwell, Cape Spear, Newfoundland, July 2012, and telephone interviews.

50. To form dense smog: The word *smog* — a portmanteau of *smoke* and *fog* — was coined to describe the London phenomenon.

50. "When one of the thick, yellowish compounds": "Pea-Soup Fog in London, New York's Worst Fog Does Not Approach It, A Dirty Yellowish Compound Which Makes Itself Felt Everywhere and by Everybody," *The New York Times* (December 29, 1889).

Bleak House (1853) begins with Charles Dickens's famous evocation of London fog: "Fog everywhere. Fog up the river, where it flows among green aits and meadows; fog down the river, where it rolls defiled among the tiers of shipping and the waterside pollutions of a great (and dirty) city. Fog on the Essex marshes, fog on the Kentish heights. Fog creeping into the cabooses of collier-brigs; fog lying out on the yards, and hovering in the rigging of great ships; fog drooping on the gunwales of barges and small boats. Fog in the eyes and throats of ancient Greenwich pensioners, wheezing by the firesides of their wards; fog in the stem and bowl of the afternoon pipe of the wrathful skipper, down in his close cabin; fog cruelly pinching the toes and fingers of his shivering little 'prentice boy on deck. Chance people on the bridges peeping over the parapets into a nether sky of fog, with fog all round them, as if they were up in a balloon, and hanging in the misty clouds."

50. Crime surged: "London Fog Tie-up Lasts for 3rd Day," *The New York Times* (December 8, 1952).

50. "With visibility near zero late last night": "Thieves get $56,000 as Fog Grips London," *The New York Times* (January 31, 1959).

50. An airplane overshot the runway: "Excursions Plane Crash Kills 28; 2 Survive in London Fog Disaster," *The New York Times* (November 1, 1950).

50. Ambulances had to be accompanied: "London Fog Tie-up Lasts for 3rd Day," *The New York Times* (December 8, 1952).

50. At least one funeral procession: "London Has a Fog so Dense Funeral Procession Is Lost," *The New York Times* (December 19, 1929).

50. In nearly windless conditions: Sue Black, Eilidh Ferguso, *Forensic Anthropology*, (Boca Raton: CRC Press, 2011), 245.

54. "When we're tied up in port": Telephone interview with Paul Bowering, June 2012.

55. "The temperature drops": Telephone interview with Captain David Fowler, June 2012.

60. "They're listening…": Sound waves — vibrations transmitted through solid, liquid or gas — are subject to *reflection:* they may bounce off solid objects. Sound can be *refracted:* a sound wave's direction gets diverted as it travels through space. Sound is also affected by temperature. Usually, in the troposphere — the part of the atmosphere closest to earth — temperature drops as altitude increases. Since sound moves faster in warm air, the speed of sound also decreases with altitude. According to Daniel Russell, professor of acoustics at Pennsylvania State University, "This means that for a sound wave traveling close to the ground, the part of the wave closest to the ground is traveling the fastest, and the part of the wave farthest above the ground is traveling the slowest. As a result, the wave changes direction and bends upwards. This can create a 'shadow zone' region into which the sound wave cannot penetrate. A person standing in the shadow zone will not hear the sound even though he/she might be able to see the source." (Daniel A. Russell, "Acoustics and Vibration Animations," acs.psu.edu.)

Acoustician Charles Ross has analyzed the role of sound shadows in the command decisions and outcomes of Civil War battles. "Before electrical and wireless communications became common on the tactical level, the sound of battle was often the quickest and most efficient method by which a commander could judge the course of a battle." When the sounds of battle were distorted by atmospheric effects, leaders could blunder into disastrous miscalculations. According to Ross, an acoustic shadow foiled a decisive Confederate victory at the Battle of Seven Pines, in Virginia, in 1862, leading to the wounding of Confederate General Joseph Johnston and the rise of Robert E. Lee. Johnston had planned a three-pronged attack against Union General George McClellan's men, but as the battle raged, he was cocooned in a pocket of silence that deceived him into thinking the fighting had yet to begin.

Fog, which was reported the morning of the Confederate attack on Seven Pines, is often an indication of a temperature inversion — a reversal of the normal decrease of temperature with altitude.

Charles Ross: "As the upper part of our [sound] wave enters such a region, it will speed up and turn the entire wave back towards the earth....The net result is that someone far from the source may be able to hear it better than someone close by. Stranger still, if the downward refracted wave reflects off the ground with sufficient intensity it can rise again and repeat the cycle. This can lead to a 'bull's eye' pattern of rings of audibility and inaudibility around the source. These rings can be many miles in width."

After the Seven Pines battle, Confederate General Joseph Johnston wrote: "Owing to some peculiar condition of the atmosphere the sound of the musketry did not reach us. I consequently deferred the signal for General Smith's advance until about 4 o'clock." In other words, until too late.

(See: Charles Ross, *Civil War Acoustic Shadows*, Shippenberg: White Mane Press, 2001, 24, 61–82. See also: "Outdoor Sound Propagation in the U.S. Civil War," Charles D. Ross, *Echoes*, Volume 9, No. 1, Winter 1999.)

61. "It disorientates you": In his book *Traffic*, Tom Vanderbilt describes the effect of fog on automobile driving:

"When fog rolls in on the highway, the result is often a huge, multicar chain-reaction crash....Obviously, it is harder to see in a fog. But the real problem may be that it is *even more difficult to see than we think it is.* The reason is that our perception of speed is affected by contrast....In fog, the contrast of cars, not to mention the surrounding landscape, is reduced. Everything around us appears to be moving more slowly than it is, and we seem to be moving more slowly through the landscape....Ironically, drivers may feel more comfortable staying closer to the vehicle ahead of them — so that they do not 'lose' them in the fog — but given the perceptual confusion, this is exactly the wrong move."

(Tom Vanderbilt, *Traffic: Why We Drive the Way We Do* [New York: Alfred A. Knopf, 2008], 99.)

62. "The strongest ships ever built": "Loss of the Arctic: Collision Between the Steamer and a Propeller of Cape Race, Probably Loss of Two to Three Hundred Lives," *The New York Times* (October 12, 1854). This article cites an earlier *NYT* article (from 1850).

62. Considered the finest: David W. Shaw, *The Sea Shall Embrace Them* (New York: Free Press, 2002), 30.

62. The "fastest steamer afloat": "Sinking and Abandoned," James Dugan, *The New York Times* (November 26, 1961).

62. Description of *Arctic* interior: Shaw, 41.

62. Description of the competition to be fastest across the Atlantic: Shaw, 45–50.

62. A week later...the *Arctic* was sailing: See: George H. Burns's statement in "Additional Particulars" sidebar, *The New York Times* (October 11, 1854) and map in Shaw, 99.

62-3. Quotes from Peter McCabe, Francis Dorian, James Smith, George Burns, James Carnegan, Thomas Stinson: In the days following the *Arctic* catastrophe, *The New York Times* published survivor statements and witness testimony. I've included an edited selection. For the complete statements see: "Loss of the Arctic: Collision Between the Steamer and a Propeller off Cape Race," *The New York Times* (October 12, 1854) and "The

Arctic: Important Details, Narrative of Capt. Luce, Dreadful Scenes on the Wreck," *The New York Times* (October 17, 1954).

62. Description of the last moments before the *Arctic* sinking: Shaw, 145.

63. **Of** the 408 people aboard: There is uncertainty about the precise number of people aboard the *Arctic* at the time of the disaster. In *The Sea Shall Embrace Them*, David W. Shaw explains that certain figures may not have included the total number of crewmembers or family members accompanying them (Shaw, 205).

CHAPTER 5: WIND

70–71. "Between the Keys of Florida and Africa": Telephone interview with Diana Nyad, December 2012.

72. Attempting to swim from Cuba to Florida: In 2013, Diana Nyad completed the Cuba–Florida swim. It was her fifth attempt.

72. "Absolutely unafraid of pain": Diana Nyad, *Other Shores* (New York: Random House, 1978), 71.

73. "The *scirocco*...has been blamed": Peter Ackroyd, *Venice* (New York: Nan Talese/ Doubleday, 2009), 24.

73. "It was one of those hot dry Santa Anas": Raymond Chandler, *Red Wind* (Cleveland: World Publishing Co., 1946).

73. "Day and night you could hear the Föhn howl": Herman Hesse, *Peter Camenzind*, (1904), Translated by Michael Roloff. (New York: Farrar, Straus, and Giroux, Inc., 1969), 191–92.

74. The term "doldrums": In his "Rime of the Ancient Mariner" Samuel Taylor Coleridge described the doldrums:
>   "Day after day, day after day,
>   We stuck, nor breath nor motion;
>   As idle as a painted ship
>   Upon a painted ocean."

78. "Gambling, lying, and stealing": Diana Nyad, "Father's Day," *The Score*, KCRW, (July 18, 2005).

79. "One night": Diana Nyad, "The Ups and Downs of Life with a Con Artist," *Newsweek* (July 31, 2005).

79. "The winds that howl from every quarter": Homer, *The Odyssey*, translated by Robert Fagles (New York: Penguin Classics, 1996), 231.

79. "A fatal plan...All the winds burst out": Ibid, 232–33.

80. Every Muslim is bound by the five pillars of Islam to make the hajj: Muslim tradition allows exceptions for those in poor health, and for those too poor to afford the journey.

81. Expansions and renovation projects: Given the historical and religious significance of the site, the changes are controversial. Preservationists object to proposed changes at the Grand Mosque that include the dismantling of historic structures. Critics have decried the commercialization of the area around the site and compared new developments, which include hotels, luxury apartments, upscale chain boutiques, and the accompanying destruction of historic sites and egalitarian access to mountain views, to the commercialism and kitsch of Las Vegas. One of many recent articles and opinion pieces addressing these issues is Ziauddin Sardar's "The Destruction of Mecca," *The New York Times* (September 30, 2014).

82. "There isn't enough ventilation": Telephone interviews with Anton Davies, 2013 and 2014.

CHAPTER 6: HEAT

91. Scientists believe climate change...is contributing: Max A. Moritz et al., "Climate Change and Disruptions to Global Fire Activity," *Ecosphere*, volume 3, Issue 6 (June 2012).

91. "We are seeing more fire activity": Felicity Ogilvie, "Bushfires intensifying as they feed climate change, scientist warns," *The World Today*, Australian Broadcasting Channel Online (April 24, 2009).

91. "We are seeing fire behaviors which are unusual": Telephone interview with David Bowman, March 2014.

93. Hottest year on record...fewest rainy days: These facts come from multiple sources including Ehud Zion Waldoks, "2010 was hottest year in Israel's recorded history," *Jerusalem Post* (January 3, 2011).

93. Help from the international community: Ethan Bronner, "Suspects Held as Deadly Fire Rages in Israel for Third Day," *The New York Times* (December 4, 2010).

93. "This is a special type of battle": Anshel Pfeffer, Barak Ravid, and Ilan Lior, "Major Carmel Wildfire Sources Have Been Doused, Firefighters Say," *Haaretz* (December 5, 2010).

97. Fire in Bhutan: Yonten Dargye, "A Brief Overview of Fire Disaster Management in Bhutan," National Library, Bhutan (2003). See also: Kencho Wangmo, "A Case Study on Forest Fire Situation in Trashigang, Bhutan," *Sherub Doeme: The Research Journal of Sherubtse College* (2012).

97. "Children playing with matchsticks": "Event Report: Forest/Wild Fire in Bhutan," *The Hungarian National Association of Radio Distress-Signalling and Infocommunications Emergency and Disaster Information Service* (January 25, 2013).

97. Threatening people and wildlife: "Forest Fire," Department of Forests and Park Services, Ministry of Agriculture and Forests, Royal Government of Bhutan (2009).

100. "Fire is one element which the Black Kite": David Hollands, *Eagles, Hawks, and Falcons of Australia* (Melbourne: Thomas Nelson, 1984), 36.

100. Aborigines used fire: Stephen J. Pyne, *Burning Bush: A Fire History of Australia* (New York: Henry Holt and Company, 1991).

100. "Kangaroos, wallabies, and wombats": Ibid, 32.

100. Victoria was in its thirteenth year of drought: See: Kevin Tolhurst, "Report on the Physical Nature of the Victorian Fires Occurring on the 7th of February," *2009 Victorian Bushfires Royal Commission* (Parliament of Victoria, Australia, 2009). See also: "Conditions on the Day," *The 2009 Victorian Bushfires Royal Commission*, Final Report, Volume IV (Parliament of Victoria, Australia, 2009).

100. "Worst day in the history of the state": Marc Moncrief, "Worst day in History," *The Age* (February 6, 2009).

100. Smoke was seen rising: See photos included in: Kevin Tolhurst, "Report on the Physical Nature of the Victorian Fires Occurring on the 7th of February."

101. "And what happens on the wind change": "Inside the Firestorm," Australian Broadcasting Channel (February 7, 2010).

101. "It was a hurricane": Jim Baruta, Ibid. See also: Jim Baruta's Witness Statement, *2009 Victoria Bushfires Royal Commission*.

101. Every firefighter in town lost his home: Interview with Glen Fiske, "Inside the Firestorm."

101. "Everything was on fire": Daryl Roderick Hull, Witness Statement, *2009 Victorian Bushfires Royal Commission*.

109. "We face the increased risk": Kate Galbraith "Wildfires and Climate Change," *The New York Times* (September 4, 2013).

109. The fire season will be longer and the air smokier: Xu Yue, Loretta Mickley et al., "Ensemble projections of wildfire activity and carbonaceous aerosol concentrations over the western United States in the mid-21st century," *Atmospheric Environment*, volume 77 (2013). Also: Telephone interview and email correspondence with Loretta Mickley, April 2014.

109. Even Siberia is on fire: What I call Siberia here refers to the land area historically and commonly referred to as Siberia (also called "North Asia") and not to the borders of the smaller Siberian Federal District, created by presidential decree in 2000.

109. Temperatures throughout Russia set record highs: "Wildfires and Russian Bureaucracy: Perfect Combination," *Pravda.ru*, English edition (August 3, 2010).

109. More than 50 people dead and the loss of a quarter of Russia's grain harvest: "Satellite images show wildfires hugging Lake Baikal as army use drones to monitor 2013 blazes," *The Siberian Times* (May 11, 2013).

109. 100 meters per minute: "Wildfires and Russian Bureaucracy: Perfect Combination," *Pravda.ru*, English edition (August 3, 2010).

109. 2012 was worse in terms of land area affected. See: "State of Emergency Declared Due to Fires in Eastern Regions," *The St. Petersburg Times* (June 18, 2012).

109. "The wildfire situation…is abnormal": "As Wildfires Rage, the Russian Government Heads East to Battle the Crisis," *The Siberian Times* (August 6, 2012).

CHAPTER 8: DOMINION

130. "Wonderful natural beauty": Jon Chol Ju, "Fascinating Frostwork," *Rodong.rep.kp* (December 1, 2010).

130. Winds were stronger, waves higher: "Unforgettable Last Days of Kim Jong Il's Life," *KCNA* (December 21, 2011).

130. Mount Paektu: Soviet documents place Kim Jong Il's birth in Siberia, not Mount Paektu.

130. "Loud roar": "KCNA Detailed Report on Mourning Period for Kim Jong Il," *KCNA* (December 30, 2011).

130. "Unusual glow tinging the sky": *Rodong.rep.kp* (December 25, 2011).

130. "Kim Jong Il was the heaven-born man," *Rodong.rep.kp* (December 30, 2011).

135. An 1894 issue of *National Geographic*: Mark W. Harrington, "Weather Making, Ancient and Modern," *National Geographic*, Volume 6 (April 25, 1894), 35–62.

136. Drought, floods, and famine also ravaged Asia: Fagan, 50.

136. The Little Ice Age: Scholars differ in dating the period identified as the "Little Ice Age," a term coined by glacial geologist François Matthes in 1939. Brian Fagan's *The Little Ice Age* traces evidence of cooling in Greenland and the Arctic to approximately 1200, with colder temperatures creeping into Europe around 1300. (Fagan, *The Little Ice Age: How Climate Made History: 1300-1850* [New York: Basic Books, 2000].) Other scholars prefer a more limited time frame, from the late 17th to mid 19th centuries.

136. "A major player": Fagan, 28.

136. A million accused witches…put to death: Emily Oster, "Witchcraft, Weather and Economic Growth in Renaissance Europe," *Journal of Economic Perspectives*, Volume 18, Number 1 (Winter 2004), 216.

136. "Many persons of both sexes…give themselves over to devils": Protestant and secular courts also prosecuted "witches." See: Teresa Kwiatkowska's "The Light was Retreating Before Darkness: Tales of the Witch Hunt and Climate Change," *Medievalia 42* (2010) and Wolfgang Berhinger's *Witches and Witch-Hunts: A Global History* (Malden, MA: 2004).

138. "It'll bring about terrorist bombs": Sentinel Staff, "Orlando Rainbow Flags Bring New Attack," *Orlando Sentinel* (August 7, 1998).

138. John McTernan…blamed: "Superstorm Sandy and many more disasters that have been blamed on the gay community," *The Guardian* (October 30, 2012).

138. Rabbi Noson Leiter…said: Brian Tashman, "Religious Rabbi Blames Sandy on Gays, Marriage Equality," *Right Wing Watch* (October 31, 3012).

138. Doubling of "witch" murders: Edward Miguel, "Poverty and Witch Killing," *Review of Economic Studies* (2005), 1153–72.

139. "It would be a mistake to think": Telephone interview with Estelle Trengrove, February 2012. See also: Trengove, E., Jandrell, I. R., "Lightning and witchcraft in southern Africa,"

2011 Asia Pacific International Conference on Lightning, Chengdu, China (November 2011).

139. "When we sat down, one of them said": Telephone interview with Estelle Trengrove, February 2012.

142. "We dress for the weather": Edmond Mathez, *Climate Change* (New York: Columbia University Press, 2009), 279.

142. "Warming of the climate system is unequivocal": IPCC, 2013: Summary for Policymakers. In: *Climate Change 2013: The Physical Science Basis. Contribution of Working Group I to the Fifth Assessment Report of the Intergovernmental Panel on Climate Change.* Edited by Stocker, T.F., D. Qin, G. K. Plattner, M. Tignor, S.K. Allen, J. Boschung, A. Nauels, Y. Xia, V. Bex, and P.M. Midgley (Cambridge University Press: Cambridge, United Kingdom, and New York, 2013).

142. The consequences...include: There are countless studies that make similar assertions, including the 2014 National Climate Assessment, which states the United States faces "increasingly frequent and intense extreme heat, which causes heat-related illnesses and deaths and, over time, worsens drought and wildfire risks, and intensifies air pollution; increasingly frequent extreme precipitation and associated flooding that can lead to injuries and increases in marine and freshwater-borne disease; and rising sea levels that intensify coastal flooding and storm surge." Jerry M. Melillo, Terese (T.C.) Richmond, and Gary W. Yohe, eds., *Climate Change Impacts in the United States: The Third National Climate Assessment* (Washington, DC: U.S. Global Change Research Program, 2014), 15.

142. The Pentagon's 2010 Quadrennial Defense Review: "Quadrennial Defense Review Report," (Washington DC: United States Department of Defense, February 2010), 84–94.

145. Geoengineering strategies are generally divided into two categories: According to scientist David Keith, the term "geoengineering" is not ideal. "First of all you have to divide these two things that are sometimes called geoengineering but I think they have nothing to do with each other. One is the changing the amount of sunlight or solar radiation management....I think there's no more connection between solar radiation management and removing carbon than there is between either of those and the other things we might do about climate change like cutting emissions or adaptation or conservation. So I think it's not a question o[f] which is better or worse, but in terms of th[e] architecture of the technology or the policy concerns around the technology, I just don't think there's any connection between them. So I think it's a little unfortunate we use the same word for both."

145. "One approach is called solar radiation management": Telephone interview with Nathan Myhrvold, April 2012.

145. "The scientific equivalent of a porn habit": Jeff Goodell, *How to Cool the Planet* (Boston: Houghton Mifflin Harcourt, 2010), 13.

147. "You've got a national park": Interview with Richard Pearson, New York, NY, May 2012.

Despite the challenges to even protected lands, Richard Pearson believes "there are good reasons to think that protected areas will remain our best bet for conserving biodiversity over the coming century. By reducing nonclimatic threats, parks and preserves are able to maintain ecosystems that have a diversity of species and healthi[ly] sized populations. As we've seen, diverse ecosystems are more resilient to climate change." (Pearson, *Driven to Extinction: The Impact of Climate Change on Biodiversity* [New York: Sterling, 2011], 210.)

148–49. The round table depicted on pages 148–49 is imaginary. The quotes are culled from separate interviews with Nathan Myhrvold, Emma Marris, Alan Robock, and David Keith, as well as from various readings, noted belo[w]

148. "We are already running the whole earth": Emma Marris, *Rambunctious Garden* (New York: Bloomsbury, 2011), 2.

148. "I'm worried about it being developed as a weapon": Telephone interview with Alan Robock, July 2012.

148. "Among the many likely consequences": Elizabeth Kolbert, "Hosed," *The New Yorker* (November 16, 2009).

149. "I could see a country": Telephone interview with Nathan Myhrvold, April 2012.

149. "It's not the end of nature": David Keith quoted in Goodell, *How to Cool the Planet*, 45. See also: Thomas Homer-Dixon and David Keith, "Blocking the Sky to Save the Earth," *The New York Times* (September 19, 2008).

149. "Even though we might really not want to interfere": Telephone interview with Emma Marris, January 2014.

CHAPTER 9: WAR

154. "They would just stand around": Seymour Hersh, "Rainmaking Is Used As Weapon by U.S.," *The New York Times* (July 3, 1972).

155. "The agency got an Air America Beechcraft": Ibid.

155. Vincent Schaefer: Bruce Lambert, "Vincent J. Schaefer, 87, Is Dead; Chemist Who First Seeded Clouds," *The New York Times* (July 28, 1993).

155. A GE promotional film: "Thinking Outside the Cold Box: How a Nobel Prize Winner and Kurt Vonnegut's Brother Made White Christmas on Demand," GE Reports, December 27, 2011. The footage described here is viewable at: www.gereports.com/thinking-outside-the-cold-box/ (The film is not dated, but refers to seeding the first cloud "last November" which would seem to situate it in 1948.)

155. Press release: Ibid.

155. "Weather control can be as powerful a war weapon as the atom bomb": "Weather Control Called 'Weapon,'" *The New York Times* (December 10, 1950).

155. "We seeded the area": Seymour Hersh, "Rainmaking Is Used As Weapon by U.S."

155. "The first confirmed use of meteorological warfare": Ibid. See also: In his book *Fixing the Sky*, James Fleming discusses earlier efforts to use weather as a weapon: cloud seeding in Korea in 1950 and French rainmaking in Vietnam in 1954. (James Fleming, *Fixing the Sky* [New York: Columbia University Press, 2010], 182.)

155. The project's focus shifted: Seymour Hersh, "Weather as Weapon of War," *The New York Times* (July 9, 1972).

155. "We were trying to arrange the weather pattern to suit our convenience": Seymour Hersh, "Rainmaking Is Used As Weapon by U.S."

155. The program was kept secret: During 1974 Congressional testimony, Deputy Assistant Secretary of Defense Dennis J. Doolin admitted that even he first learned of the cloud-seeding effort in Jack Anderson's 1971 *Washington Post* column.

156. "The object of the cloud seeding in Vietnam": Interview with Ben Livingston, Midland, TX, July 2013, and by telephone in 2013 and 2014.

156. The project aimed to "increase rainfall...": "Weather Modification," Top Secret hearing, Washington, DC: United States Senate, *Subcommittee on Oceans and International Environment of the Committee on Foreign Relations* (March 20, 1974, made public May 19, 1974).

156. Mounted in rows on the wing of an airplane: The mounting position of cartridges can vary.

156. "Back to Thailand or wherever": According to Livingston, "In '66, we flew out of Da Nang, and '67 flew out of Udorn, Thailand."

158. "Air Force rainmakers, operating secretly": Jack Anderson, "Air Force Turns Rainmaker in Laos," *The Washington Post* (March 18, 1971).

158. "Legal status of clouds": P.K. Menon, "Modifying the Weather: A Stormy Issue," Letter to the Editor, *The New York Times* (July 10, 1972).

158. "Weapon of mass destruction": Paul Bock, "Outlaw the Martial Rainmakers," Letter to the Editor, *The New York Times* (July 18, 1972).

159. Soyster and Doolin were asked about the effectiveness: It is implied in the back and forth of the transcript that one reason for secrecy may have been the perceived ineffectiveness of the project.

159. State Department officials objected: Seymour Hersh, "Rainmaking Is Used as Weapon by U.S."

160. "Weather modification can provide battlespace dominance to a degree never before imagined": Col Tamzy J. House et al., "Weather as a Force Multiplier: Owning the Weather in 2025" (August 1996).

160. A scenario from the future: Writer Edward Bellamy believed human management of weather was a necessary component of an ideal society. The protagonist of Bellamy's 19th century bestseller *Looking Backward*, Julian West, falls asleep under hypnosis in Boston in 1887 and wakes up in an utopian Boston in the year 2000. Through the rest of the novel, West is guided around the new millennium by the man who discovers him — a physician named Dr. Leete — as well as the physician's alluring daughter, Edith, whose "delicately tinted complexion" and "abounding physical vitality" make her a particular comfort to West as he adapts to life in the future. West describes the class struggles and inequities of the late 1800s. The Leetes welcome him into a world of social harmony and shared plenty. The weather is never a problem. West describes an evening on the town:

"A heavy rainstorm came up during the day, and I had concluded that the condition of the streets would be such that my hosts would have to give up the idea of going out to dinner, although the dining-hall I had understood to be quite near. I was much surprised when at the dinner hour the ladies appeared prepared to go out, but without either rubbers or umbrellas.

"The mystery was explained when we found ourselves on the street, for a continuous waterproof covering had been let down so as to enclose the sidewalk and turn it into a well-lighted and perfectly dry corridor, which was filled with a stream of ladies and gentlemen dressed for dinner. At the corners the entire open space was similarly roofed in. Edith Leete, with whom I walked, seemed much interested in learning what appeared to be entirely new to her, that in the storm weather the streets of the Boston of my day had been impassable, except to persons protected by umbrellas, boots, and heavy clothing. 'Were sidewalk coverings not used a all?' she asked.

"They were used, I explained, but in a scattered and utterly unsystematic way, being private enterprises. She said to me that at the present time all the streets were provided against inclement weather in the manner I saw, the apparatus being rolle out of the way when it was unnecessary. She intimated that it would be considered an extraordinary imbecility to permit the weather to have any effect on the social movements of the people."

Edward Bellamy, *Looking Backward* (Cambridge: Houghton Mifflin, 1887).

163. Thailand has a Bureau of Royal Rainmaking and Agricultural Aviation: "Special Report: The Roles of the Bureau of Royal Rainmaking and Agricultural Aviation," *Thai Financial Post* (March 1, 2013).

163. The Beijing Meteorological Bureau announced: Jonathon Watts, "China's largest cloud seeding assault aims to stop rain on the national parade," *The Guardian* (September 23, 2009).

163. Indonesia's Agency for the Assessment and Application of Technology: "BPPT to Use Cloud Seeding to Minimize Flood Risk in Jakarta," *Jakarta Globe* (January 25, 2013).

163. Scientists still debate: A 2014 *Scientific American* article reported that new data collection technology and more sophisticated methods of analysis bolster claims of cloud seeding's effectiveness: "New satellite and

radar evidence and more powerful computer models have lent qualified credibility to the practice of silver iodide cloud seeding." Dan Baum, "Summon the Rain," *Scientific American* (June 2014).

164. "Our mission is to defang, tame, wear down, and finally destroy": Waylon A. (Ben) Livingston, *Dr. Lively's Ultimatum* (New York: iUniverse, Inc., 2004), 159.

165. "In an instant of intense light": Waylon A. (Ben) Livingston, *Dr. Lively's Ultimatum*, 248.

CHAPTER 10: PROFIT

175. "I went to the woods": Henry David Thoreau, *Walden* (1854) (New York: Penguin Books, 1983), 135.

176. "Every winter the liquid and trembling surface": Thoreau, 330–31.

177. "Wild and ruinous": Gavin Weightman, quoting from Tudor's diary, *The Frozen-Water Trade* (New York: Hyperion, 2003), 30.

177. "No joke": Ibid, 37.

178. "Celestial dews": Thoreau, 227.

178. "The beholder measures the depth": Ibid, 233.

179. "A hundred Irishmen": Ibid, 343–44.

179. "A great emerald": Ibid, 345.

179. "On the passage to the East Indies": James Parton, *Captains of Industry, or, Men of Business Who Did Something Besides Making Money* (Boston: Houghton, Mifflin & Co, 1884), 156–62.

179. "Thus it appears": Thoreau, 346.

180. On ancient history of the ice trade: Elizabeth David, *Harvest of the Cold Months* (New York: Viking, 1995).

180. "A sack full of snow": Fernand Braudel, *The Mediterranean and the Mediterranean world in the age of Philip II*, Volume 1, (Berkeley: University of California Press, 1995), 28–29.

180. Queen Victoria…placed buckets of ice: Paul Brown, "Queen Victoria's Cooling System," *The Guardian* (July 17, 2011).

181. The Enron Corporation made the first weather derivative deal: Loren Fox, *Enron: The Rise and Fall* (New York: John Wiley & Sons, 2003), 133.

181. A $12 billion dollar business: "2011 Weather Risk Derivative Survey," *Weather Risk Management Association*, PriceWaterhouse Cooper (2011). (I rounded up from the survey's figure of $11.8 billion.)

181. Weather's impact on the U.S. economy: Jeffrey K. Lazo et al. "U.S. Economic Sensitivity to Weather Variability," *American Meteorological Society* (June 2011), 709–20.

181. More on weather's economic impact: John A. Dutton, "Opportunities and Priorities in a New Era for Weather and Climate Services," *American Meteorological Society* (September 2002), 1306. See also: John Dutton, "Weather and Climate Sensitive GDP Components," 1999, Pennsylvania State University (2001).

182. "We have a supermarket client": Interview with Frederick Fox, Berwyn, PA, August 2012, and by telephone in 2012 and 2013.

184. "I am not without a prospect that my woodlot": Ralph L. Rusk, *The Letters of Ralph Waldo Emerson*, Volume 3 (New York: Columbia University Press, 1939), 383.

185. On the rise of mechanical refrigeration: Oscar Edward Anderson, *Refrigeration in America* (Princeton: Princeton University Press, 1953), 86–102. On rise and fall of the ice trade: Joseph C. Jones, Jr. *America's Icemen* (Olathe, Kansas: Jobeco Books, 1984), 154, 159.

185. "Intestinal germs": Weightman, 241.

185. "Unless there is cold weather and plenty of it": "Ice Famine Threatens Unless Cold Sets In," *The New York Times* (February 2, 1906).

CHAPTER 11: PLEASURE

188. John Ruskin: Cited in John Lubbock, *The Use of Life*, (New York: MacMillan and Co., 1895), 69.

189. Craigslist: I retrieved the postings "If the hurricane gonna destroy nyc" and "Just met at Soho Evacuation Center" shortly after they were posted on Craigslist. They are no longer accessible. The ad "How hot would it be" was posted on Buzzfeed: "8 People Looking For Sex (And Love) During Hurricane Sandy," Anna North, *Buzzfeed* (Oct. 28, 2012).

192. "You take off all your clothes": Interview with Mark Norell, New York, NY, April 2012.

193. "There is such whimsies on the frozen ice": The website thames.me.uk gives the following citation for this poem: "Printed by M. Haly, and J. Millet, and sold by Robert Waltor, at the Globe on the North-side of St. Pauls-Church, near that end towards Ludgate; Where you may have all sorts and sizes of Maps, Coppy-Books, and Prints, not only English, but Italian, French, and Dutch. And by John Seller in the West-side of the Royal Exchange. 1684."

196. "The voice of the announcer": Adam Nicholson, "Whipping up a storm over the BBC shipping forecast sacking," *The Guardian* (September 15, 2009).

197. "It snowed all week," Truman Capote, "Miriam," *Mademoiselle* (June 1945).

200. "They now had to run for it": Charles Cowden Clarke, "Adam the Gardener," *The Monthly Repository*, Volume 8 (London: Effingham Wilson, 1834), 103.

200. Mineralogists...coined a word for the smell: I.J. Bear and R.G. Thomas, "The Nature of Argillaceous Odour," *Nature* (March 7, 1964).

CHAPTER 12: FORECASTING

208-9. Charles Golub...his daughter: Charles Golub was my grandfather. Robin (Redniss) is my mother.

210. Facts about the Worcester tornado: John M. O'Toole, *Tornado! 84 Minutes, 94 Lives,* (Worcester: Data Books, 1993). Until the Joplin tornado in May 2011, the Worcester tornado was the deadliest in U.S. history.

210. The Weather Bureau at Logan Airport: A study by the National Research Council found that local weather reconnaissance planes as well as meteorologists at the G.E. labs in upstate New York also anticipated the possibility of a tornado in the area, but also stopped short of alerting the public. See: William Chittock, *The Worcester Tornado* (Bristol, RI: Self-published pamphlet, 2003), 12-13.

210. "Nasty": Interview with Jud Hale, Dublin, NH, November 2011. See also: Judson Hale, *The Best of The Old Farmer's Almanac: The First 200 Years* (New York: Random House, 1992), 46.

211. "How to get through life": Richard Anders, "Almanacs," americanantiquarian.org/almanacs.htm

214. "Mild" or "wet" or "frosty": Judson Hale, *The Best of The Old Farmer's Almanac*, 43-44.

214. "You request the actual number of snowflakes": Robb Sagendorph, *Old Farmer's Almanac* (Dublin, NH: Yankee Publishing, 1949) cited by Hale, *The Best of The Old Farmer's Almanac*, 43.

214. "He begins with a series of cycles": "Old Faithful Goes Out on a Limb," *Life* (November 18, 1966).

214. "This isn't science": Ibid.

214. "Refined and enhanced": "How We Predict Weather," This statement appears in every issue of the *OFA*.

214. "There is, I am almost convinced": Robb Sagendorph, "My Life with the Old Farmer's Almanac," *American Legion Magazine* (January 1965), 26.

216. Lincoln asked the witness to read an almanac entry: Other almanacs also lay claim to this story, but according to Jud Hale, "We're the only almanac that says, for August 29 — the night of the murder: 'Moon rides low.' You look at other almanacs

that are out that year, and they don't say anything for August 29."

Norman Rockwell depicted the courtroom scene in "Lincoln for the Defense." In the painting, Armstrong is seated; his head is down, his hands shacked and fingers interlaced as if in prayer. Lincoln dominates the foreground of the long, vertical composition, dressed in white. His face and his right fist are clenched. In his left hand he holds a pair of spectacles and the 1857 *Almanac*.

216. About the substitution of weather "indications" for weather "forecasts," Jud Hale says, "It was semantics. We didn't change the information. In those days — in 1944 — there was no serious weather forecasting."

218. Stones…stones: Jud Hale: "A friend of mine went to the very spot where Alexander the Great did something or other. She got some stones for me."

220. "The *Old Farmer's Almanac* claims to be right": Interview with Harold Brooks, Norman, *OK*, November 2011, and telephone interviews, 2012.

221. Edward Lorenz: Kenneth Chang, "Edward N. Lorenz, a Meteorologist and a Father of Chaos Theory, Dies at 90," *The New York Times* (April 17, 2008).

221. A decade earlier: For a detailed description of these events and more about Lorenz and his work, see James Gleick's *Chaos: The Making of a New Science* (New York: Vintage, 1987).

221. "Long-range weather forecasting must be doomed": Ibid.

221. "Since we do not know exactly how many butterflies": Edward Lorenz, *The Essence of Chaos* (Seattle: University of Washington Press, 1995), 182.

221. "A perfect forecast": Kenneth Chang, "Edward N. Lorenz, a Meteorologist and a Father of Chaos Theory, Dies at 90."

221. "What Is a Good Forecast?": Allan H. Murphy, "What Is a Good Forecast? An Essay on the Nature of Goodness in Weather Forecasting," *American Meteorological Society* (June 1993).

222. "There's a fuzziness you have to live with": Interview with Greg Carbin, Norman, *OK*, November 2011, and telephone interviews, 2012, 2013, 2014.

222. "Wet bias": Nate Silver, *The Signal and the Noise* (New York: Penguin Press, 2012), 135.

222. "People want a yes or a no": Interview with Rick Smith, Norman, *OK*, November 2011.

222. On folk weather predictors: Harold Brooks: "Most folklore things are based upon lots of observations. And the real question becomes: Do you understand the circumstances in which those observations were made, and how do they apply to you? 'Red sky at night [sailor's delight, Red sky in morning, sailor's warning'] is a classic. It is essentially looking at where is cloud cover in the morning, at sunrise, and in the afternoon. It works well when weather systems are moving west to east, which is what happens in mid-latitudes, so what you're seeing in the morning, if you got lots of cloud cover coming in from the west, is storms approaching. If you've got a clear sunset and there's lots of cloud cover off to the east when you're getting red sky in the evening, everything's clearing, so it's going to be nice. Well, if you now take the converse possibility, and you put yourself on the east coast of Florida and you've got red sky in the evening — okay, you've got clouds to your east. It's September. What that actually has just told you is that there is a hurricane's approaching. So that's a really bad. Because you aren't in the zone where a system would be west to east. So that's the thing: a lot of the folklore things are based upon a large number of observations, and when you're in an abnormal situation they can be horribly wrong."

222. "Forecasts…acquire value": Allan H. Murphy, "What Is a Good Forecast? An Essay on the Nature of Goodness in Weather Forecasting," *American Meteorological Society* (June 1993).

The Allied forecast for June 6, 1944 —
D-Day — offers a paradoxical example
of what makes a "good" forecast. Harold
Brooks: "If you go back and look at the
forecast from D-Day, the Allied forecast
was that conditions will be good enough to
invade — the big question was wave height,
how many landing craft would get lost. The
Germans' forecast was that the waves would
be too high: it would be foolish to invade.
Both sides acted like their forecasts were
correct. It turns out the German forecasts
were probably more accurate. The waves
were higher than what the Allies considered
to be acceptable for invading. So the
Germans were unprepared for the invasion,
because they thought, 'no one's going to
invade when the weather's this bad.' [German
field marshal Erwin] Rommel had gone
home to see his wife for her birthday. He
was the only person other than Hitler who
could actually mobilize some of the defense
systems. If the Germans would've gotten
a worse forecast, they might've actually
been better prepared for the invasion. If
the Allies would've gotten a more accurate
forecast, they might've decided, 'Oh, it's not
worth invading. We're going to lose too many
folks.' The Allies should be very happy that
they got a less accurate forecast."

In the case of the D-Day forecast, low
quality (accuracy) meant high value (benefit
to user) — at least for the Allied Forces.

222. "A methodology that looks at relationships":
Telephone interview with Michael Steinberg,
December 2012.

224. The Dublin Community Church: According to
the *OFA* website, "Built in 1852, the church
gained notoriety during the Hurricane of
'38, when winds snapped the steeple off the
building, spun it around, and plunged it back
through the roof of the church."

NOTE ON THE ART

239. "Instruments in the processes of inquiry
into the natural world": Susan Dackerman,
*Prints and the Pursuit of Knowledge in
Early Modern Europe* (Cambridge: Harvard
Art Museums, 2011), 20.

NOTE ON THE TYPE

241. Qaneq…"falling snow": This spelling and
definition come from Boas, as cited by
Krupnik and Müller-Wille: Igor Krupnik
and Ludger Müller-Wille, "Frank Boas and
Inuktitut Terminology for Ice and Snow:
From the Emergence of the Field to the
'Great Eskimo Vocabulary Hoax,'" *SIKU:
Knowing Our Ice*, Dordrecht: *Springer
Science + Business Media* (2010), 384.

241. "Trivialization of the complexity"…
"responsible scholarship": Laura Martin, "Eskimo
Words for Snow: A Case Study in the Genesis
and Decay of an Anthropological Example,"
American Anthropologist, New Series, Volume
88, Number 2 (June 1986), 421.

241. "Nine, forty-eight, a hundred, two hundred,
who cares?"…"The truth is": Geoffrey K.
Pullman, *The Great Eskimo Vocabulary Hoax
and Other Irreverent Essays on the Study
of Language* (Chicago: University of Chicago
Press, 1991), 164…160.

241. Cultural anthropologist Igor Krupnik: Phone
interview with Igor Krupnik, July 2014. See
also: Igor Krupnik and Ludger Müller-Wille,
table 16.3, 392–93.

ACKNOWLEDGMENTS

This book would not exist without the support of many people.

I am particularly indebted to my editor Susan Kamil, to my literary agent Elyse Cheney, to Lewis Bernard and the American Museum of Natural History, and to the Solomon R. Guggenheim Foundation.

Tamara Connolly, Jackie Hahn, and Duncan Tonatiuh helped me with production and design. Paul Mullowney and Paul Taylor turned my drawings into prints. Greg Carbin, Jenifer & Dane Clark, Mary Ann Cooper, and Ron Holle fact-checked the book's scientific passages. Alexa Tsoulis-Reay did additional fact-checking. (Any mistakes that remain are, of course, my own.)

Enormous thanks to all those who submitted to interviews, many of whose names appear in the book.

My deepest gratitude also to:

Omar Ali, Ted Allen, Dave Andra, Tom Baione, Mark Benner, Julio Betancourt, Jamie Boettcher, Nadine Bourgeois, Harold Brooks, Gerry Cantwell, Emma Caruso, Marie d'Origny, Bella Desai, Benjamin Dreyer, Eleana Duarte, Richard Elman, Gina Eosco, David Ferriero, Barbara Fillon, Mike Foster, Sam Freilich, Ellen Futter, Anne Gaines, Jennifer Garza, Malcolm Gladwell, Liz Goldwyn, J.J. Gourley, Amy Gray, Eve Gruntfest, Steven Guarnaccia, Judson Hale, Joshua Hammerman, Pam Heinselman, Charlotte Herscher, Janet Howe, Alex Jacobs, Justin Jampol, Gillian Kane, Ben Katchor, David Keith, Daniel Kevles, Kim Klockow, Nora Krug, Jim LaDue, Todd Lambrix, Davie Lerner, Ben Livingston, Lenaya Lynch, Leigh Marchant, Sally Marvin, Richard McGuire, Carolyn Meers, Stephen Metcalf, Kaela Myers, Tess Nellis, Alana Newhouse, Mark Norell, Loren Noveck, Richard Pearson, Tom Perry, Abigail Pope, Liz Quoetone, Christopher Raxworthy, Lily Redniss, Seth Redniss, Rick & Robin Redniss, Mia Reitmeyer, Susan Grant Rosen, Marc Rosen, Russ Schneider, Erin Sheehy, Mark Siddall, Sandra Sjursen, Rick Smith, Michael Steinberg, Dave Stensrud, Janice Stillman, Jean Strouse, Keli Tarp, Stewart Thorndike, Joel Towers, Sven Travis, Molly Turpin, David Wettergreen, Andy Wood, Teresa Zoro.

This book is dedicated to my family: Jody Rosen, Sasha Rosen, and Theo Rosen.

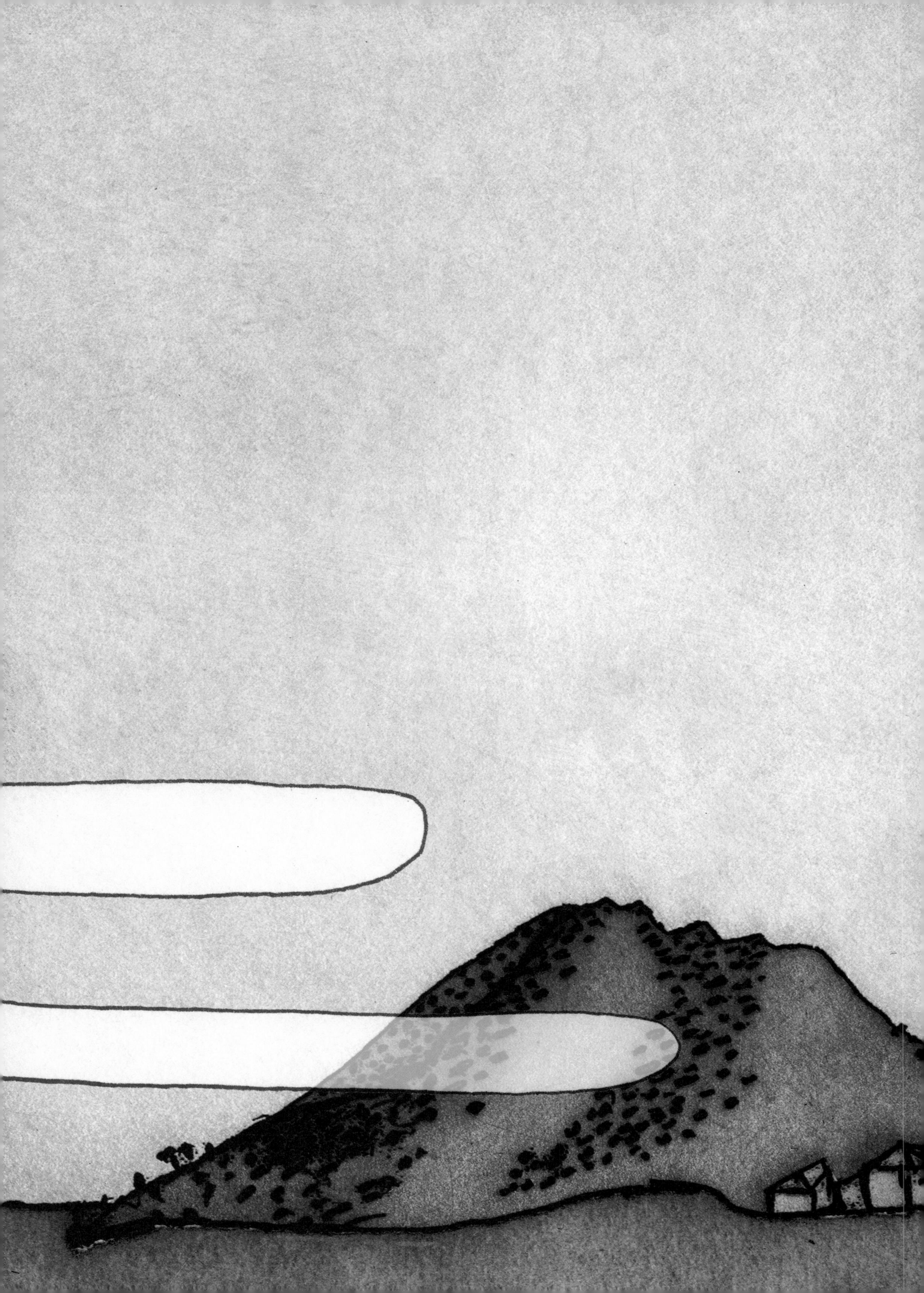

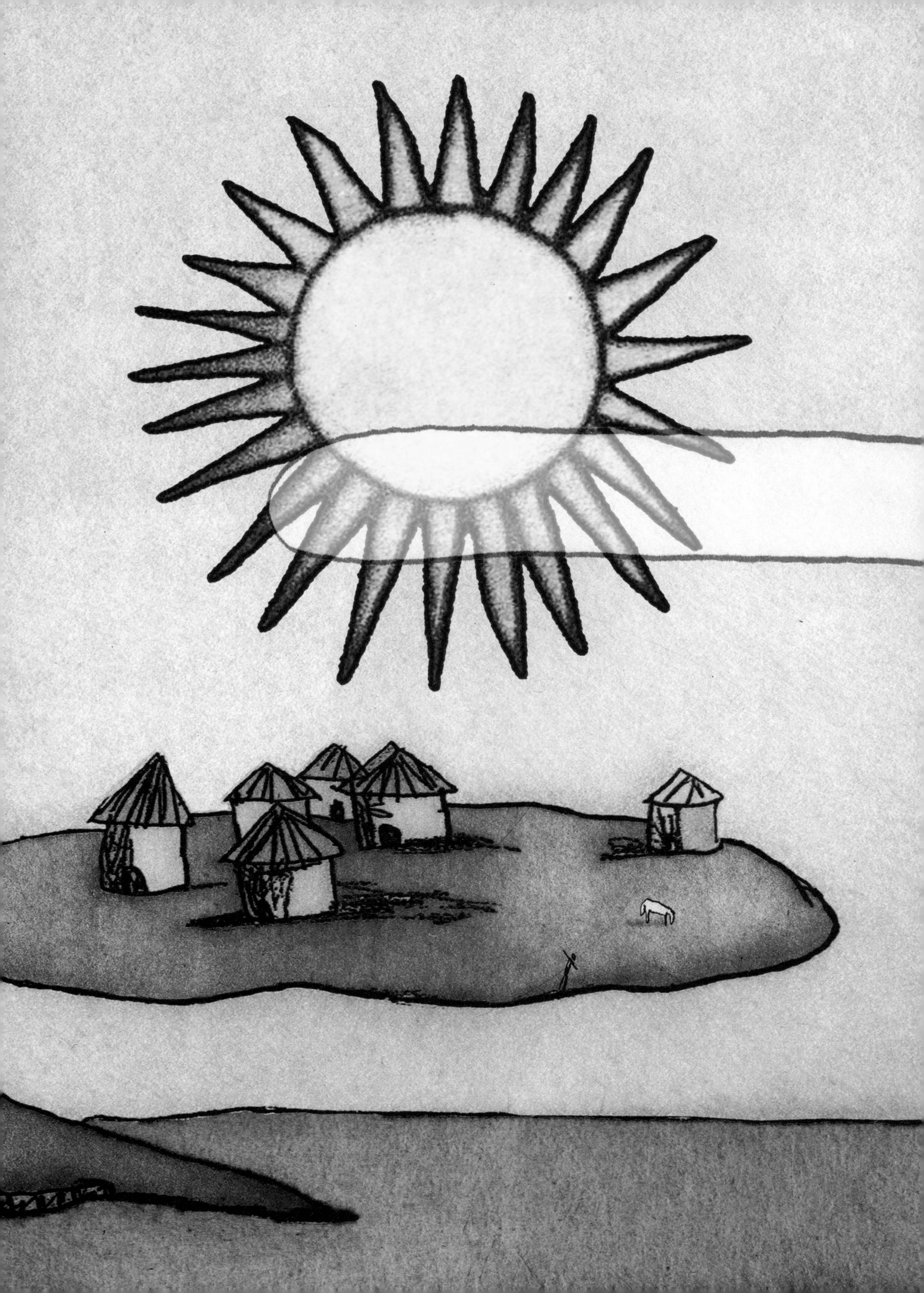

Published in the United States by Random House, an imprint and division of Penguin Random House LLC, New York.

RANDOM HOUSE and the HOUSE colophon are registered trademarks of Penguin Random House LLC.

Author photo by Abigail Pope.

Excerpt from PETER CAMENZIND by Hermann Hesse, translated by Michael Roloff. Copyright © 1969, renewed 1998 by Farrar, Straus and Giroux, Inc. Reprinted by permission of Farrar, Straus and Giroux, LLC.

ISBN 978-0-8129-9317-2
eBook ISBN 978-0-679-64472-3

Printed in China on acid-free paper

randomhousebooks.com

987654321

First Edition

Lauren Redniss is the author of *Century Girl: 100 Years in the Life of Doris Eaton Travis, Last Living Star of the Ziegfeld Follies* and *Radioactive: Marie & Pierre Curie, A Tale of Love and Fallout*, a finalist for the National Book Award. She teaches at Parsons The New School for Design.